Circuit Analysis I
with MATLAB® Applications

Steven T. Karris

Orchard Publications, Fremont, California
www.orchardpublications.com

Circuit Analysis I with MATLAB® Applications

Direct all inquiries to Orchard Publications, 39510 Paseo Padre Parkway, Fremont, California 94538, U.S.A. URL: http://www.orchardpublications.com

Product and corporate names are trademarks or registered trademarks of the MathWorks®, Inc., and Microsoft® Corporation. They are used only for identification and explanation, without intent to infringe.

Library of Congress Cataloging-in-Publication Data

Catalog record is available from the Library of Congress   LCCN 2002096946

ISBN 0-9709511-2-4

Disclaimer

The author has made every effort to make this text as complete and accurate as possible, but no warranty is implied. The author and publisher shall have neither liability nor responsibility to any person or entity with respect to any loss or damages arising from the information contained in this text.

This book was created electronically using Adobe Framemaker®.

Preface

This text is an introduction to the basic principles of electrical engineering. It is the outgrowth of lecture notes prepared by this author while teaching for the electrical engineering and computer engineering departments at San José State University, DeAnza college, and the College of San Mateo, all in California. Many of the examples and problems are based on the author's industrial experience. It can be used as a primary text or supplementary text. It is also ideal for self-study.

This book is intended for students of college grade, both community colleges and universities. It presumes knowledge of first year differential and integral calculus and physics. While some knowledge of differential equations would be helpful, it is not absolutely necessary. Chapters 9 and 10 include step-by-step procedures for the solutions of simple differential equations used in the derivation of the natural and forces responses. Appendices B and C provide a thorough review of complex numbers and matrices respectively.

There are several textbooks on the subject that have been used for years. The material of this book is not new, and this author claims no originality of its content. This book was written to fit the needs of the average student. Moreover, it is not restricted to computer oriented circuit analysis. While it is true that there is a great demand for electrical and computer engineers, especially in the internet field, the demand also exists for power engineers to work in electric utility companies, and facility engineers to work in the industrial areas.

Circuit analysis is comprised of numerous topics. It would be impractical to include all related topics in a single text. This book, *Circuit Analysis I with MATLAB® Applications*, contains the standard subject matter of electrical engineering. Accordingly, it is intended as a first course in circuits and the material can be covered in one semester or two quarters. A sequel, *Circuit Analysis II with MATLAB® Applications*, is intended for use in a subsequent semester or two subsequent quarters.

It is not necessary that the reader has previous knowledge of MATLAB®. The material of this text can be learned without MATLAB. However, this author highly recommends that the reader studies this material in conjunction with the inexpensive MATLAB Student Version package that is available at most college and university bookstores. Appendix A of this text provides a practical introduction to MATLAB. As shown on the front cover, a system of equations with complex coefficients can be solved with MATLAB very accurately and rapidly. MATLAB will be invaluable in later studies such as the design of analog and digital filters.

In addition to several problems provided at the end of each chapter, this text includes multiple-choice questions to test and enhance the reader's knowledge of this subject. Moreover, answers to these questions and detailed solutions of all problems are provided at the end of each chapter. The rationale is to encourage the reader to solve all problems and check his effort for correct solutions and appropriate steps in obtaining the correct solution. And since this text was written to serve as a self-study or supplementary textbook, it provides the reader with a resource to test his knowledge.

The author has accumulated many additional problems for homework assignment and these are available to those instructors who adopt this text either as primary or supplementary text, and prefer to assign problems without the solutions. He also has accumulated many sample exams.

Like any other new book, this text may contain some grammar and typographical errors. Accordingly, all feedback for errors, advice and comments will be most welcomed and greatly appreciated.

Orchard Publications
Fremont, California

Contents

Contents

Chapter 3

Nodal and Mesh Equations - Circuit Theorems

Chapter 4

Introduction to Operational Amplifiers

Chapter 5

Inductance and Capacitance

Chapter 6

Sinusoidal Circuit Analysis

Chapter 7

Phasor Circuit Analysis

Contents

Chapter 8

Average and RMS Values, Complex Power, and Instruments

Chapter 9

Natural Response

Chapter 10

Forced and Total Response in RL and RC Circuits

Appendix A

Introduction to MATLAB®

Appendix B

A Review of Complex Numbers

Contents

Appendix C

Matrices and Determinants

Chapter 1

Basic Concepts and Definitions

T his chapter begins with the basic definitions in electric circuit analysis. It introduces the concepts and conventions used in introductory circuit analysis, the unit and quantities used in circuit analysis, and includes several practical examples to illustrate these concepts.

1.1 The Coulomb

Two identically charged (both positive or both negative) particles possess a charge of one *coulomb* when being separated by one meter in a vacuum, repel each other with a force of $10^{-7}c^2$ newton where $c = velocity\ of\ light \approx 3 \times 10^8\ m/s$. The definition of coulomb is illustrated in Figure 1.1.

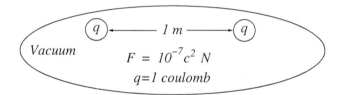

Figure 1.1. Definition of the coulomb

The coulomb, abbreviated as C, is the fundamental unit of charge. In terms of this unit, the charge of an electron is $1.6 \times 10^{-19}\ C$ and one negative coulomb is equal to 6.24×10^{18} electrons. Charge, positive or negative, is denoted by the letter q or Q.

1.2 Electric Current and Ampere

Electric *current* i at a specified point and flowing in a specified direction is defined as the instantaneous rate at which net positive charge is moving past this point in that specified direction, that is,

$$i = \frac{dq}{dt} = \lim_{\Delta t \to 0} \frac{\Delta q}{\Delta t} \qquad (1.1)$$

The unit of current is the *ampere* abbreviated as A and corresponds to charge q moving at the *rate* of one coulomb per second. In other words,

$$1\ ampere = \frac{1\ coulomb}{1\ second} \qquad (1.2)$$

Chapter 1 Basic Concepts and Definitions

Note: Although it is known that current flow results from electron motion, it is customary to think of current as the motion of positive charge; this is known as *conventional current flow.*

To find an expression of the charge q in terms of the current i, let us consider the charge q transferred from some reference time t_0 to some future time t. Then, since

$$i = \frac{dq}{dt}$$

the charge q is

$$q\Big|_{t_0}^{t} = \int_{t_0}^{t} i\,dt$$

or

$$q(t) - q(t_0) = \int_{t_0}^{t} i\,dt$$

or

$$q(t) = \int_{t_0}^{t} i\,dt + q(t_0) \tag{1.3}$$

Example 1.1

For the waveform of current i shown in Figure 1.2, compute the total charge q transferred between

a. $t = 0$ and $t = 3\ s$

b. $t = 0$ and $t = 9\ s$

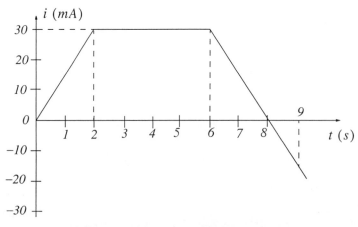

Figure 1.2. Waveform for Example 1.1

Solution:

We know that

$$q\big|_{t=0}^{t} = \int_{0}^{t} i\,dt = Area\big|_{0}^{t}$$

Then, by calculating the areas, we find that:

a. For $0 < t < 2$ s, area $= \frac{1}{2} \times (2 \times 30 \text{ mA}) = 30$ mC
   For $2 < t < 3$ s, area $= 1 \times 30 = 30$ mC
   Therefore, for $0 < t < 3$ s, total charge $=$ total area $= 30$ mC $+ 30$ mC $= 60$ mC.

b. For $0 < t < 2$ s, area $= \frac{1}{2} \times (2 \times 30 \text{ mA}) = 30$ mC
   For $2 < t < 6$ s, area $= 4 \times 30 = 120$ mC
   For $6 < t < 8$ s, area $= \frac{1}{2} \times (2 \times 30 \text{ mA}) = 30$ mC
   For $8 < t < 9$ s, we observe that the slope of the straight line for $t > 6$ s is -30 mA $/ 2$ s, or -15 mA $/$ s. Then, for $8 < t < 9$ s, area $= \frac{1}{2} \times \{1 \times (-15)\} = -7.5$ mC. Therefore, for $0 < t < 9$ s, total charge $=$ total area $= 30 + 120 + 30 - 7.5 = 172.5$ mC.

Convention: We denote the current i by placing an arrow with the numerical value of the current next to the device in which the current flows. For example, the designation shown in Figure 1.3 indicates either a current of 2 A is flowing from left to right, or that a current of -2 A is moving from right to left.

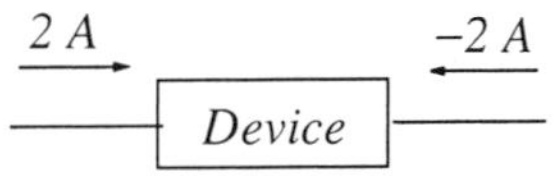

Figure 1.3. Direction of conventional current flow

Caution: The arrow may or may not indicate the *actual* conventional current flow. We will see later in Chapters *2* and *3* that in some circuits (to be defined shortly), the actual direction of the current cannot be determined by inspection. In such a case, we assume a direction with an arrow for said current i; then, if the current with the assumed direction turns out to be negative, we conclude that the actual direction of the current flow is opposite to the direction of the arrow. Obviously, reversing the direction reverses the algebraic sign of the current as shown in Figure 1.3.

In the case of time-varying currents which change direction from time-to-time, it is convenient to think or consider the instantaneous current, that is, the direction of the current which flows at some particular instant. As before, we assume a direction by placing an arrow next to the device in which the current flows, and if a negative value for the current i is obtained, we conclude that the actual direction is opposite of that of the arrow.

1.3 Two Terminal Devices

In this text we will only consider two-terminal devices. In a two-terminal device the current entering one terminal is the same as the current leaving the other terminal[*] as shown in Figure 1.4.

Figure 1.4. *Current entering and leaving a two-terminal device*

Let us assume that a constant value current (commonly known as *Direct Current* and abbreviated as DC) enters terminal *A* and leaves the device through terminal *B* in Figure 1.4. The passage of current (or charge) through the device requires some expenditure of energy, and thus we say that a *potential difference* or *voltage* exists "across" the device. This voltage across the terminals of the device is a measure of the work required to move the current (or charge) through the device.

Example 1.2

In a two-terminal device, a current $i(t) = 20\cos 100\pi t \; mA$ enters the left (first) terminal.

a. What is the amount of current which enters that terminal in the time interval $-10 \le t \le 20 \; ms$?

b. What is the current at $t = 40 \; ms$?

c. What is the charge q at $t = 5 \; ms$ given that $q(0) = 0$?

Solution:

a.

$$i\Big|_{t_0}^{t} = 20\cos 100\pi t \Big|_{-10 \times 10^{-3}}^{20 \times 10^{-3}} = 20\cos 100\pi(20 \times 10^{-3}) - 20\cos 100\pi(-10 \times 10^{-3})$$

$$= 20\cos 2\pi - 20\cos(-\pi) = 40 \; mA$$

b.

$$i\Big|_{t = 0.4 \; ms} = 20\cos 100\pi t \Big|_{t = 0.4 \; ms} = 20\cos 40\pi = 20 \; mA$$

c.

$$q(t) = \int_{0}^{5 \times 10^{-3}} i \, dt + q(0) = \int_{0}^{5 \times 10^{-3}} 20\cos 100\pi t \, dt + 0$$

$$= \frac{0.2}{\pi}\sin 100\pi t \Big|_{0}^{5 \times 10^{-3}} = \frac{0.2}{\pi}\sin\frac{\pi}{2} - 0 = \frac{0.2}{\pi} \; C$$

---

[*] *We will see in Chapter 5 that a two terminal device known as capacitor is capable of storing energy.*

1.4 Voltage (Potential Difference)

The voltage (potential difference) across a two-terminal device is defined as the work required to move a positive charge of one coulomb from one terminal of the device to the other terminal.

The unit of voltage is the *volt* (abbreviated as V or v) and it is defined as

$$1 \ volt = \frac{1 \ joule}{1 \ coulomb} \tag{1.4}$$

Convention: We denote the voltage v by a plus (+) minus (–) pair. For example, in Figure 1.5, we say that terminal A is *10 V* positive with respect to terminal B or there is a potential difference of *10 V* between points A and B. We can also say that there is a *voltage drop* of *10 V* in going from point A to point B. Alternately, we can say that there is a *voltage rise* of *10 V* in going from B to A.

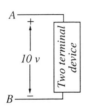

Figure 1.5. Illustration of voltage polarity for a two-terminal device

Caution: The (+) and (–) pair may or may not indicate the actual voltage drop or voltage rise. As in the case with the current, in some circuits the actual polarity cannot be determined by inspection. In such a case, again we assume a voltage reference polarity for the voltage; if this reference polarity turns out to be negative, this means that the potential at the (+) sign terminal is at a lower potential than the potential at the (–) sign terminal.

In the case of time-varying voltages which change (+) and (–) polarity from time-to-time, it is convenient to think the *instantaneous voltage*, that is, the voltage reference polarity at some particular instance. As before, we assume a voltage reference polarity by placing (+) and (–) polarity signs at the terminals of the device, and if a negative value of the voltage is obtained, we conclude that the actual polarity is opposite to that of the assumed reference polarity. We must remember that reversing the reference polarity reverses the algebraic sign of the voltage as shown in Figure 1.6.

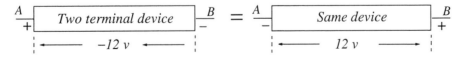

Figure 1.6. Alternate ways of denoting voltage polarity in a two-terminal device

Chapter 1 Basic Concepts and Definitions

Example 1.3

The $i - v$ (current-voltage) relation of a non-linear electrical device is given by

$$i(t) = 0.1(e^{0.2 \sin 3t} - 1) \tag{10.5}$$

a. Use MATLAB® [*] to sketch this function for the interval $0 \le t \le 10$ s

b. Use the MATLAB **quad** function to find the charge at $t = 5$ s given that $q(0) = 0$

Solution:

a. We use the following code to sketch $i(t)$.

```
t=0: 0.1: 10;
it=0.1.*(exp(0.2.*sin(3.*t))–1);
plot(t,it), grid, xlabel('time in sec.'), ylabel('current in amp.')
```

The plot for $i(t)$ is shown in Figure 1.7.

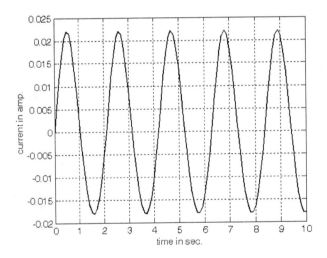

Figure 1.7. Plot of $i(t)$ for Example 1.3

b. The charge $q(t)$ is the integral of the current $i(t)$, that is,

$$q(t) = \int_{t_0}^{t_1} i(t)dt = 0.1 \int_{0}^{t_1} (e^{0.2 \sin 3t} - 1)dt \tag{1.6}$$

---

[*] *MATLAB and SIMULINK are registered marks of The MathWorks, Inc., 3 Apple Hill Drive, Natick, MA, 01760, www.mathworks.com. An introduction to MATLAB is given in Appendix A.*

Circuit Analysis I with MATLAB Applications
Orchard Publications

We will use the MATLAB **int(f,a,b)** integration function where **f** is a symbolic expression, and **a** and **b** are the lower and upper limits of integration respectively.

Note

When MATLAB cannot find a solution, it returns a warning. For this example, MATLAB returns the following message when integration is attempted with the symbolic expression of (1.6).

```
t=sym('t');
s=int(0.1*(exp(0.2*sin(3*t))−1),0,10)
```

When this code is executed, MATLAB displays the following message:

```
Warning: Explicit integral could not be found.
In C:\MATLAB 12\toolbox\symbolic\@sym\int.m at line 58

s = int(1/10*exp(1/5*sin(3*t))-1/10,t = 0. . 10)
```

We will use numerical integration with Simpson's rule. MATLAB has two quadrature functions for performing numerical integration, the **quad**[*] and **quad8**. The description of these can be seen by typing help quad or help quad8. Both of these functions use *adaptive quadrature methods*; this means that these methods can handle irregularities such as singularities. When such irregularities occur, MATLAB displays a warning message but still provides an answer.

For this example, we will use the **quad** function. It has the syntax **q=quad('f',a,b,tol)**, and performs an integration to a relative error **tol** which we must specify. If **tol** is omitted, it is understood to be the standard tolerance of 10^{-3}. The string **'f'** is the name of a user defined function, and **a** and **b** are the lower and upper limits of integration respectively.

First, we need to create and save a function m-file. We define it as shown below, and we save it as CA_1_Ex_1_3.m. This is a mnemonic for Circuit Analysis I, Example 1.3.

```
function t = fcn_example_1_3(t); t = 0.1*(exp(0.2*sin(3*t))-1);
```

With this file saved as CA_1_Ex_1_3.m, we write and execute the following code.

```
charge=quad('CA_1_Ex_1_3',0,5)
```

and MATLAB returns

```
charge =
    0.0170
```

---

[*] *For a detailed discussion on numerical analysis and the MATLAB functions* **quad** *and* **quad8**, *the reader may refer to Numerical Analysis Using MATLAB® and Spreadsheets by this author, Orchard Publications, ISBN 0-9709511-1-6.*

1.5 Power and Energy

Power p is the rate at which *energy* (or *work*) W is expended. That is,

$$\boxed{Power = p = \frac{dW}{dt}}$$

(1.7)

Absorbed power is proportional both to the current and the voltage needed to transfer one coulomb through the device. The unit of power is the *watt*. Then,

$$Power = p = volts \times amperes = vi = \frac{joul}{coul} \times \frac{coul}{sec} = \frac{joul}{sec} = watts$$

(1.8)

and

$$\boxed{1 \; watt = 1 \; volt \times 1 \; ampere}$$

(1.9)

Passive Sign Convention: Consider the two-terminal device shown in Figure 1.8.

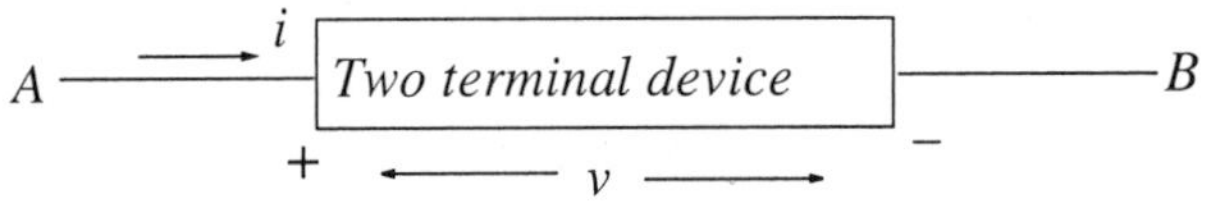

$$A \xrightarrow{\quad i \quad} \boxed{Two \; terminal \; device} \longrightarrow B$$

Figure 1.8. Illustration of the passive sign convention

In Figure 1.8, terminal A is v volts positive with respect to terminal B and current i enters the device through the positive terminal A. In this case, we satisfy the *passive sign convention* and $power = p = vi$ is said to be *absorbed* by the device.

The passive sign convention states that if the arrow representing the current i and the $(+)$ $(-)$ pair are placed at the device terminals in such a way that the current enters the device terminal marked with the $(+)$ sign, and if both the arrow and the sign pair are labeled with the appropriate algebraic quantities, the power absorbed or delivered to the device can be expressed as $p = vi$. If the numerical value of this product is positive, we say that the device is *absorbing power* which is equivalent to saying that power is delivered to the device. If, on the other hand, the numerical value of the product $p = vi$ is negative, we say that the device delivers power to some other device. The passive sign convention is illustrated with the examples in Figures 1.9 and 1.10.

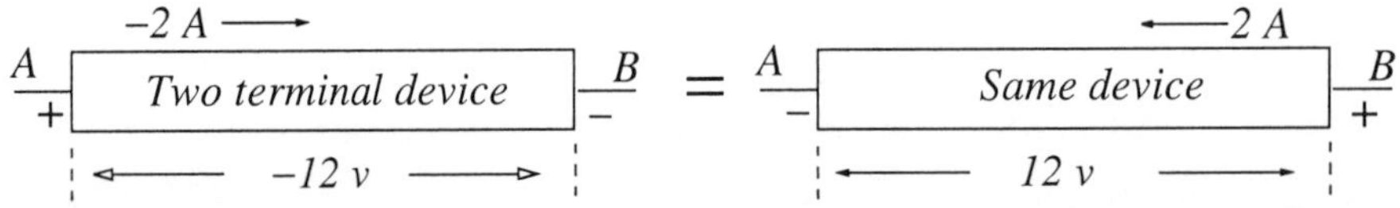

$$Power = p = (-12)(-2) = 24 \; w \qquad Power = p = (12)(2) = 24 \; w$$

Figure 1.9. Examples where power is absorbed by a two-terminal device

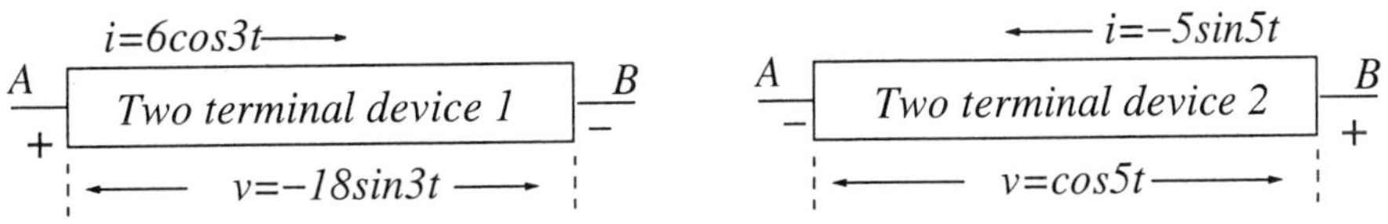

$$i = 6cos3t \longrightarrow$$

A | Two terminal device 1 | B

$$v = -18sin3t \longrightarrow$$

$$p = (-18sin3t)(6cos3t) = -54sin6t \ w$$

$$\longleftarrow i = -5sin5t$$

A | Two terminal device 2 | B

$$\longleftarrow v = cos5t \longrightarrow$$

$$p = (cos5t)(-5sin5t) = -2.5sin10t \ w$$

Figure 1.10. Examples where power is delivered to a two-terminal device

In Figure 1.9, power is absorbed by the device, whereas in Figure 1.10, power is delivered to the device.

Example 1.4

It is assumed a 12-volt automotive battery is completely discharged and at some reference time $t = 0$, is connected to a battery charger to trickle charge it for the next 8 hours. It is also assumed that the charging rate is

$$i(t) = \begin{cases} 8e^{-t/3600} \ A & 0 \leq t \leq 8 \ hr \\ 0 & otherwise \end{cases}$$

For this 8-hour interval compute:

a. the total charge delivered to the battery

b. the *maximum* power (in watts) absorbed by the battery

c. the total energy (in joules) supplied

d. the *average* power (in watts) absorbed by the battery

Solution:

The current entering the positive terminal of the battery is the decaying exponential shown in Figure 1.11 where the time has been converted to seconds.

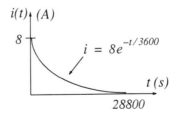

Figure 1.11. Decaying exponential for Example 1.4

Then,

a.

$$q\Big|_{t=0}^{15000} = \int_0^{15000} idt = \int_0^{28800} 8e^{-t/3600}dt = \frac{8}{-1/3600}e^{-t/3600}\Big|_0^{28800}$$

$$= -8 \times 3600(e^{-8}-1) \approx 28800 \ C \ or \ 28.8 \ kC$$

b.

$$i_{max} = 8 \ A \ (occurs \ at \ t=0)$$

Therefore,

$$p_{max} = vi_{max} = 12 \times 8 = 96 \ w$$

c.

$$W = \int pdt = \int_0^{28800} vidt = \int_0^{28800} 12 \times 8e^{-t/3600}dt = \frac{96}{-1/3600}e^{-t/3600}\Big|_0^{28800}$$

$$= 3.456 \times 10^5(1-e^{-8}) \approx 345.6 \ KJ.$$

d.

$$P_{ave} = \frac{1}{T}\int_0^T pdt = \frac{1}{28800}\int_0^{28800} 12 \times 8e^{-t/3600}dt = \frac{345.6 \times 10^3}{28.8 \times 10^3} = 12 \ w.$$

Example 1.5

The power absorbed by a non-linear device is $p = 9(e^{0.16t^2}-1)$. If $v = 3(e^{0.4t}+1)$, how much charge goes through this device in two seconds?

Solution:

The power is

$$p = vi, \ i = \frac{p}{v} = \frac{9(e^{0.16t^2}-1)}{3(e^{0.4t}+1)} = \frac{9(e^{0.4t}+1)(e^{0.4t}-1)}{3(e^{0.4t}+1)} = 3(e^{0.4t}-1) \ A$$

then, the charge for 2 seconds is

$$q\Big|_{t_0}^t = \int_{t_0}^t idt = 3\int_0^2 (e^{0.4t}-1)dt = \frac{3}{0.4}e^{0.4t}\Big|_0^2 - 3t\Big|_0^2 = 7.5(e^{0.8}-1)-6 = 3.19 \ C$$

The two-terminal devices which we will be concerned with in this text are shown in Figure 1.12.

Linear devices are those in which there is a linear relationship between the voltage across that device and the current that flows through that device. Diodes and Transistors are *non-linear* devices, that is, their voltage-current relationship is non-linear. These will not be discussed in this text. A simple circuit with a diode is presented in Chapter 3.

Independent and Dependent Sources

v or v(t)

Ideal Independent Voltage Source – Maintains same voltage regardless of the amount of current that flows through it. Its value is either constant (DC) or sinusoidal (AC).

i or i(t)

Ideal Independent Current Source – Maintains same current regardless of the voltage that appears across its terminals. Its value is either constant (DC) or sinusoidal (AC).

$k_1 v$ or $k_2 i$

Dependent Voltage Source – Its value depends on another voltage or current elsewhere in the circuit. Here, k_1 is a constant and k_2 is a resistance as defined in linear devices below. When denoted as $k_1 v$ it is referred to as *voltage controlled voltage source*, and when denoted as $k_2 i$ it is referred to *as current controlled voltage source.*

$k_3 i$ or $k_4 v$

Dependent Current Source – Its value depends on another current or voltage elsewhere in the circuit. Here, k_3 is a constant and k_4 is a conductance as defined in linear devices below. When denoted as $k_3 i$ it is referred to as *current controlled current source* and when denoted as $k_4 v$ it is referred to as *voltage controlled current source.*

Linear Devices

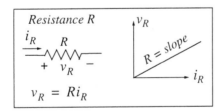

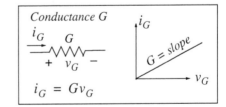

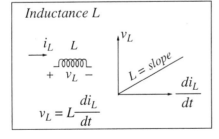

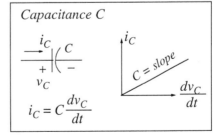

Figure 1.12. Voltage and current sources and linear devices

1.6 Active and Passive Devices

Independent and dependent voltage and current sources are *active devices*; they normally (but not always) deliver power to some external device. Resistors, inductors and capacitors are *passive devices*; they normally receive (absorb) power from an active device.

1.7 Circuits and Networks

A *network* is the interconnection of two or more simple devices as shown in Figure 1.13.

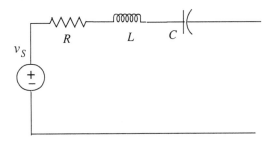

Figure 1.13. A network but not a circuit

A *circuit* is a network which contains at least one closed path. Thus every circuit is a network but not all networks are circuits. An example is shown in Figure 1.14.

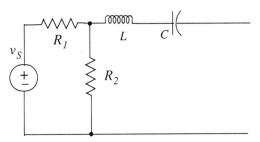

Figure 1.14. A network and a circuit

1.8 Active and Passive Networks

Active Network is a network which contains at least one active device (voltage or current source).

Passive Network is a network which does not contain any active device.

1.9 Necessary Conditions for Current Flow

There are two conditions which are necessary to set up and maintain a flow of current in a network or circuit. These are:

1. There must be a voltage source (potential difference) present to provide the electrical work which will force current to flow.

2. The circuit must be closed.

These conditions are illustrated in Figures 1.15 through 1.17.

Figure 1.15 shows a network which contains a voltage source but it is not closed and therefore, current will not flow.

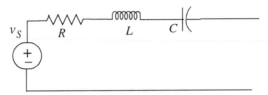

Figure 1.15. A network in which there is no current flow

Figure 1.16 shows a closed circuit but there is no voltage present to provide the electrical work for current to flow.

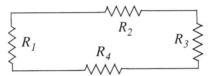

Figure 1.16. A closed circuit in which there is no current flow

Figure 1.17 shows a voltage source present and the circuit is closed. Therefore, both conditions are satisfied and current will flow.

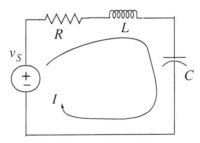

Figure 1.17. A circuit in which current flows

1.10 International System of Units

The *International System of Units* (abbreviated *SI* in all languages) was adopted by the General Conference on Weights and Measures in 1960. It is used extensively by the international scientific community. It was formerly known as the *Metric System*. The basic units of the *SI* system are listed in Table 1.1.

TABLE 1.1 SI Base Units

Unit of	Name	Abbreviation
Length	Metre	m
Mass	Kilogram	kg
Time	Second	s
Electric Current	Ampere	A
Temperature	Degrees Kelvin	°K
Amount of Substance	Mole	mol
Luminous Intensity	Candela	cd
Plane Angle	Radian	rad
Solid Angle	Steradian	sr

The *SI* uses larger and smaller units by various powers of 10 known as *standard prefixes*. The *common prefixes* are listed in Table 1.2 and the less frequently in Table 1.3. Table 1.4 shows some conversion factors between the *SI* and the English system. Table 1.5 shows typical temperature values in degrees Fahrenheit and the equivalent temperature values in degrees Celsius and degrees Kelvin. Other units used in physical sciences and electronics are derived from the SI base units and the most common are listed in Table 1.6.

TABLE 1.2 Most Commonly Used SI Prefixes

Value	Prefix	Symbol	Example
10^9	Giga	G	12 GHz (Gigahertz) = 12×10^9 Hz
10^6	Mega	M	*25 MΩ (Megaohms) = 25×10^6 Ω (ohms)*
10^3	Kilo	K	13.2 KV (Kilovolts) = 13.2×10^3 volts
10^{-2}	centi	c	2.8 cm (centimeters) = 2.8×10^{-2} meter
10^{-3}	milli	m	4 mH (millihenries) = 4×10^{-3} henry
10^{-6}	micro	μ	6 μw (microwatts) = 6×10^{-6} watt
10^{-9}	nano	n	2 ns (nanoseconds) = 2×10^{-9} second
10^{-12}	pico	p	3 pF (picofarads) = 3×10^{-12} Farad

TABLE 1.3 Less Frequently Used SI Prefixes

Value	Prefix	Symbol	Example
10^{18}	*Exa*	*E*	*1 Em (Exameter) = 10^{18} meters*
10^{15}	*Peta*	*P*	*5 Pyrs (Petayears) = 5×10^{15} years*
10^{12}	*Tera*	*T*	*3 T\$ (Teradollars) = 3×10^{12} dollars*
10^{-15}	*femto*	*f*	*7 fA (femtoamperes) = 7×10^{-15} ampere*
10^{-18}	*atto*	*a*	*9 aC (attocoulombs) = 9×10^{-18} coulomb*

TABLE 1.4 Conversion Factors

1 in. (inch)	*2.54 cm (centimeters)*
1 mi. (mile)	*1.609 Km (Kilometers)*
1 lb. (pound)	*0.4536 Kg (Kilograms)*
1 qt. (quart)	*946 cm^3 (cubic centimeters)*
1 cm (centimeter)	*0.3937 in. (inch)*
1 Km (Kilometer)	*0.6214 mi. (mile)*
1 Kg (Kilogram)	*2.2046 lbs (pounds)*
1 lt. (liter) = 1000 cm^3	*1.057 quarts*
1 Å (Angstrom)	*10^{-10} meter*
1 mm (micron)	*10^{-6} meter*

TABLE 1.5 Temperature Scale Equivalents

°F	°C	°K
−523.4	−273	0
32	0	273
0	−17.8	255.2
77	25	298
98.6	37	310
212	100	373

TABLE 1.6 SI Derived Units

Unit of	Name	Formula
Force	Newton (N)	$N = kg \cdot m/s^2$
Pressure or Stress	Pascal (Pa)	$Pa = N/m^2$
Work or Energy	Joule (J)	$J = N \cdot m$
Power	Watt (W)	$W = J/s$
Voltage	Volt (V)	$V = W/A$
Resistance	Ohm (Ω)	$\Omega = V/A$
Conductance	Siemens (S) or (Ω^{-1})	$S = A/V$
Capacitance	Farad (F)	$F = A \cdot s/V$
Inductance	Henry (H)	$H = V \cdot s/A$
Frequency	Hertz (Hz)	$Hz = 1/s$
Quantity of Electricity	Coulomb (C)	$C = A \cdot s$
Magnetic Flux	Weber (Wb)	$Wb = V \cdot s$
Magnetic Flux Density	Tesla (T)	$T = Wb/m^2$
Luminous Flux	Lumen (lm)	$lm = cd \cdot sr$
Illuminance	Lux (lx)	$lx = lm/m^2$
Radioactivity	Becquerel (Bq)	$Bq = s^{-1}$
Radiation Dose	Gray (Gy)	$S = J/kg$
Volume	Litre (L)	$L = m^3 \times 10^{-3}$

1.11 Sources of Energy

The principal sources of energy are from chemical processes (coal, fuel oil, natural gas, wood etc.) and from mechanical forms (water falls, wind, etc.). Other sources include nuclear and solar energy.

Example 1.6

A certain type of wood used in the generation of electric energy and we can get 12,000 BTUs from a pound (lb) of that wood when burned. Suppose that a computer system that includes a monitor, a printer, and other peripherals absorbs an average power of 500 w gets its energy from that burned wood and it is turned on for 8 hours. It is known that 1 BTU is equivalent to 778.3 ft-lb of energy, and 1 joule is equivalent to 0.7376 ft-lb.

Compute:

a. the energy consumption during this 8-hour interval

b. the cost for this energy consumption if the rate is $0.15 per kw-hr

c. the amount of wood in lbs burned during this time interval.

Solution:

a. Energy consumption for 8 hours is

$$Energy \ W = P_{ave}t = 500 \ w \ \times 8 \ hrs \ \times \frac{3600 \ s}{1 \ hr} = 14.4 \ Mjoules$$

b. Since $1 \ kilowatt - hour = 3.6 \times 10^6 \ joules$,

$$Cost = \frac{\$0.15}{kw - hr} \times \frac{1 \ kw - hr}{3.6 \times 10^6 \ joules} \times 14.4 \times 10^6 = \$0.60$$

c. Wood burned in 8 hours,

$$14.4 \times 10^6 \ joules \times 0.7376 \frac{ft-lb}{joule} \times \frac{1 \ BTU}{778.3 \ ft-lb} \times \frac{1 \ lb}{12000 \ BTU} = 1.137 \ lb$$

1.12 Summary

- Two identically charged (both positive or both negative) particles possess a charge of one coulomb when being separated by one meter in a vacuum, repel each other with a force of $10^{-7}c^2$ newton where $c = velocity \ of \ light \approx 3 \times 10^8 \ m/s$. Thus, the force with which two electrically charged bodies attract or repel one another depends on the product of the charges (in coulombs) in both objects, and also on the distance between the objects. If the polarities are the same (negative/negative or positive/positive), the so-called coulomb force is repulsive; if the polarities are opposite (negative/positive or positive/negative), the force is attractive. For any two charged bodies, the coulomb force decreases in proportion to the square of the distance between their charge centers.

- Electric current is defined as the instantaneous rate at which net positive charge is moving past this point in that specified direction, that is,

$$i = \frac{dq}{dt} = \lim_{\Delta t \to 0} \frac{\Delta q}{\Delta t}$$

- The unit of current is the ampere, abbreviated as A, and corresponds to charge q moving at the rate of one coulomb per second.

- In a two-terminal device the current entering one terminal is the same as the current leaving the other terminal.

- The voltage (potential difference) across a two-terminal device is defined as the work required to move a positive charge of one coulomb from one terminal of the device to the other terminal.

- The unit of voltage is the volt (abbreviated as V or v) and it is defined as

$$1 \ volt \ = \ \frac{1 \ joule}{1 \ coulomb}$$

- Power p is the rate at which *energy* (or *work*) W is expended. That is,

$$Power \ = \ p \ = \ \frac{dW}{dt}$$

- Absorbed power is proportional both to the current and the voltage needed to transfer one coulomb through the device. The unit of power is the watt and

$$1 \ watt \ = \ 1 \ volt \times 1 \ ampere$$

- The passive sign convention states that if the arrow representing the current i and the plus (+) minus (−) pair are placed at the device terminals in such a way that the current enters the device terminal marked with the plus (+) sign, and if both the arrow and the sign pair are labeled with the appropriate algebraic quantities, the power absorbed or delivered to the device can be expressed as $p \ = \ vi$. If the numerical value of this product is positive, we say that the device is absorbing power which is equivalent to saying that power is delivered to the device. If, on the other hand, the numerical value of the product $p \ = \ vi$ is negative, we say that the device delivers power to some other device.

- An ideal independent voltage source maintains the same voltage regardless of the amount of current that flows through it.

- An ideal independent current source maintains the same current regardless of the amount of voltage that appears across its terminals.

- The value of an dependent voltage source depends on another voltage or current elsewhere in the circuit.

- The value of an dependent current source depends on another current or voltage elsewhere in the circuit.

- Ideal voltage and current sources are just mathematical models. We will discuss practical voltage and current sources in Chapter 3.

- Independent and Dependent voltage and current sources are active devices; they normally (but not always) deliver power to some external device.

- Resistors, inductors, and capacitors are passive devices; they normally receive (absorb) power from an active device.

- A network is the interconnection of two or more simple devices.

- A circuit is a network which contains at least one closed path. Thus every circuit is a network but not all networks are circuits.

- An active network is a network which contains at least one active device (voltage or current source).

- A passive network is a network which does not contain any active device.

- To set up and maintain a flow of current in a network or circuit there must be a voltage source (potential difference) present to provide the electrical work which will force current to flow and the circuit must be closed.

- Linear devices are those in which there is a linear relationship between the voltage across that device and the current that flows through that device.

- The International System of Units is used extensively by the international scientific community. It was formerly known as the Metric System.

- The principal sources of energy are from chemical processes (coal, fuel oil, natural gas, wood etc.) and from mechanical forms (water falls, wind, etc.). Other sources include nuclear and solar energy.

1.13 Exercises

Multiple choice

1. The unit of charge is the

   A. ampere

   B. volt

   C. watt

   D. coulomb

   E. none of the above

2. The unit of current is the

   A. ampere

   B. coulomb

   C. watt

   D. joule

   E. none of the above

3. The unit of electric power is the

   A. ampere

   B. coulomb

   C. watt

   D. joule

   E. none of the above

4. The unit of energy is the

   A. ampere

   B. volt

   C. watt

   D. joule

   E. none of the above

5. Power is

   A. the integral of energy

B. the derivative of energy

C. current times some constant k

D. voltage times some constant k

E. none of the above

6. Active voltage and current sources

   A. always deliver power to other external devices

   B. normally deliver power to other external devices

   C. neither deliver or absorb power to or from other devices

   D. are just mathematical models

   E. none of the above

7. An ideal independent voltage source

   A. maintains the same voltage regardless of the amount of current that flows through it

   B. maintains the same current regardless of the voltage rating of that voltage source

   C. always delivers the same amount of power to other devices

   D. is a source where both voltage and current can be variable

   E. none of the above

8. An ideal independent current source

   A. maintains the same voltage regardless of the amount of current that flows through it

   B. maintains the same current regardless of the voltage that appears across its terminals

   C. always delivers the same amount of power to other devices

   D. is a source where both voltage and current can be variable

   E. none of the above

9. The value of a dependent voltage source can be denoted as

   A. kV where k is a conductance value

   B. kI where k is a resistance value

   C. kV where k is an inductance value

   D. kI where k is a capacitance value

E. none of the above

10. The value of a dependent current source can be denoted as

A. kV where k is a conductance value

B. kI where k is a resistance value

C. kV where k is an inductance value

D. kI where k is a capacitance value

E. none of the above

Problems

1. A two terminal device consumes energy as shown by the waveform of Figure 1.18 below, and the current through this device is $i(t) = 2\cos 4000\pi t \ A$. Find the voltage across this device at t = 0.5, 1.5, 4.75 and 6.5 ms. Answers: *2.5 V, 0 V, 2.5 V, –2.5 V*

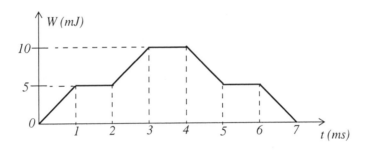

Figure 1.18. Waveform for Problem 1

2. A household light bulb is rated 75 watts at 120 volts. Compute the number of electrons per second that flow through this bulb when it is connected to a 120 volt source.

Answer: 3.9×10^{18} *electrons/s*

3. An airplane, whose total mass is *50,000* metric tons, reaches a height of 32,808 feet in 20 minutes after takeoff.

a. Compute the potential energy that the airplane has gained at this height. Answer: *1, 736 MJ*

b. If this energy could be converted to electric energy with a conversion loss of 10%, how much would this energy be worth at $0.15 per kilowatt-hour? Answer: *$65.10*

c. If this energy were converted into electric energy during the period of 20 minutes, what average number of kilowatts would be generated? Answer: *1, 450 Kw*

4. The power input to a television station transmitter is 125 kw and the output is 100 kw which is transmitted as radio frequency power. The remaining 25 kw of power is converted into heat.

   a. How many BTUs per hour does this transmitter release as heat? *1 BTU = 1054.8 J*

     Answer: *85,234 BTU/hr*

   b. How many electron-volts per second is this heat equivalent to?

$$1 \; electron-volt \; = \; 1.6 \times 10^{-19} \; J \quad \text{Answer: } 1.56 \times 10^{23} \frac{electron-volts}{sec.}$$

1.14 Answers to Exercises

Dear Reader:

The remaining pages on this chapter contain answers to the multiple-choice questions and solutions to the exercises.

You must, for your benefit, make an honest effort to answer the multiple-choice questions and solve the problems without first looking at the solutions that follow. It is recommended that first you go through and answer those you feel that you know. For the multiple-choice questions and problems that you are uncertain, review this chapter and try again. If your answers to the problems do not agree with those provided, look over your procedures for inconsistencies and computational errors. Refer to the solutions as a last resort and rework those problems at a later date.

You should follow this practice with the multiple-choice and problems on all chapters of this book.

Chapter 1 Basic Concepts and Definitions

Multiple choice

1. D

2. A

3. C

4. D

5. B

6. B

7. A

8. B

9. B

10. A

Problems

1.

$$v = \frac{p}{i} = \frac{dW/dt}{i} = \frac{slope}{i}$$

a.

$$slope\Big|_{0}^{1\ ms} = \frac{5\ mJ}{1\ ms} = 5\ J/s$$

$$v\Big|_{t\,=\,0.5\ ms} = \frac{5\ J/s}{2\cos 4000\pi(0.5\times 10^{-3})\ A} = \frac{5\ J/s}{2\cos 2\pi\ A} = \frac{5\ J/s}{2\ A} = 2.5\ V$$

b.

$$slope\Big|_{1}^{2\ ms} = 0$$

$$v\Big|_{t\,=\,1.5\ ms} = \frac{0}{i} = 0\ V$$

c.

$$slope\Big|_{4}^{5\ ms} = \frac{-5\ mJ}{1\ ms} = -5\ J/s$$

$$v\Big|_{t\,=\,4.75\ ms} = \frac{-5\ J/s}{2\cos 4000\pi(4.75\times 10^{-3})\ A} = \frac{-5\ J/s}{2\cos 19\pi\ A} = \frac{-5\ J/s}{2\cos\pi\ A} = \frac{-5\ J/s}{-2\ A} = 2.5\ V$$

d.

$$slope\Big|_6^{7 \ ms} = \frac{-5 \ mJ}{1 \ ms} = -5 \ J/s$$

$$v\Big|_{t \ = \ 6.5 \ ms} = \frac{-5 \ J/s}{2\cos4000\pi(6.5 \times 10^{-3}) \ A} = \frac{-5 \ J/s}{2\cos26\pi \ A} = \frac{-5 \ J/s}{2 \ A} = -2.5 \ V$$

2.

$$i = \frac{p}{v} = \frac{75 \ w}{120 \ V} = \frac{5}{8} \ A$$

$$q = \int_{t_0}^t i dt$$

$$q\Big|_{t \ = \ 1 \ s} = \int_0^{1 \ s} \frac{5}{8} dt = \frac{5}{8} t \Big|_0^{1 \ s} = \frac{5}{8} \ C/s$$

$$\frac{5}{8} \ C/s \times \frac{6.24 \times 10^{18} \ electrons}{1 \ C} = 3.9 \times 10^{18} \ electrons/s$$

3.

$$W_p = W_k = \frac{1}{2} m v^2$$

where $m = mass \ in \ kg$ and $v = velocity \ in \ meters/sec.$

$$33,808 \ ft \times \frac{0.3048 \ m}{ft} = 10,000 \ m = 10 \ Km$$

$$20 \ minutes \times \frac{60 \ sec.}{min} = 1,200 \ sec.$$

$$v = \frac{10,000 \ m}{1,200 \ sec.} = \frac{25}{3} \ m/s$$

$$50,000 \ metric \ tons \times \frac{1,000 \ Kg}{metric \ ton} = 5 \times 10^7 \ Kg$$

Then,

a.

$$W_p = W_k = \frac{1}{2}(5 \times 10^7)\left(\frac{25}{3}\right)^2 = 173.61 \times 10^7 \ J \approx 1,736 \ MJ$$

b.

$$1 \; joule \; = \; 1 \; watt\text{-}sec$$

$$1,736 \times 10^6 J \times \frac{1 \; watt\text{-}sec}{1 \; joule} \times \frac{1 \; Kw}{1,000 \; w} \times \frac{1 \; hr}{3,600 \; sec.} \; = \; 482.22 \; Kw\text{-}hr$$

and with 10% conversion loss, the useful energy is

$$482.22 \times 0.9 \; = \; 482.22 \times 0.9 \; = \; 434 \; Kw\text{-}hr$$

$$Cost \; of \; Energy \; = \; \frac{\$0.15}{Kw\text{-}hr} \times 434 \; Kw\text{-}hr \; = \; \$65.10$$

c.

$$P_{ave} \; = \; \frac{W}{t} \; = \; \frac{1,736 \; MJ}{20 \; min \times \dfrac{60 \; sec}{min}} \; = \; 1.45 \; Mw \; = \; 1450 \; Kw$$

4.

a.

$$1 \; BTU \; = \; 1054.8 \; J$$

$$25,000 \; watts \times \frac{1 \; joule/sec.}{watt} \times \frac{1 \; BTU}{1054.8 \; J} \times \frac{3600 \; sec.}{1 \; hr} \; = \; 85,234 \; BTU/hr$$

b.

$$1 \; electron - volt \; = \; 1.6 \times 10^{-19} \; J$$

$$\frac{1 \; electron - volt}{sec.} \; = \; \frac{1.6 \times 10^{-19} \; J}{sec.} \; = \; 1.6 \times 10^{-19} \; watt$$

$$25,000 \; watts \times \frac{1 \; electron - volt/ \; sec.}{1.6 \times 10^{-19} \; watt} \; = \; 1.56 \times 10^{23} \frac{electron - volts}{sec.}$$

Chapter 2

Analysis of Simple Circuits

T his chapter defines constant and instantaneous values, Ohm's law, and Kirchhoff's Current and Voltage laws. Series and parallel circuits are also defined and nodal, mesh, and loop analyses are introduced. Combinations of voltage and current sources and resistance combinations are discussed, and the voltage and current division formulas are derived.

2.1 Conventions

We will use lower case letters such as v, i, and p to denote *instantaneous values* of voltage, current, and power respectively, and we will use subscripts to denote specific voltages, currents, resistances, etc. For example, v_S and i_S will be used to denote voltage and current sources respectively. Notations like v_{R1} and i_{R2} will be used to denote the voltage across resistance R_1 and the current through resistance R_2 respectively. Other notations like v_A or v_1 will represent the voltage (potential difference) between point A or point 1 *with respect* to some arbitrarily chosen reference point taken as "zero" volts or "ground".

The designations v_{AB} or v_{12} will be used to denote the voltage between point A or point 1 *with respect* to point B or 2 respectively. We will denote voltages as $v(t)$ and $i(t)$ whenever we wish to emphasize that these quantities are time dependent. Thus, sinusoidal (AC) voltages and currents will be denoted as $v(t)$ and $i(t)$ respectively. Phasor quantities, to be introduced in Chapter 6, will be represented with bold capital letters, V for phasor voltage and I for phasor current.

2.2 Ohm's Law

We recall from Chapter 1 that *resistance* R is a *constant* that relates the voltage and the current as:

$$\boxed{v_R = Ri_R} \tag{2.1}$$

This relation is known as *Ohm's law.*

The unit of resistance is the *Ohm* and its symbol is the Greek capital letter Ω. One ohm is the resistance of a conductor such that a constant current of one ampere through it produces a voltage of one volt between its ends. Thus,

$$\boxed{1\ \Omega = \frac{1\ V}{1\ A}} \tag{2.2}$$

Chapter 2 Analysis of Simple Circuits

Physically, a *resistor* is a device that opposes current flow. Resistors are used as a *current limiting devices* and *as voltage dividers*.

In the previous chapter we defined *conductance* G as the constant that relates the current and the voltage as

$$\boxed{i_G = Gv_G}\tag{2.3}$$

This is another form of Ohm's law since by letting $i_G = i_R$ and $v_G = v_R$, we get

$$\boxed{G = \frac{1}{R}}\tag{2.4}$$

The unit of conductance is the *siemens* or *mho* (ohm spelled backwards) and its symbol is S or Ω^{-1} Thus,

$$\boxed{1\,\Omega^{-1} = \frac{1\,A}{1\,V}}\tag{2.5}$$

Resistances (or conductances) are commonly used to define an "open circuit" or a "short circuit". An *open circuit* is an adjective describing the "open space" between a pair of terminals, and can be thought of as an "infinite resistance" or "zero conductance". In contrast, a *short circuit* is an adjective describing the connection of a pair of terminals by a piece of wire of "infinite conductance" or a piece of wire of "zero" resistance.

The current through an "open circuit" is always zero but the voltage across the open circuit terminals may or may not be zero. Likewise, the voltage across a short circuit terminals is always zero but the current through it may or may not be zero. The open and short circuit concepts and their equivalent resistances or conductances are shown in Figure 2.1.

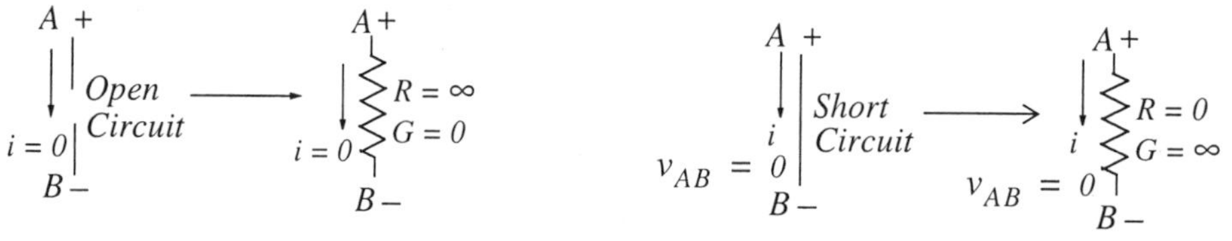

Figure 2.1. The concepts of open and short circuits

The fact that current does not flow through an open circuit and that zero voltage exists across the terminals of a short circuit, can also be observed from the expressions $v_R = Ri_R$ and $i_G = Gv_G$.

That is, since $G = \frac{1}{R}$, infinite R means zero G and zero R means infinite G. Then, for a finite voltage, say v_G, and an open circuit,

$$\lim_{G \to 0} i_G = \lim_{G \to 0} Gv_G = 0 \qquad (2.6)$$

Likewise, for a finite current, say i_R, and a short circuit,

$$\lim_{R \to 0} v_R = \lim_{R \to 0} Ri_R = 0 \qquad (2.7)$$

Reminder:

We must remember that the expressions

$$v_R = Ri_R$$

and

$$i_G = Gv_G$$

are true only when the passive sign convention is observed. This is consistent with our classification of R and G being passive devices and thus $v_R = Ri_R$ implies the current direction and voltage polarity are as shown in Figure 2.2.

Figure 2.2. Voltage polarity and current direction in accordance to passive sign convention

But if the voltage polarities and current directions are as shown in Figure 2.3, then,

$$v_R = -Ri_R \qquad (2.8)$$

Figure 2.3. Voltage polarity and current direction not in accordance to passive sign convention

Note: "Negative resistance," as shown in (2.8), can be thought of as being a math model that supplies energy.

2.3 Power Absorbed by a Resistor

A resistor, being a passive device, absorbs power. This absorbed power can be found from Ohm's law, that is,

$$v_R = Ri_R$$

and the power relation

$$p_R = v_R i_R$$

Then,

$$p_R = v_R i_R = (Ri_R) i_R = Ri_R^2 = v_R \left(\frac{v_R}{R}\right) = \frac{v_R^2}{R} \qquad (2.9)$$

The voltage, current, resistance and power relations are arranged in the pie chart shown in Figure 2.4.

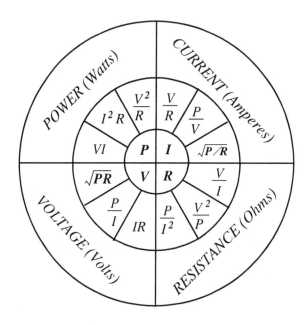

Figure 2.4. Pie chart for showing relations among voltage, current, resistance, and power

Note:

A resistor, besides its resistance rating (ohms) has a power rating in watts commonly referred to as the *wattage* of the resistor. Common resistor wattage values are ¼ watt, ½ watt, 1 watt, 2 watts, 5 watts and so on. This topic will be discussed in Section 2.16.

2.4 Energy Dissipated in a Resistor

A resistor, by its own nature, dissipates energy in the form of heat; it never stores energy. The energy dissipated in a resistor during a time interval, say from t_1 to t_2, is given by the integral of the instantaneous power p_R. Thus,

$$W_{R\ diss} = \int_{t_1}^{t_2} p_R \, dt \qquad (2.10)$$

If the power is constant, say P, then (2.10) reduces to

$$W_{R\ diss} = Pt \tag{2.11}$$

Alternately, if the energy is known, we can find the power by taking the derivative of the energy, that is,

$$p_R = \frac{d}{dt}W_{R\ diss} \tag{2.12}$$

Reminder:

When using all formulas, we must express the quantities involved in their primary units. For instance in (2.11) above, the energy is in joules when the power is in *watts* and the time is in *seconds*.

2.5 Nodes, Branches, Loops and Meshes

Definition 2.1

A *node* is the common point at which two or more devices (passive or active) are connected. An example of a node is shown in Figure 2.5.

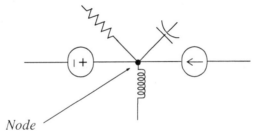

Figure 2.5. Definition of node

Definition 2.2

A *branch* is a simple path composed of one single device as shown in Figure 2.6.

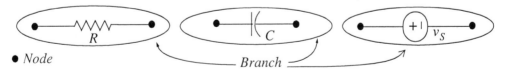

Figure 2.6. Definition of branch

Definition 2.3

A *loop* is a closed path formed by the interconnection of simple devices. For example, the network shown in Figure 2.7 is a loop.

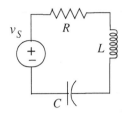

Figure 2.7. Definition of a loop

Definition 2.4

A *mesh* is a loop which does not enclose any other loops. For example, in the circuit shown in Figure 2.8, *ABEF* is both a loop and a mesh, but *ABCDEF* is a loop but not a mesh.

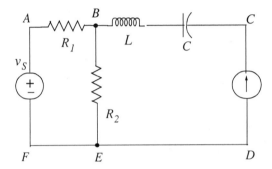

Figure 2.8. Example showing the difference between mesh and loop

2.6 Kirchhoff's Current Law (KCL)

KCL states that *the algebraic sum of all currents leaving (or entering) a node is equal to zero*. For example, in Figure 2.9, if we assign a plus (+) sign to the currents *leaving* the node, we must assign a minus (−) sign to the currents *entering* the node. Then by KCL,

$$-i_1 - i_2 + i_3 + i_4 = 0 \tag{2.13}$$

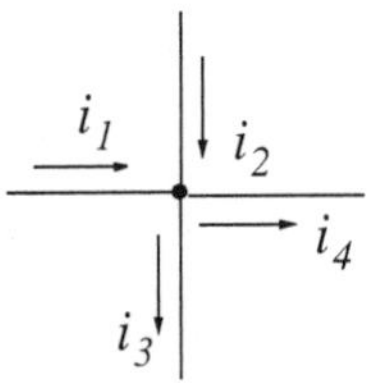

Figure 2.9. Node to illustrate KCL

But if we assign a plus (+) sign to the currents *entering* the node and minus (−) sign to the currents *leaving* the node, then by KCL,

$$i_1 + i_2 - i_3 - i_4 = 0 \tag{2.14}$$

or

$$-i_1 - i_2 + i_3 + i_4 = 0 \qquad (2.15)$$

We observe that (2.13) and (2.15) are the same; therefore, it does not matter which we choose as plus (+).

Convention:

In our subsequent discussion we will assign plus (+) signs to the currents *leaving* the node.

2.7 Kirchhoff's Voltage Law (KVL)

KVL states that *the algebraic sum of the voltage drops (voltages from + to −) or voltage rises (voltages from − to +) around any closed path (mesh or loop) in a circuit is equal to zero.* For example, in the circuit shown in Figure 2.10, voltages v_1, v_2, v_3, and v_4 represent the voltages across devices 1, 2, 3, and 4 respectively, and have the polarities shown.

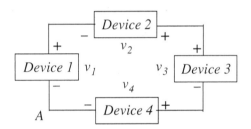

Figure 2.10. Circuit to illustrate KVL

Now, if we assign a (+) sign to the voltage drops, we must assign a (−) sign to the voltage rises. Then, by KVL starting at node *A* and going *clockwise* we get:

$$-v_1 - v_2 + v_3 + v_4 = 0 \qquad (2.16)$$

or going *counterclockwise*, we get:

$$-v_4 - v_3 + v_2 + v_1 = 0 \qquad (2.17)$$

Alternately, if we assign a (+) sign to the voltage rises, we must assign a (−) sign to the voltage drops. Then, by KVL starting again at node A and going *clockwise* we get:

$$v_1 + v_2 - v_3 - v_4 = 0 \qquad (2.18)$$

or going *counterclockwise*, we get:

$$v_4 + v_3 - v_2 - v_1 = 0 \qquad (2.19)$$

We observe that expressions (2.16) through (2.19) are the same.

Convention:

In our subsequent discussion we will assign plus (+) signs to voltage drops.

Definition 2.5

Two or more devices are said to be *connected in series* if and only if the same current flows through them. For example, in the circuit of Figure 2.11, the same current i flows through the voltage source, the resistance, the inductance and the capacitance. Accordingly, this is classified as a series circuit.

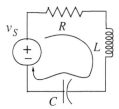

Figure 2.11. A simple series circuit

Definition 2.6

Two or more devices are said to be *connected in parallel* if and only if the same voltage exists across each of the devices. For example, in the circuit of Figure 2.12, the same voltage v_{AB} exists across the current source, the conductance, the inductance, and the capacitance and therefore it is classified as a parallel circuit

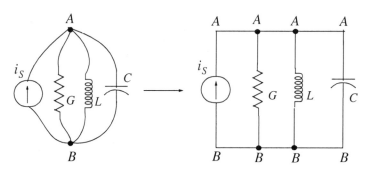

Figure 2.12. A simple parallel circuit

Convention:

In our subsequent discussion we will adopt the conventional current flow, i.e., the current that flows from a higher (+) to a lower (−) potential. For example, if in Figure 2.13 we are given the indicated polarity,

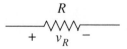

Figure 2.13. Device with established voltage polarity

then, the current arrow will be pointing to the right direction as shown in Figure 2.14.

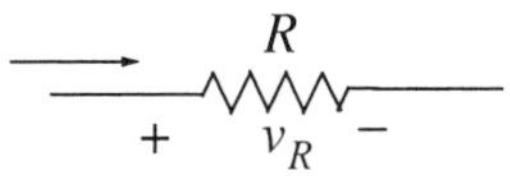

$$\xrightarrow{} \underset{+ \quad v_R \quad -}{-\!\!\!/\!\!\!\bigvee\!\!\!\bigvee\!\!\!\bigvee\!\!\!-}^{R}$$

Figure 2.14. Direction of conventional current flow in device with established voltage polarity

Alternately, if current flows in an assumed specific direction through a device thus producing a voltage, we will assign a (+) sign at the terminal of the device at which the current enters. For example, if we are given this designation a device in which the current direction has been established as shown in Figure 2.15,

$$\xrightarrow{} -\!\!\!/\!\!\!\bigvee\!\!\!\bigvee\!\!\!\bigvee\!\!\!-^{R}$$

Figure 2.15. Device with established conventional current direction

then we assign (+) and (−) as shown in Figure 2.16.

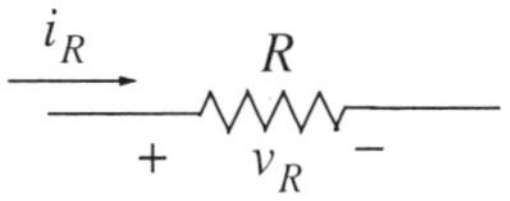

$$\overset{i_R}{\xrightarrow{}} \underset{+ \quad v_R \quad -}{-\!\!\!/\!\!\!\bigvee\!\!\!\bigvee\!\!\!\bigvee\!\!\!-}^{R}$$

Figure 2.16. Voltage polarity in a device with established conventional current flow

Note: Active devices, such as voltage and current sources, have their voltage polarity and current direction respectively, established as part of their notation. The current through and the voltage across these devices can easily be determined *if these devices deliver power to the rest of the circuit.* Thus with the voltage polarity as given in the circuit of Figure 2.17 (a), we assign a clockwise direction to the current as shown in Figure 2.17 (b). This is consistent with the passive sign convention since we have assumed that the voltage source delivers power to the rest of the circuit.

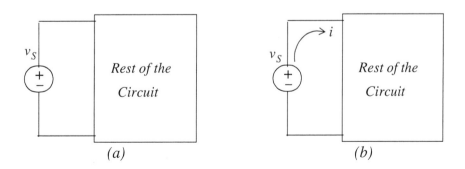

Figure 2.17. Direction of conventional current flow produced by voltage sources

Likewise, in the circuit of Figure 2.18 (a) below, the direction of the current source is clockwise, and assuming that this source delivers power to the rest of the circuit, we assign the voltage polarity shown in Figure 2.18 (b) to be consistent with the passive sign convention.

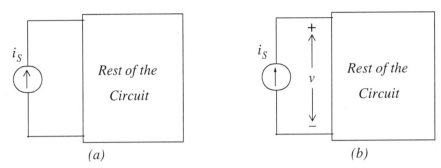

(a)   *(b)*

Figure 2.18. Voltage polarity across current sources

The following facts were discussed in the previous chapter but they are repeated here for emphasis.

There are two conditions required to setup and maintain the flow of an electric current:

1. *There must be some voltage (potential difference) to provide the energy (work) which will force electric current to flow in a specific direction in accordance with the conventional current flow (from a higher to a lower potential).*

2. *There must be a continuous (closed) external path for current to flow around this path (mesh or loop).*

The external path is usually made of two parts: (a) the *metallic wires* and (b) the *load* to which the electric power is to be delivered in order to accomplish some useful purpose or effect. The load may be a resistive, an inductive, or a capacitive circuit, or a combination of these.

2.8 Single Mesh Circuit Analysis

We will use the following example to develop a step-by-step procedure for analyzing (finding current, voltage drops and power) in a circuit with a single mesh.

Example 2.1

For the series circuit shown in Figure 2.19, we want to find:

a. The current i which flows through each device

b. The voltage drop across each resistor

c. The power absorbed or delivered by each device

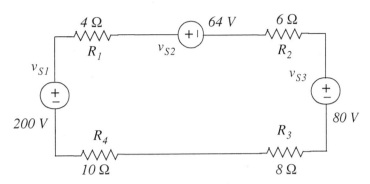

Figure 2.19. Circuit for Example 2.1

Solution:

a. Step 1: We do not know which voltage source(s) deliver power to the other sources, so let us assume that the current i flows in the clockwise direction[*] as shown in Figure 2.20.

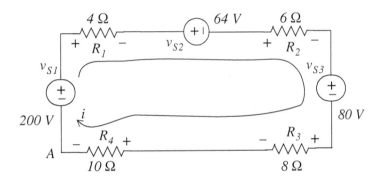

Figure 2.20. Circuit for Example 2.1 with assumed current direction

Step 2: We assign (+) and (−) polarities at each resistor's terminal in accordance with the established passive sign convention.

Step 3: By application of KVL and the adopted conventions, starting at node A and going clockwise, we get:

$$-v_{S1} + v_{R1} + v_{S2} + v_{R2} + v_{S3} + v_{R3} + v_{R4} = 0 \qquad (2.20)$$

and by Ohm's law,

$$v_{R1} = R_1 i \qquad v_{R2} = R_2 i \qquad v_{R3} = R_3 i \qquad v_{R4} = R_4 i$$

Then, by substitution of given values into (2.20), we get

---

* *Henceforth, the current direction will be assumed to be that of the conventional current flow.*

$$-200 + 4i + 64 + 6i + 80 + 8i + 10i = 0$$

or

$$28i = 56$$

or

$$i = 2 \ A \tag{2.21}$$

b. Knowing the current i from part (a), we can now compute the voltage drop across each resistor using Ohm's law $v = Ri$.

$$v_{R1} = 4 \times 2 = 8 \ \text{V} \qquad v_{R2} = 6 \times 2 = 12 \ \text{V}$$
$$v_{R3} = 8 \times 2 = 16 \ \text{V} \qquad v_{R4} = 10 \times 2 = 20 \ \text{V} \tag{2.22}$$

c. The power absorbed (or delivered) by each device can be found from the power relation $p = vi$. Then, the power absorbed by each resistor is

$$p_{R1} = 8 \times 2 = 16 \ \text{w} \qquad p_{R2} = 12 \times 2 = 24 \ \text{w}$$
$$p_{R3} = 16 \times 2 = 32 \ \text{w} \qquad p_{R4} = 20 \times 2 = 40 \ \text{w} \tag{2.23}$$

and the power delivered (or absorbed) by each voltage source is

$$p_{V_{S1}} = -200 \times 2 = -400 \ \text{w} \qquad p_{V_{S2}} = 64 \times 2 = 128 \ \text{w} \qquad p_{V_{S3}} = 80 \times 2 = 160 \ \text{w} \tag{2.24}$$

From (2.24), we observe that the 200 volt source absorbs -400 watts of power. This means that this source delivers (supplies) 400 watts to the rest of the circuit. However, the other two voltage sources receive (absorb) power from the 200 volt source. Table 2.1 shows that the conservation of energy principle is satisfied since the total absorbed power is equal to the power delivered.

Example 2.2

Repeat Example 2.1 with the assumption that the current i flows counterclockwise.

Solution:

We denote the current as i' (i prime) for this example. Then, starting at Node A and going counterclockwise, the voltage drops across each resistor are as indicated in Figure 2.21.

Repeating Steps 2 and 3 of Example 2.1, we get:

$$v_{R4} + v_{R3} - v_{S3} + v_{R2} - v_{S2} + v_{R1} + v_{S1} = 0 \tag{2.25}$$

Next, by Ohm's law,

$$v_{R1} = R_1 i' \qquad v_{R2} = R_2 i' \qquad v_{R3} = R_3 i' \qquad v_{R4} = R_4 i'$$

TABLE 2.1 *Power delivered or absorbed by each device on the circuit of Figure 2.19*

Device	*Power Delivered (watts)*	*Power Absorbed (watts)*
200 V Source	400	
64 V Source		128
80 V Source		160
4 Ω Resistor		16
6 Ω Resistor		24
8 Ω Resistor		32
10 Ω Resistor		40
Total	400	400

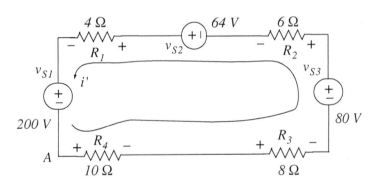

Figure 2.21. Circuit for Example 2.2

By substitution of given values, we get

$$200 + 4i' - 64 + 6i' - 80 + 8i' + 10i' = 0$$

or

$$28i' = -56$$

or

$$i' = -2A \qquad (2.26)$$

Comparing (2.21) with (2.26) we see that $i' = -i$ as expected.

Definition 2.7

A single node-pair circuit is one in which any number of simple elements are connected between the same pair of nodes. For example, the circuit of Figure 2.22 (a), which is more conveniently shown as Figure 2.22 (b), is a single node-pair circuit.

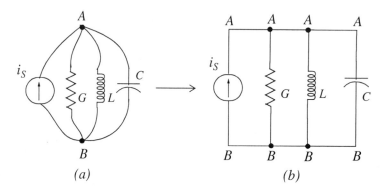

Figure 2.22. Circuit with a single node-pair

2.9 Single Node-Pair Circuit Analysis

We will use the following example to develop a step-by-step procedure for analyzing (finding currents, voltage drop and power) in a circuit with a single node-pair.

Example 2.3

For the parallel circuit shown in Figure 2.23, find:

a. The voltage drop across each device

b. The current i which flows through each conductance

c. The power absorbed or delivered by each device

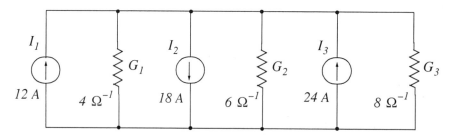

Figure 2.23. Circuit for Example 2.3

Solution:

a. Step 1: We denote the single node-pair with the letters A and B as shown in Figure 2.24. It is important to observe that the same voltage (or potential difference) exists across each device. Node B is chosen as our reference node and it is convenient to assume that this reference node is at zero potential (ground) as indicated by the symbol $\perp\!\!\!=$

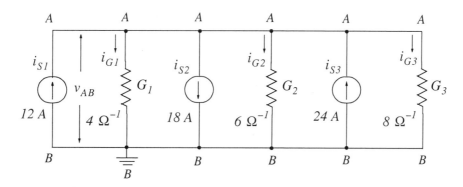

Figure 2.24. Circuit for Example 2.3 with assumed current directions

Step 2: We assign currents through each of the conductances G_1, G_2, and G_3 in accordance with the conventional current flow. These currents are shown as i_{G1}, i_{G2}, and i_{G3}.

Step 3: By application of KCL and in accordance with our established convention, we choose node A which is the plus (+) reference point and we form the algebraic sum of the currents leaving (or entering) this node. Then, with plus (+) assigned to the currents leaving this node and with minus (−) entering this node we get

$$-i_{S1} + i_{G1} + i_{S2} + i_{G2} - i_{S3} + i_{G3} = 0 \tag{2.27}$$

and since

$$i_{G1} = G_1 v_{AB} \quad i_{G2} = G_2 v_{AB} \quad i_{G3} = G_3 v_{AB} \tag{2.28}$$

by substitution into (2.27),

$$-i_{S1} + G_1 v_{AB} + i_{S2} + G_2 v_{AB} - i_{S3} + G_3 v_{AB} = 0 \tag{2.29}$$

Solving for v_{AB}, we get

$$v_{AB} = \frac{i_{S1} - i_{S2} + i_{S3}}{G_1 + G_2 + G_3} \tag{2.30}$$

and by substitution of the given values, we get

$$v_{AB} = \frac{12 - 18 + 24}{4 + 6 + 8} \tag{2.31}$$

or

$$v_{AB} = 1 \ V \tag{2.32}$$

b. From (2.28),

$$i_{G1} = 4 \qquad i_{G2} = 6 \qquad i_{G3} = 8 \tag{2.33}$$

and we observe that with these values, (2.27) is satisfied.

c. The power absorbed (or delivered) by each device can be found from the power relation $p = vi$. Then, the power absorbed by each conductance is

$$p_{G1} = 1 \times 4 = 4 \ \text{w}$$
$$p_{G2} = 1 \times 6 = 6 \ \text{w} \tag{2.34}$$
$$p_{G3} = 1 \times 8 = 8 \ \text{w}$$

and the power delivered (or absorbed) by each current source is

$$p_{I1} = 1 \times (-12) = -12 \ \text{w}$$
$$p_{I2} = 1 \times 18 = 18 \ \text{w} \tag{2.35}$$
$$p_{I3} = 1 \times (-24) = -24 \ \text{w}$$

From (2.35) we observe that the *12 A* and *24 A* current sources absorb *–12 w* and *–24 w* respectively. This means that these sources deliver (supply) a total of *36 w* to the rest of the circuit. The *18 A* source absorbs power.

Table 2.2 shows that the conservation of energy principle is satisfied since the absorbed power is equal to the power delivered.

2.10 Voltage and Current Source Combinations

Definition 2.8

Two or more voltage sources connected in series are said to be *series aiding* when the plus (+) terminal of any one voltage source is connected to the minus (–) terminal of another, or when the minus (–) terminal of any one voltage source is connected to the plus (+) terminal of another.

Two or more series aiding voltage sources may be replaced by an equivalent voltage source whose value is the algebraic sum of the individual voltage sources as shown in Figure 2.25.

TABLE 2.2 *Power delivered or absorbed by each device of Figure 2.23*

Device	Power Delivered (watts)	Power Absorbed (watts)
12 A Source	12	
18 A Source		18
24 A Source	24	
4 Ω^{-1} Conductance		4
6 Ω^{-1} Conductance		6
8 Ω^{-1} Conductance		8
Total	36	36

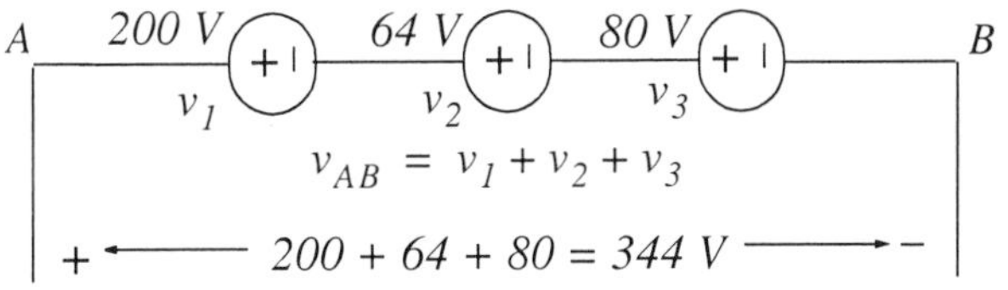

Figure 2.25. Addition of voltage sources in series when all have same polarity

A good example of combining voltage sources as series aiding is when we connect several AA size batteries each rated at *1.5 v* to power up a hand calculator, or a small flashlight.

Definition 2.9

Two or more voltage sources connected in series are said to be *series opposing* when the plus (+) terminal of one voltage source is connected to the plus (+) terminal of the other voltage source or when the minus (−) of one voltage source is connected to the minus (−) terminal of the other voltage source. Two series opposing voltage sources may be replaced by an equivalent voltage source whose value is the algebraic difference of the individual voltage sources as shown in Figure 2.26.

Definition 2.10

Two or more current sources connected in parallel are said to be *parallel aiding* when the arrows indicating the direction of the current flow have the same direction. They can be combined into a single current source as shown in Figure 2.27.

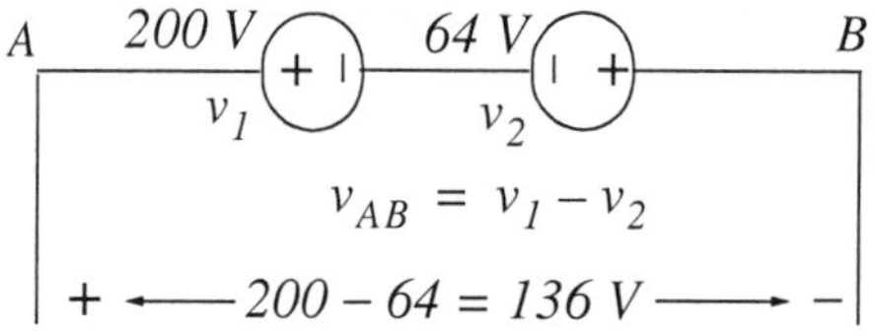

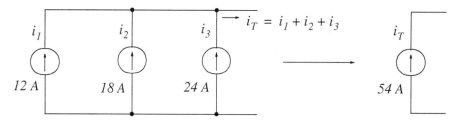

Figure 2.26. Addition of voltage sources in series when they have different polarity

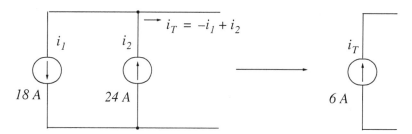

Figure 2.27. Addition of current sources in parallel when all have same direction

Definition 2.11

Two or more current sources connected in parallel are said to be *parallel opposing* when the arrows indicating the direction of the current flow have opposite direction. They can be replaced by an equivalent current source whose value is the algebraic difference of the individual current sources as shown in Figure 2.28.

Figure 2.28. Addition of current sources in parallel when they have opposite direction

2.11 Resistance and Conductance Combinations

Often, resistors are connected in series or in parallel. With either of these connections, series or parallel, it is possible to replace these resistors by a single resistor to simplify the computations of the voltages and currents. Figure 2.29 shows *n* resistors connected in series.

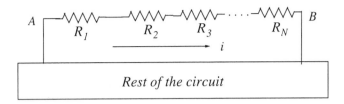

Figure 2.29. Addition of resistances in series

The combined or equivalent resistance R_{eq} is

$$R_{eq} = \frac{v_{AB}}{i} = \frac{v_{R1}}{i} + \frac{v_{R2}}{i} + \frac{v_{R3}}{i} + \dots + \frac{v_{Rn}}{i}$$

or

$$R_{eq} = R_1 + R_2 + R_3 + \dots + R_n = \sum_{k=1}^{n} R_K \tag{2.36}$$

For Resistors in Series

Example 2.4

For the circuit of Figure 2.30, find the value of the current i after combining the voltage sources to a single voltage source and the resistances to a single resistor.

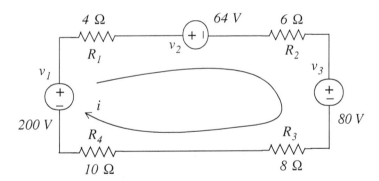

Figure 2.30. Circuit for Example 2.4

Solution:

We add the values of the voltage sources as indicated in Definitions 8 and 9, we add the resistances in accordance with (2.36), and we apply Ohm's law. Then,

$$i = \frac{\Sigma v}{\Sigma R} = \frac{200 - (64 + 80)}{28} = \frac{56}{28} = 2\ A \tag{2.37}$$

Next, we consider thecase where n resistors are connected in parallel as shown in Figure 2.31.

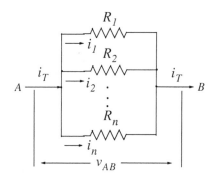

Figure 2.31. Addition of resistances in parallel

By KCL,

$$i_T = i_1 + i_2 + \dots + i_n \qquad (2.38)$$

The same voltage exists across each resistor; therefore, dividing each term of (2.38) by V_{AB}, we get

$$\frac{i_T}{v_{AB}} = \frac{i_1}{v_{AB}} + \frac{i_2}{v_{AB}} + \dots + \frac{i_n}{v_{AB}} \qquad (2.39)$$

and since $v/i = R$, then $i/v = 1/R$ and thus (2.39) can be written as

$$\frac{1}{R_{AB}} = \frac{1}{R_1} + \frac{1}{R_2} + \dots + \frac{1}{R_n}$$

or

$$\boxed{\frac{1}{R_{eq}} = \frac{1}{R_1} + \frac{1}{R_2} + \dots + \frac{1}{R_n} \atop \textit{For Resistors in Parallel}} \qquad (2.40)$$

For the special case of two parallel resistors, (2.40) reduces to

$$\frac{1}{R_{eq}} = \frac{1}{R_1} + \frac{1}{R_2}$$

or

$$\boxed{R_{eq} = R_1 \| R_2 = \frac{R_1 \cdot R_2}{R_1 + R_2}} \qquad (2.41)$$

where the designation $R_1 \| R_2$ indicates that R_1 and R_2 are in parallel.

Also, since $G = 1/R$, from (2.38),

$$G_{eq} = G_1 + G_2 + \ldots + G_n = \sum_{k=1}^{n} G_k \qquad (2.42)$$

that is, *parallel conductances combine as series resistors do.*

Example 2.5

In the circuit of Figure 2.32,

a. Replace all resistors with a single equivalent resistance R_{eq}

b. Compute the voltage v_{AB} across the current source.

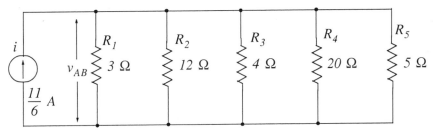

Figure 2.32. Circuit for Example 2.5

Solution:

a. We could use (2.40) to find the equivalent resistance R_{eq}. However, it is easier to form groups of two parallel resistors as shown in Figure 2.33 and use (2.41) instead.

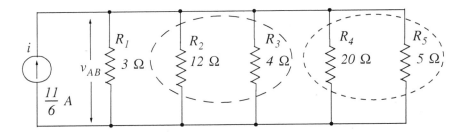

Figure 2.33. Groups of parallel combinations for the circuit of Example 2.5.

Then,

$$R_2 \| R_3 = \frac{12 \times 4}{12 + 4} = 3 \ \Omega$$

Also,

$$R_4 \| R_5 = \frac{20 \times 5}{20 + 5} = 4 \ \Omega$$

and the circuit reduces to that shown in Figure 2.34.

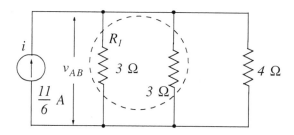

Figure 2.34. Partial reduction for the circuit of Example 2.5

Next,

$$3 \| 3 = \frac{3 \times 3}{3 + 3} = 1.5 \ \Omega$$

Finally,

$$R_{eq} = 1.5 \| 4 = \frac{1.5 \times 4}{1.5 + 4} = \frac{12}{11} \ \Omega$$

and the circuit reduces to that shown in Figure 2.35

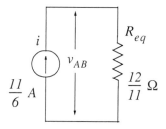

Figure 2.35. Reduction of the circuit of Example 2.5 to its simplest form

b. The voltage v_{AB} across the current source is

$$v_{AB} = IR_{eq} = \frac{11}{6} \cdot \frac{12}{11} = 2 \ V \qquad (2.43)$$

2.12 Voltage Division Expressions

In the circuit of Figure 2.36, v_S, R_1, and R_2 are known.

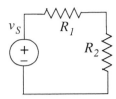

Figure 2.36. Circuit for the derivation of the voltage division expressions

For the circuit of Figure 2.36, we will derive the voltage division expressions which state that:

$$v_{R1} = \frac{R_1}{R_1 + R_2} v_S \text{ and } v_{R2} = \frac{R_2}{R_1 + R_2} v_S$$

These expressions enable us to obtain the voltage drops across the resistors in a series circuit simply by observation.

Derivation:

By Ohm's law in the circuit of Figure 2.36 where i is the current flowing through i, we get

$$v_{R1} = R_1 i \quad \text{and} \quad v_{R2} = R_2 i \tag{2.44}$$

Also,

$$(R_1 + R_2)i = v_S$$

or

$$i = \frac{v_S}{R_1 + R_2} \tag{2.45}$$

and by substitution of (2.45) into (2.44) we obtain the voltage division expressions below.

$$v_{R1} = \frac{R_1}{R_1 + R_2} v_S \quad \text{and} \quad v_{R2} = \frac{R_2}{R_1 + R_2} v_S \tag{2.46}$$

VOLTAGE DIVISION EXPRESSIONS

Example 2.6

In the network of Figure 2.37, the arrows indicate that resistors R_1 and R_2 are variable and that the power supply is set for *12 V*.

a. Compute v_{R1} and v_{R2} if R_1 and R_2 are adjusted for *7 Ω* and *5 Ω* respectively.

b. To what values should R_1 and R_2 be adjusted so that $v_{R1} = 3\ V$, $v_{R2} = 9\ V$, and $R_1 + R_2 = 12\ \Omega$?

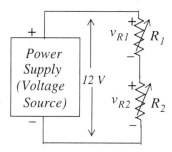

Figure 2.37. Network for Example 2.6

Solution:

a. Using the voltage division expressions of (2.46), we get

$$v_{R1} = \frac{R_1}{R_1 + R_2} v_S = \frac{7}{7 + 5} \times 12 = 7\ V$$

and

$$v_{R2} = \frac{R_2}{R_1 + R_2} v_S = \frac{5}{7 + 5} \times 12 = 5\ V$$

b. Since $v_{R1} + v_{R2} = 3 + 9 = 12\ V$, $R_1 + R_2 = 12\ \Omega$, and the voltage drops are proportional to the resistances, it follows that if we let $R_1 = 3\ \Omega$ and $R_2 = 9\ \Omega$, the voltage drops v_{R1} and v_{R2} will be $3\ V$ and $9\ V$ respectively.

2.13 Current Division Expressions

In the circuit shown in Figure 2.38, i_S, G_1, and G_2 are known.

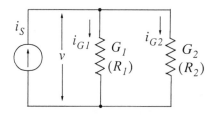

Figure 2.38. Circuit for the derivation of the current division expressions

For the circuit of Figure 2.38, we will derive the current division expressions which state that

$$i_{G1} = \frac{G_1}{G_1 + G_2} i_S \quad \text{and} \quad i_{G2} = \frac{G_2}{G_1 + G_2} i_S$$

and these expressions enable us to obtain the currents through the conductances (or resistances) in a parallel circuit simply by observation.

Derivation:

By Ohm's law for conductances, we get

$$i_{G1} = G_1 v \quad \text{and} \quad i_{G2} = G_2 v \tag{2.47}$$

Also,

$$(G_1 + G_2)v = i_S$$

or

$$v = \frac{i_S}{G_1 + G_2} \tag{2.48}$$

and by substitution of (2.48) into (2.47)

$$i_{G1} = \frac{G_1}{G_1 + G_2} i_S \quad \text{and} \quad i_{G2} = \frac{G_2}{G_1 + G_2} i_S \tag{2.49}$$

Also, since

$$R_1 = \frac{1}{G_1} \qquad R_2 = \frac{1}{G_2}$$

by substitution into (2.49) we get

$$\boxed{i_{R1} = \frac{R_2}{R_1 + R_2} i_S \quad \text{and} \quad i_{R2} = \frac{R_1}{R_1 + R_2} i_S} \tag{2.50}$$

$$\textit{CURRENT DIVISION EXPRESSIONS}$$

Example 2.7

For the circuit of figure 2.39, compute the voltage drop v

Solution:

The current source i_S divides into currents i_1 and i_2 as shown in Figure 2.40.

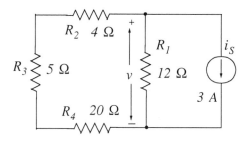

Figure 2.39. Circuit for Example 2.7

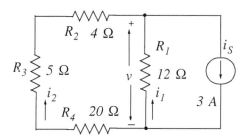

Figure 2.40. Application of current division expressions for the circuit of Example 2.7

We observe that the voltage v is the voltage across the resistor R_1. Therefore, we are only interested in current i_1. This is found by the current division expression as

$$i_1 = \frac{R_2 + R_3 + R_4}{R_1 + R_2 + R_3 + R_4} \cdot i_S = \frac{4 + 5 + 20}{12 + 4 + 5 + 20} \cdot 3 = \frac{87}{41} \ A$$

and observing the passive sign convention, the voltage v is

$$v = -i_1 R_1 = -\frac{87}{41} \cdot 12 = -\frac{1044}{41} \ V$$

or

$$v = -25.46 \ V$$

2.14 Standards for Electrical and Electronic Devices

Standardization of electronic components such as resistors, capacitors and diodes is carried out by various technical committees. In the United States, the Electronics Industries Association (EIA) and the American National Standards Institute (ANSI) have established and published several standards for electrical and electronic devices to provide interchangeability among similar products made by different manufacturers. Also, the U.S. Department of Defense or its agencies issue standards known as *Military Standards,* or simply MIL-stds. All of the aforementioned standards are updated periodically. The interested reader may find the latest revisions in the Internet or the local library.

2.15 Resistor Color Code

The Resistor Color Code is used for marking and identifying pertinent data for standard resistors. Figures 2.41 and 2.42 show the color coding scheme per EIA Standard RS-279 and MIL-STD-1285A respectively.

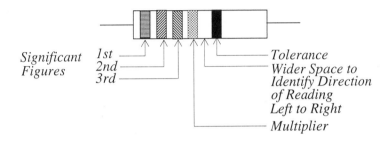

Figure 2.41. Resistor Color Code per EIA Standard RS-279

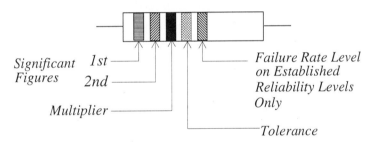

Figure 2.42. Resistor Color Code per MIL-STD-1285A

In a color coded scheme, each color represents a single digit number, or conversely, a single digit number can be represented by a particular color band as shown in Table 2.3 that is based on MIL-STD-1285A color code.

As shown in Figure 2.42, the first and second bands designate the first and second significant digits respectively, the third represents the multiplier, that is, the number by which the first two digits are multiplied, and the fourth and fifth bands, if they exist, indicate the tolerance and failure rate respectively. The *tolerance* is the maximum deviation from the specified nominal value and it is given as a percentage. The *failure rate* is the percent probability of failure in a 1000-hour time interval.

Let A and B represent the first and second significant digits and C represent the multiplier. Then the resistance value is found from the expression

$$R = (10 \times A + B) \times 10^C \qquad (2.51)$$

TABLE 2.3 Resistor values per MIL-STD-1285A

Color Code	1st & 2nd Digits	Multiplier (3rd Digit)	Tolerance (Percent)	Fail Rate (Percent)
Black	0	1		
Brown	1	10	1	1
Red	2	100	2	0.1
Orange	3	1000		0.01
Yellow	4	10000		0.001
Green	5	100000	0.5	
Blue	6	1000000	0.25	
Violet	7		0.1	
Gray	8			
White	9			
Gold		0.1	5	
Silver		0.01	10	
No Color			20	

Example 2.8

The value of a resistor is coded with the following colored band code, left to right: Brown, Green, Blue, Gold, Red. What is the value, tolerance, and probability of failure for that resistor?

Solution:

Table 2.3 yields the following data: Brown (1st significant digit) = 1, Green (2nd significant digit) = 5, and Blue (multiplier) = 1,000,000. Therefore, the nominal value of this resistor is 15,000,000 Ohms or 15 MΩ The 4th band is Gold indicating a ±5% tolerance meaning that the maximum deviation from the nominal value is 15,000,000 ±5% = 15,000,000 × ±0.05 = ±750,000 Ohms or ±0.75 MΩ That is, this resistor can have a value anywhere between 14.25 MΩ and 15.75 MΩ Since the 5th band is Red, there is a 0.1% probability that this resistor will fail after 1000 hours of operation.

2.16 Power Rating of Resistors

As it was mentioned in Section 2.2, a resistor, besides its resistance rating (ohms) has a power rating (watts) commonly referred to as the *wattage* of the resistor, and common resistor wattage values are ¼ watt, ½ watt, 1 watt, 2 watts, 5 watts and so on. To appreciate the importance of the wattage of a resistor, let us refer to the voltage divider circuit of Example 2.6, Figure 2.37 where the current is $12 \ V / 12 \ \Omega = 1 \ A$. Using the power relation $p_R = i^2 R$, we find that the wattage of the 7 Ω and

5 Ω resistors would be 7 watts and 5 watts respectively. We could also divide the 12 volt source into two voltages of 7 V and 5 V using a *7 kΩ* and a *5 kΩ* resistor. Then, with this arrangement the current would be *12 V/12 kΩ = 1 mA*. The wattage of the *7 kΩ* and *5 kΩ* resistors would then be $(10^{-3})^2 \times 7 \times 10^3 = 7 \times 10^{-3}$ W = *7 mW* and $(10^{-3})^2 \times 5 \times 10^3 = 5$ mW respectively.

2.17 Temperature Coefficient of Resistance

The resistance of any pure metal, such as copper, changes with temperature. For each degree that the temperature of a copper wire rises above *20°C* Celsius, up to about *200°C*, the resistance increases 0.393 of 1 percent of what it was at 20 degrees Celsius. Similarly, for each degree that the temperature drops below *20°C*, down to about *−50°C*, the resistance decreases 0.393 of 1 percent of what it was at *20°C*. This percentage of change in resistance is called the *Temperature Coefficient of Resistance*. In general, the resistance of any pure metal at temperature *T* in degrees Celsius is given by

$$R = R_{20}[1 + \alpha_{20}(T - 20)] \tag{2.52}$$

where R_{20} is the resistance at *20°C* and α_{20} is the temperature coefficient of resistance at *20°C*.

Example 2.9

The resistance of a long piece of copper wire is *48 Ω* at *20°C*.

a. What would the resistance be at *50°C*?

b. Construct a curve showing the relation between resistance and temperature.

Solution:

a. The temperature rise is *50 − 20 = 30* degrees Celsius and the resistance increases 0.393% for every degree rise. Therefore the resistance increases by *30 × 0.393 = 11.79%*. This represents an increase of *0.1179 × 48 Ω* in resistance or 5.66 Ω Therefore, the resistance at 50 degrees Celsius is *48 + 5.66 = 53.66 Ω*.

b. The relation of (2.52) is an equation of a straight line with *slope* = $R_{20}\alpha_{20}$. This straight line is easily constructed with the Microsoft Excel spreadsheet shown in Figure 2.43.

From Figure 2.43, we observe that the resistance reaches zero value at approximately *−235°C*.

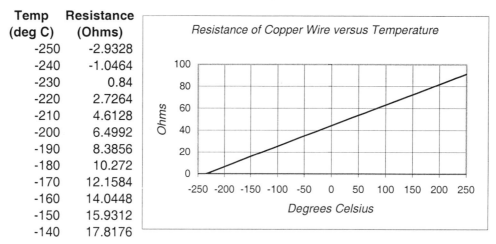

Temp (deg C)	Resistance (Ohms)
-250	-2.9328
-240	-1.0464
-230	0.84
-220	2.7264
-210	4.6128
-200	6.4992
-190	8.3856
-180	10.272
-170	12.1584
-160	14.0448
-150	15.9312
-140	17.8176

Figure 2.43. Spreadsheet for construction of equation (2.52)

2.18 Ampere Capacity of Wires

For public safety, electric power supply (mains) wiring is controlled by local, state and federal boards, primarily on the *National Electric Code* (NEC) and the *National Electric Safety Code*. Moreover, many products such as wire and cable, fuses, circuit breakers, outlet boxes and appliances are governed by *Underwriters Laboratories* (UL) Standards which approves consumer products such as motors, radios, television sets etc.

Table 2.4 shows the NEC allowable current-carrying capacities for copper conductors based on the type of insulation.

The ratings in Table 2.4 are for copper wires. The ratings for aluminum wires are typically *84%* of these values. Also, these rating are for not more than three conductors in a cable with temperature *30°C* or *86°F*. The NEC contains tables with correction factors at higher temperatures.

2.19 Current Ratings for Electronic Equipment

There are also standards for the internal wiring of electronic equipment and chassis. Table 2.5 provides recommended current ratings for copper wire based on *45°C (40°C* for wires smaller than 22 AWG. Listed also, are the circular mils and these denote the area of the cross section of each wire size. A *circular mil* is the area of a circle whose diameter is 1 mil (one-thousandth of an inch). Since the area of a circle is proportional to the square of its diameter, and the area of a circle one mil in diameter is one circular mil, the area of any circle in circular mils is the square of its diameter in mils.

A *mil-foot wire* is a wire whose length is one foot and has a cross-sectional area of one circular mil.

TABLE 2.4 Current Carrying Capacities for Copper Conductors

Size (AWG)	Copper Conductor Insulation			
	RUH(14-2) T, TW, UF	RH, RHW, RUH, (14-2) THW, THWN, XHHW	TA, TBS, SA, FEP, FEPB, RHH, THHN XHHW†	TFE‡
14	15	15	25	40
12	20	20	30	55
10	30	30	40	75
8	40	45	50	95
6	55	65	70	120
4	70	85	90	145
3	80	100	105	170
2	95	115	120	195
1	110	130	140	220
0	125	150	155	250
00	145	175	185	280
000	165	200	210	315
0000	195	230	235	370

† Dry Locations Only   ‡ Nickel or nickel-coated copper only

TABLE 2.5 *Current Ratings for Electronic Equipment and Chassis Copper Wires*

Wire Size		Nominal Resistance (Ohms/1000 ft) at 100 °C	Maximum Current (Amperes)	
AWG	Circular Mils		Wire in Free Air	Wire Confined in Insulation
32	63.2	188	0.53	0.32
30	100.5	116	0.86	0.52
28	159.8	72	1.4	0.83
26	254.1	45.2	2.2	1.3
24	404	28.4	3.5	2.1
22	642.4	22	7	5
20	10.22	13.7	11	7.5
18	1624	6.5	16	10
16	2583	5.15	22	13
14	4107	3.2	32	17
12	6530	2.02	41	23
10	10380	1.31	55	33
8	16510	0.734	73	46
6	26250	0.459	101	60
4	41740	0.29	135	80
2	66370	0.185	181	100
1	83690	0.151	211	125
0	105500	0.117	245	150
00	133100	0.092	283	175
000	167800	0.074	328	200
0000	211600	0.059	380	225

The resistance of a wire of length l can be computed by the relation

$$R = \frac{\rho l}{d^2} \tag{2.53}$$

where ρ = resistance per mil-foot, l = length of wire in feet, d = diameter of wire in mils, and R is the resistance at $20°C$.

Example 2.10

Compute the resistance per mile of a copper conductor $1/8$ inch in diameter given that the resistance per mil-foot of copper is 10.4 Ω at $20°C$.

Solution:

$$(1/8) \ in \ = \ 0.125 \ in \ = \ 125 \ mils$$

and from (2.53)

$$R \ = \ \frac{\rho l}{d^2} \ = \ \frac{10.4 \times 5280}{125^2} \ = \ 3.51 \ \Omega$$

Column 3 of Table 2.5 shows the copper wire resistance at $100°C$. Correction factors must be applied to determine the resistance at other temperatures or for other materials. For copper, the conversion equation is

$$R_T \ = \ R_{100}[1 + 0.004(T - 100)] \tag{2.54}$$

where R_T is the resistance at the desired temperature, R_{100} is the resistance at $100°C$ for copper, and T is the desired temperature.

Example 2.11

Compute the resistance of $1000 \ ft$ of size $AWG \ 12$ copper wire at $30°C$.

Solution:

From Table 2.5 we find that the resistance of $1000 \ ft$ of size $AWG \ 12$ copper wire at $100°C$ is $2.02 \ \Omega$. Then, by (2.54), the resistance of the same wire at $30°C$ is

$$R_{30°C} \ = \ 2.02[1 + 0.004(30 - 100)] \ = \ 1.45 \ \Omega$$

2.20 Copper Conductor Sizes for Interior Wiring

In the design of an interior electrical installation, the electrical contractor must consider two important factors:

a. The wiring size in each section must be selected such that the current shall not exceed the current carrying capacities as defined by the NEC tables. Therefore, the electrical contractor must accurately determine the current which each wire must carry and make a tentative selection of the size listed in Table 2.4.

b. The voltage drop throughout the electrical system must then be computed to ensure that it does not exceed certain specifications. For instance, in the lighting part of the system referred to as the *lighting load*, a variation of more than *5%* in the voltage across each lamp causes an unpleasant variation in the illumination. Also, the voltage variation in the heating and air conditioning load must not exceed *10%*.

Important! The requirements stated here are for instructional purposes only. They change from time to time. It is, therefore, imperative that the designer consults the latest publications of the applicable codes for compliance.

Example 2.12

Figure 2.44 shows a lighting load distribution diagram for an interior electric installation.

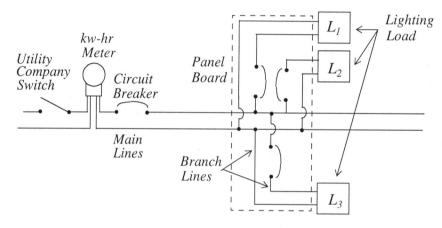

Figure 2.44. Load distribution for an interior electric installation

The panel board is 200 feet from the meter. Each of the three branches has 12 outlets for 75 w, 120 volt lamps. The *load center* is that point on the branch line at which all lighting loads may be considered to be concentrated. For this example, assume that the distance from the panel to the load center is 60 ft. Compute the size of the main lines. Use T (thermoplastic insulation) type copper conductor and base your calculations on *25°C* temperature environment.

Solution:

It is best to use a spreadsheet for the calculations so that we can compute sizes for more and different branches if need be.

The computations for Parts I and II are shown on the spreadsheet of Figure 2.45 where from the last line of Part II we see that the percent line drop is *12.29* and this is more than twice the allowable *5%* drop. With the *12.29%* voltage variation the brightness of the lamps would vary through wide ranges, depending on how many lamps were in use at one time.

A much higher voltage than the rated *120 V* would cause these lamps to glow far above their rated candle power and would either burn them immediately, or shorten their life considerably. It is therefore necessary to install larger than *12 AWG* main line. The computations in Parts III through V of the spreadsheet of Figures 2.45 and 2.46 indicate that we should not use a conductor less than size *6 AWG*.

Part I   Sizing in Accordance with NEC (Table 2.2)

Step	Description	Calculation	Value	Units
1	Meter to Panel Board Distance		200	ft
2	No. of Branches from Panel Board		3	
3	Outlets per branch		12	
4	Outlet Voltage Rating		120	V
5	Power drawn from each outlet		75	w
6	Power required by each branch	Step 3 x Step 5	900	w
7	Current drawn by each branch	Step 6 / Step 4	7.50	A
8	Required wire size for branches *	Table 2.4	14 AWG	Carries up to 15 A
9	Current required by the main line	Step 2 x Step 7	22.50	A
10	Required wire size for main line	Table 2.4	12 AWG	Carries up to 20 A

Part II Check for Voltage Drops

Step	Description	Calculation	Value	Units
11	Distance from panel to load center		60	ft
12	Number of wires in each branch		2	
13	Total length of wire in each branch	Step 11 x Step 12	120	ft
14	Specified Temperature in °C		25	°C
15	Resistance of 1000 ft 14 AWG wire at 100 °C	Table 2.5	3.20	Ω
16	Resistance of 1000 ft 14 AWG wire at Spec. Temp	Equation (2.54)	2.24	Ω
17	Resistance of actual length of wire	(Step 13 x Step 16) / 1000	0.269	Ω
18	Voltage drop in each branch	Step 7 x Step 17	2.02	V
19	Number of wires in each main		2	
20	Total length of wire in main line	Step 1 x Step 19	400	ft
21	Resistance of 1000 ft 12 AWG wire at 100 °C	Table 2.5	2.02	Ω
22	Resistance of 1000 ft 12 AWG wire at Spec. Temp	Equation (2.54)	1.41	Ω
23	Resistance of actual length of wire	(Step 20 x Step 22) /1000	0.57	Ω
24	Voltage drop in main line	Step 9 x Step 23	12.73	V
25	Voltage drop from meter to load center	Step 18 + Step 24	14.74	V
26	Percent line drop	(Step 25 / Step 4) x 100	12.29	%

* No size smaller can be installed; smaller sizes have insufficient mechanical strength

Figure 2.45. Spreadsheet for Example 2.12, Parts I and II

Part III Recalculation Using 10 AWG Conductor Size

27	Next larger conductor size for main line	10 AWG	Carries up to 30 A
28	Resistance of 1000 ft 10 AWG wire at 100 °C	Table 2.4	1.310 Ω
29	Resistance of 1000 ft 10 AWG wire at Spec. Temp	Table 2.5	0.917 Ω
30	Resistance of actual length of wire	Equation (2.54)	0.367 Ω
31	Voltage drop in main line	(Step 20 x Step 29)/1000	8.25 V
32	Voltage drop from meter to load center	Step 9 x Step 30	10.27 V
		Step 18 + Step 24	
33	Percent line drop	(Step 32 / Step 4) x 100	8.56 %

Part IV Recalculation Using 8 AWG Conductor Size

34	Next larger conductor size for main line	8 AWG	Carries up to 40 A
35	Resistance of 1000 ft 8 AWG wire at 100 °C	Table 2.4	0.734 Ω
36	Resistance of 1000 ft 8 AWG wire at Spec. Temp	Table 2.5	0.514 Ω
37	Resistance of actual length of wire	Equation (2.54)	0.206 Ω
38	Voltage drop in main line	(Step 20 x Step 36)/1000	4.62 V
39	Voltage drop from meter to load center	Step 9 x Step 37	6.64 V
		Step 18 + Step 38	
40	Percent line drop	(Step 39 / Step 4) x 100	5.53 %

Part V Recalculation Using 6 AWG Conductor Size

41	Next larger conductor size for main line	6 AWG	Carries up to 55 A
42	Resistance of 1000 ft 6 AWG wire at 100 °C	Table 2.4	0.459 Ω
43	Resistance of 1000 ft 6 AWG wire at Spec. Temp	Table 2.5	0.321 Ω
44	Resistance of actual length of wire	Equation (2.54)	0.129 Ω
45	Voltage drop in main line	(Step 20 x Step 43)/1000	2.89 V
46	Voltage drop from meter to load center	Step 9 x Step 44	4.91 V
		Step 18 + Step 45	
47	Percent line drop	(Step 46 / Step 4) x 100	4.09 %

Figure 2.46. Spreadsheet for Example 2.12, Parts III, IV, and V

2.21 Summary

- Ohm's Law states that the voltage across a device is proportional to the current through that device and the resistance is the constant of proportionality.

- Open circuit refers to an open branch (defined below) in a network. It can be thought of as a resistor with infinite resistance (or zero conductance). The voltage across the terminals of an open may have a finite value or may be zero whereas the current is always zero.

- Short circuit refers to a branch (defined below) in a network that contains no device between its terminals, that is, a piece of wire with zero resistance. The voltage across the terminals of a short is always zero whereas the current may have a finite value or may be zero.

- A resistor absorbs power.

- A resistor does not store energy. The energy is dissipated in the form of heat.

- A node is a common point where one end of two or more devices are connected.

- A branch is part of a network that contains a device and its nodes.

- A mesh is a closed path that does not contain other closed paths

- A loop contains two or more closed paths.

- Kirchoff's Current Law (KCL) states that the algebraic sum of the currents entering (or leaving) a node is zero.

- Kirchoff's Voltage Law (KVL) states that the algebraic sum of the voltage drops (or voltage rises) around a closed mesh or loop is zero.

- Two or more devices are said to be connected in series if and only if the same current flows through them.

- Two or more devices are said to be connected in parallel if and only if the same voltage exists across their terminals.

- A series circuit with a single mesh can be easily analyzed by KVL.

- A parallel circuit with a single node pair can be easily analyzed by KCL.

- If two or more voltage sources are in series, they can be replaced by a single voltage source with the proper polarity.

- If two or more current sources are in parallel, they can be replaced by a single current source with the proper current direction.

- If two or more resistors are connected in series, they can be replaced by an equivalent resistance whose value is

$$R_{eq} = R_1 + R_2 + R_3 + \ldots + R_n = \sum_{k=1}^{n} R_K$$

- If two or more resistors are connected in parallel, they can be replaced by an equivalent resistance whose value is

$$\frac{1}{R_{eq}} = \frac{1}{R_1} + \frac{1}{R_2} + \ldots + \frac{1}{R_n}$$

- For the special case of two parallel resistors, the equivalent resistance is found from the relation

$$R_{eq} = R_1 \| R_2 = \frac{R_1 \cdot R_2}{R_1 + R_2}$$

- Conductances connected in series combine as resistors in parallel do.

- Conductances connected in parallel combine as resistors in series do.

- For the simple series circuit below

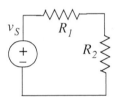

the voltage division expressions state that:

$$v_{R1} = \frac{R_1}{R_1 + R_2} v_S \text{ and } v_{R2} = \frac{R_2}{R_1 + R_2} v_S$$

- For the simple parallel circuit below

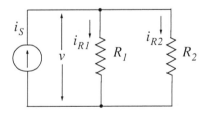

the current division expressions state that:

$$i_{R1} = \frac{R_2}{R_1 + R_2} i_S \text{ and } i_{R2} = \frac{R_1}{R_1 + R_2} i_S$$

- In the United States, the Electronics Industries Association (EIA) and the American National Standards Institute (ANSI) have established and published several standards for electrical and electronic devices to provide interchangeability among similar products made by different manufacturers.

- The resistor color code is used for marking and identifying pertinent data for standard resistors. Two standards are the EIA Standard RS-279 and MIL-STD-1285A.

- Besides their resistance value, resistors have a power rating.

- The resistance of a wire increases with increased temperature and decreases with decreased temperature.

- The current ratings for wires and electronic equipment are established by national standards and codes.

2.22 Exercises

Multiple Choice

1. Ohm's Law states that

    A. the conductance is the reciprocal of resistance

    B. the resistance is the slope of the straight line in a voltage versus current plot

    C. the resistance is the sum of the voltages across all the devices in a closed path divided by the sum of the currents through all the devices in the closed path

    D. the sum of the resistances around a closed loop is zero

    E. none of the above

2. Kirchoff's Current Law (KCL) states that

    A. the sum of the currents in a closed path is zero

    B. the current that flows through a device is inversely proportional to the voltage across that device

    C. the sum of the currents through all the devices in a closed path is equal to the sum of the voltages across all the devices

    D. the sum of the currents entering a node is equal to the sum of the currents leaving that node

    E. none of the above

3. Kirchoff's Voltage Law (KCL) states that

    A. the voltage across a device is directly proportional to the current through that device

    B. the voltage across a device is inversely proportional to the current through that device

    C. the sum of the voltages across all the devices in a closed path is equal to the sum of the currents through all the devices

    D. the sum of the voltages in a node is equal to the sum of the currents at that node

    E. none of the above

4. For the three resistors connected as shown on the network of Figure 2.47, the equivalent resistance R_{AB} is computed with the formula

    A. $R_{AB} = \sqrt{R_1 + R_2 + R_3}$

    B. $R_{AB} = \sqrt{R_1^2 + R_2^2 + R_3^2}$

C. $R_{AB} = \dfrac{R_1 R_2 R_3}{R_1 + R_2 + R_3}$

D. $R_{AB} = \sqrt{\dfrac{R_1 R_2 R_3}{R_1 + R_2 + R_3}}$

E. none of the above

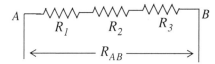

Figure 2.47. Network for Question 4

5. For the three conductances connected as shown on the network of Figure 2.48, the equivalent conductance G_{AB} is computed with the formula

A. $G_{AB} = \sqrt{G_1 + G_2 + G_3}$

B. $G_{AB} = \sqrt{G_1^2 + G_2^2 + G_3^2}$

C. $G_{AB} = \dfrac{G_1 G_2 G_3}{G_1 + G_2 + G_3}$

D. $\dfrac{1}{G_{AB}} = \dfrac{1}{G_1} + \dfrac{1}{G_2} + \dfrac{1}{G_3}$

E. none of the above

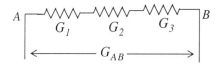

Figure 2.48. Network for Question 5

6. For the three resistances connected as shown on the network of Figure 2.49, the equivalent conductance G_{AB} is

A. $21\ \Omega^{-1}$

B. $1.5\ \Omega^{-1}$

C. $2/3 \ \Omega^{-1}$

D. $144/19 \ \Omega^{-1}$

E. none of the above

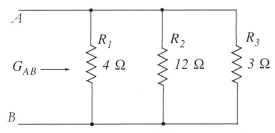

Figure 2.49. Network for Question 6

7. In the network shown in Figure 2.50, when $R = 4 \ \Omega$, the voltage $v_R = 6 \ V$. When $R = 0 \ \Omega$, $i_R = 2 \ A$. When $R = \infty$, v_R is

A. *6 V*

B. *24 V*

C. *8 V*

D. *16 V*

E. none of the above

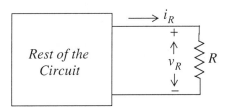

Figure 2.50. Network for Question 7

8. The node voltages shown in the partial network of Figure 2.51 are relative to some reference node not shown. The value of the voltage v_X is

A. *–6 V*

B. *16 V*

C. *0 V*

D. *10 V*

E. none of the above

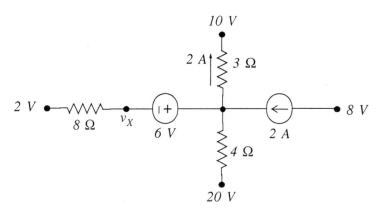

Figure 2.51. Network for Question 8

9. For the network of Figure 2.52 the value of the voltage *v* is

A. *8 V*

B. *2 V*

C. *–2 V*

D. *–8 V*

E. none of the above

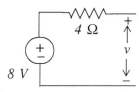

Figure 2.52. Network for Question 9

10. For the circuit of Figure 2.53 the value of the current *i* is

A. *2 A*

B. *0 A*

C. *∞ A*

D. *1 A*

E. none of the above

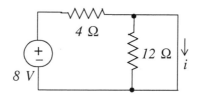

Figure 2.53. Network for Question 10

Problems

1. In the circuit of Figure 2.54, the voltage source and both resistors are variable.

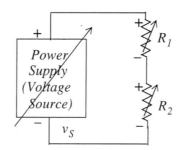

Figure 2.54. Circuit for Problem 1

a. With $v_S = 120$ V, $R_1 = 70 \, \Omega$, and $R_2 = 50 \, \Omega$, compute the power absorbed by R_2.
Answer: *50 w*

b. With $v_S = 120$ V and $R_1 = 0 \, \Omega$, to what value should R_2 be adjusted so that the power absorbed by it will be 200 w? Answer: *72 Ω*

c. With $R_1 = 0 \, \Omega$ and $R_2 = 100 \, \Omega$, to what value should v_S be adjusted to so that the power absorbed by R_2 will be 100 w? Answer: *100 V*

2. In the circuit of Figure 2.55, R_{LOAD} represents the load of that circuit.

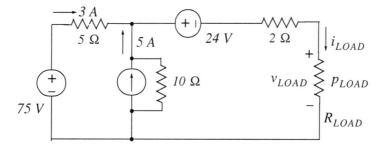

Figure 2.55. Circuit for Problem 2

Compute:

a. i_{LOAD} Answer: *8 A*

b. v_{LOAD} Answer: *20 V*

c. p_{LOAD} Answer: *160 w*

3. For the circuit of Figure 2.56, compute the power supplied or absorbed by each device.

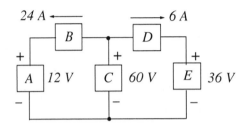

Figure 2.56. Circuit for Problem 3

Answers: $p_A = 288 \ w$, $p_B = 1152 \ w$, $p_C = -1800 \ w$, $p_D = 144 \ w$, $p_E = 216 \ w$

4. In the circuit of Figure 2.57, compute the power delivered or absorbed by the dependent voltage source.

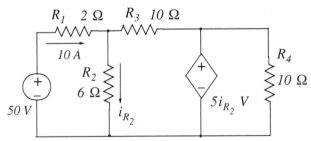

Figure 2.57. Circuit for Problem 4

Answer: *62.5 w*

5. In the network of Figure 2.58, each resistor is 10 Ω. Compute the equivalent resistance R_{eq}.

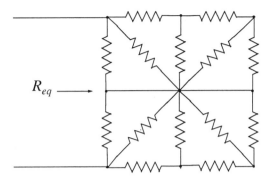

Figure 2.58. Circuit for Problem 5

Answer: *360/21 Ω*

6. In the network of Figure 2.59, $R_1 = 10 \, \Omega$ and $R_2 = 20 \, \Omega$. Compute the current i supplied by the 15 V source.

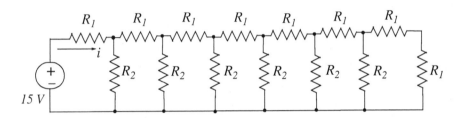

Figure 2.59. Circuit for Problem 6

Hint: Start at the right end and by series and parallel combinations of the resistors, reduce the circuit to a simple series circuit. This method is known as *analysis by network reduction*.

Answer: *0.75 A*

7. In the circuit of Figure 2.60, use the voltage division expression to compute v_X and v_Y.

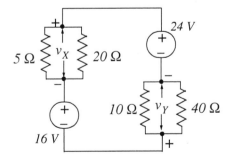

Figure 2.60. Circuit for Problem 7

Answers: $v_X = 8/3 \, V$, $v_Y = -16/3 \, V$

8. In the circuit of Figure 2.61, use the current division expression to compute i_X and i_Y.

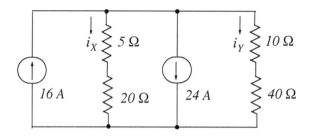

Figure 2.61. Circuit for Problem 8

Answers: $i_X = -16/3 \ V$, $i_Y = -8/3 \ V$

9. A *transformer* consists of two separate *coils* (inductors) wound around an *iron core* as shown in Figure 2.62. There are many turns in both the primary and secondary coils but, for simplicity, only few are shown. It is known that the primary coil has a resistance of 5.48 Ω at 20 degrees Celsius. After two hours of operation, it is found that the primary coil resistance has risen to 6.32 Ω Compute the temperature rise of this coil.

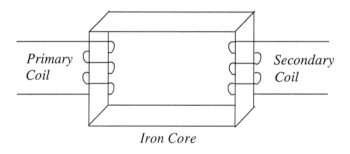

Figure 2.62. Circuit for Problem 9

Answer: *36°C*

10. A new facility is to be constructed at a site which is 1.5 miles away from the nearest electric utility company substation. The electrical contractor and the utility company have made load calculations, and decided that the main lines from the substation to the facility will require several copper conductors in parallel. Each of these conductors must have insulation type THHN and must carry a maximum current of 220 A in a *20°C* temperature environment.

a. Compute the voltage drop on each of these conductors from the substation to the facility when they carry the maximum required current of 220 A in a *20°C* temperature environment.
Answer: *70 V*

b. The power absorbed by each conductor under the conditions stated above.

Answer: *15.4 Kw*

c. The power absorbed per square cm of the surface area of each conductor under the conditions stated above.

Answer: $0.02 \ w/cm^2$

2.23 Answers to Exercises

Multiple Choice

1. B

2. D

3. E

4. E

5. D

6. C

7. B When $R = 4 \, \Omega$, the voltage $v_R = 6 \, V$. Therefore, $i_R = 6/4 = 1.5 \, A$. Also, when $R = 0 \, \Omega$,

$i_R = 2 \, A$, and thus $v_R = 0$ (short circuit). When $R = \infty$, $i_R = 0$ but v_R has a finite value

and it is denoted as $v_{R = \infty}$ in the figure below. Now, we observe that the triangles abc and dbe

are similar. Then $\dfrac{be}{bc} = \dfrac{de}{ac}$ or $\dfrac{2.0 - 1.5}{2.0} = \dfrac{6}{v_{R = \infty}}$ and thus $v_{R = \infty} = 24 \, V$

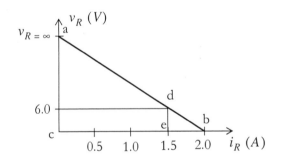

8. D We denote the voltage at the common node as v_A shown on the figure of the next page. Then,

from the branch that contains the $3 \, \Omega$ resistor, we observe that $\dfrac{v_A - 10}{3} = 2$ or $v_A = 16$ and

thus $v_X = -6 + 16 = 10 \, V$

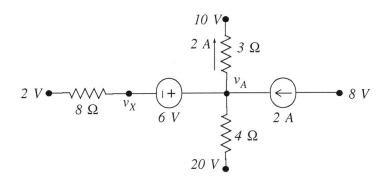

9. A This is an open circuit and therefore no current flows through the resistor. Accordingly, there is no voltage drop across the resistor and thus $v = 8\ V$.

10. A The $12\ \Omega$ resistor is shorted out by the short on the right side of the circuit and thus the only resistance in the circuit is the $4\ \Omega$ resistor.

Problems

1. a. With $v_S = 120\ V$, $R_1 = 70\ \Omega$, and $R_2 = 50\ \Omega$, the circuit is as shown below.

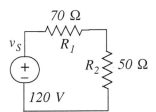

Using the voltage division expression, we get

$$v_{R_2} = \frac{50}{70 + 50} \times 120 = 50\ V$$

Then,

$$p_{R_2} = \frac{v_{R_2}^2}{R_2} = \frac{50^2}{50} = 50\ w$$

b. With $v_S = 120\ V$ and $R_1 = 0\ \Omega$, the circuit is as shown below.

We observe that

$$v_{R_2} = v_s = 120V$$

and

$$\frac{v_{R_2}^2}{R_2} = 200 \ w$$

or

$$\frac{120^2}{R_2} = 200 \ w$$

or

$$R_2 = \frac{120^2}{200} = 72 \ \Omega$$

c. With $R_1 = 0 \ \Omega$ and $R_2 = 100 \ \Omega$, the circuit is as shown below.

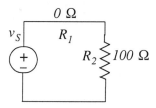

Then,

$$\frac{v_S^2}{R_2} = 100 \ w$$

or

$$\frac{v_S^2}{100} = 100 \ w$$

or

$$v_S^2 = 100 \times 100 = 10,000$$

or

$$v_S = \sqrt{10,000} = 100 \ V$$

2. a. Application of KCL at node A of the circuit below yields

$$i_{LOAD} = 3 + 5 = 8 \ A$$

b. Application of KVL around Mesh 1 yields

$$-75 + 3(5) + v_{AB} = 0$$

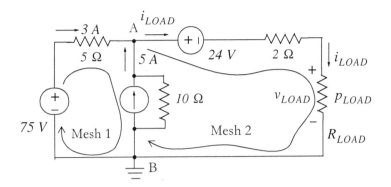

or

$$v_{AB} = 60 \ V$$

Application of KVL around Mesh 2 yields

$$-v_{AB} + 24 + 2i_{LOAD} + v_{LOAD} = 0$$

or

$$-60 + 24 + 2 \times 8 + v_{LOAD} = 0$$

or

$$v_{LOAD} = 20 \ V$$

c.

$$p_{LOAD} = v_{LOAD} \times i_{LOAD} = 20 \times 8 = 160 \ w \ (absorbed \ power)$$

3. Where not shown, we assign plus (+) and minus (−) polarities and current directions in accordance with the passive sign convention as shown below.

We observe that $i_A = i_B$ and $i_E = i_D$. Also, by KCL at Node X

$$i_C = i_B + i_D = 24 + 6 = 30 \ A$$

Then,

$$p_A = v_A i_A = 12 \times 24 = 288 \ w \quad (absorbed)$$

$$p_E = v_E i_E = 36 \times 6 = 216 \ w \quad (absorbed)$$

$$p_C = v_C(-i_C) = 60 \times (-30) = -1800 \ w \quad (supplied)$$

By KVL

$$v_A + v_B = v_C$$

or

$$v_B = v_C - v_A = 60 - 12 = 48 \ V$$

and thus

$$p_B = v_B i_B = 48 \times 24 = 1152 \ w \quad (absorbed)$$

Also by KVL

$$v_D + v_E = v_C$$

or

$$v_D = v_C - v_E = 60 - 36 = 24 \ V$$

and thus

$$p_D = v_D i_D = 24 \times 6 = 144 \ w \quad (absorbed)$$

Check: We must show that *Power supplied = Power absorbed*

$$p_C = p_A + p_B + p_C + p_D = 288 + 216 + 1152 + 144 = 1800 \ w$$

4. We assign voltages and currents v_{R_2}, v_{R_4}, i_{R_3}, i_{R_4}, and i_D as shown in the circuit below.

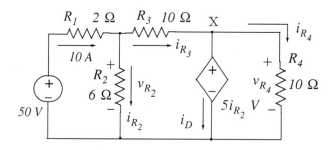

Figure 2.63. Circuit for Exercise 4

By KVL,

$$v_{R_2} = 50 - 2 \times 10 = 30 \ V$$

and by Ohm's law,

$$i_{R_2} = \frac{v_{R_2}}{R_2} = \frac{30}{6} = 5 \ A$$

Therefore, the value of the dependent voltage source is

$$5i_{R_2} = 5 \times 5 = 25 \ V$$

and

$$v_{R_4} = 5i_{R_2} = 25 \ V$$

Then,

$$i_{R_4} = \frac{v_{R_4}}{R_4} = \frac{25}{10} = 2.5 \ A$$

By KCL at Node X

$$i_D = i_{R_3} - i_{R_4}$$

where

$$i_{R_3} = 10 - i_{R_2} = 10 - 5 = 5 \ A$$

and thus

$$i_D = i_{R_3} - i_{R_4} = 5 - 2.5 = 2.5 \ A$$

with the indicated direction through the dependent source. Therefore,

$$p_D = 5i_{R_2}i_D = 25 \times 2.5 = 62.5 \ w \ (absorbed)$$

5. The simplification procedure starts with the resistors in parallel which are indicated below.

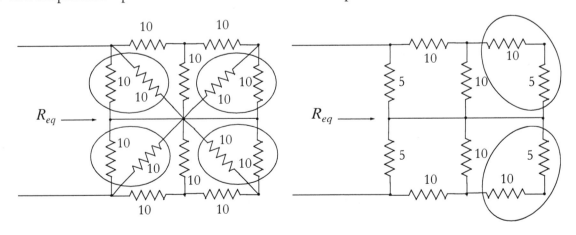

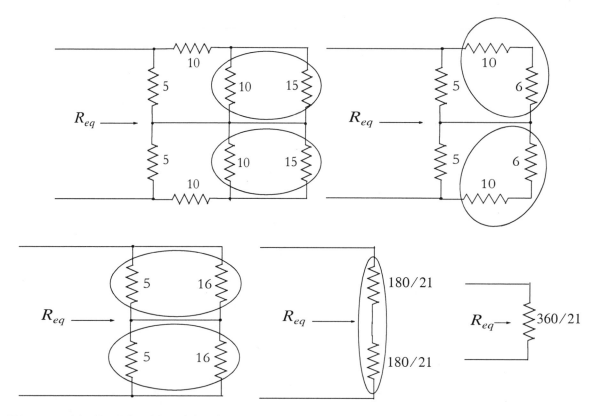

6. We start with the right side of the circuit where the last two resistors are in series as shown below.

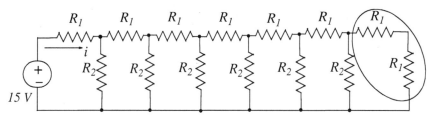

Then,

$$R_1 + R_1 = 10 + 10 = 20 \ \Omega$$

Next,

$$20 \parallel 20 = 10 \ \Omega$$

$$10 + 10 = 20 \ \Omega$$

and so on. Finally, addition of the left most resistor with its series equivalent yields

$$10 + 10 = 20 \ \Omega$$

and thus

$$i = 15/20 = 0.75 \ A$$

7. We first simplify the given circuit by replacing the parallel resistors by their equivalents. Thus,

$$5 \parallel 20 = \frac{5 \times 20}{5 + 20} = 4 \ \Omega$$

and

$$10 \parallel 40 = \frac{10 \times 40}{10 + 40} = 8 \ \Omega$$

The voltage sources are in series opposing connection and they can be replaced by a single voltage source with value $24 - 16 = 8 \ V$. The simplified circuit is shown below.

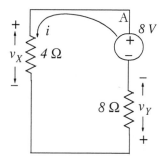

Now, by the voltage division expression,

$$v_X = \frac{4}{4 + 8} \times 8 = \frac{8}{3} \ V$$

and

$$-v_Y = \frac{8}{4 + 8} \times 8 = \frac{16}{3} \ V$$

or

$$v_Y = -\frac{16}{3} \ V$$

Check: By application of KVL starting at point A and going counterclockwise, we get

$$v_X + (-v_Y) - 8 = \frac{8}{3} + \frac{16}{3} - 8 = 0$$

8. We first simplify the given circuit by replacing the series resistors by their equivalents. Thus,

$$5 + 20 = 25 \ \Omega$$

and

$$10 + 40 = 50 \ \Omega$$

The current sources are in parallel opposing connection and they can be replaced by a single current source with value $24 - 16 = 8 \ A$. The simplified circuit is shown below.

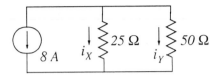

By the current division expression,

$$i_X = \frac{50}{25 + 50} \times (-8) = -\frac{16}{3} \ A$$

and

$$i_Y = \frac{25}{25 + 50} \times (-8) = -\frac{8}{3} \ A$$

Check: By application of KCL starting at point A and going counterclockwise, we get

$$8 + i_X + i_Y = 8 - \frac{16}{3} - \frac{8}{3} = 0$$

9. We construct the resistance versus temperature plot shown below.

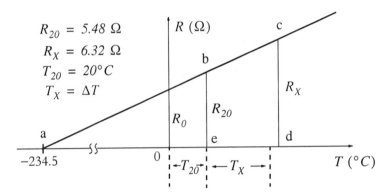

From the similar triangles acd and abe, we get

$$\frac{R_X}{R_{20}} = \frac{234.5 + T_{20} + T_X}{234.5 + T_{20}} = \frac{234.5 + 20 + T_X}{234.5 + 20} = \frac{254.5 + T_X}{254.5}$$

or

$$\Delta T = T_X = \frac{R_X}{R_{20}} \times 254.5 - 254.5 = \frac{6.32}{5.48} \times 254.5 - 254.5 = 36°C$$

10. a. From Table 2.4 we find that the cable size must be 0000 AWG and this can carry up to $235 \ A$. Also, from Table 2.5 we find that the resistance of this conductor is $0.059 \ \Omega / 1000 \ ft$ at $100°C$. Then, the resistance of this conductor that is 1.5 miles long is

$$0.059 \frac{\Omega}{1000ft} \times \frac{5280}{1 \; mile} \times 1.5 \; miles = 0.4673 \; \Omega \; at \; 100°C$$

To find the resistance of this cable at $20°C$, we use the relation of (2.54). Thus,

$$R_{20} = R_{100}[1 + 0.004(20 - 100)] = 0.4673(1 - 0.32) = 0.3178 \; \Omega$$

and the voltage drop on each of these conductors is

$$v = iR = 220 \times 0.3178 = 70 \; V$$

b. The power absorbed by each conductor is

$$p = vi = 70 \times 220 = 15,400 \; w = 15.4 \; Kw$$

c. Table 2.5 gives wire sizes in circular mils. We recall that a circular mil is the area of a circle whose diameter is $0.001 \; in$. To find the diameter in cm, we perform the following conversion:

$$1 \; circular \; mil = \frac{\pi}{4}d^2 = \frac{\pi}{4}(0.001)^2 = 7.854 \times 10^{-7} \; in^2$$

$$= 7.854 \times 10^{-7} \; in^2 \times \frac{(2.54 \; cm)^2}{in^2} = 5.067 \times 10^{-6} \; cm^2$$

From Table 2.5 we find that the cross section of a $0000 \; AWG$ cable is 211,600 circular mils. Then, the cross-section of this cable in cm^2 is

$$211,600 \; circular \; mils \times \frac{5.067 \times 10^{-6} \; cm^2}{circular \; mil} = 1.072 \; cm^2$$

Therefore, the cable diameter in cm is

$$d = \sqrt{1.072} = 1.035 \; cm$$

The cross-section circumference of the cable is

$$\pi d = \pi \times 1.035 = 3.253 \; cm$$

and the surface area of the cable is

$$Surface \; area = \pi dl = 3.253 \; cm \times 1.5 \; miles \times \frac{1.609 \; Km}{1 \; mile} \times \frac{10^5 \; cm}{1 \; Km} = 7.851 \times 10^5 \; cm^2$$

Then, the power absorbed per cm^2 is

$$P_{cm^2} = \frac{Total \; power}{cm^2} = \frac{15,400 \; w}{7.851 \times 10^5 \; cm^2} = 0.02 \; w/cm^2$$

NOTES

Chapter 3

Nodal and Mesh Equations - Circuit Theorems

T his chapter begins with nodal, loop and mesh equations and how they are applied to the solution of circuits containing two or more node-pairs and two or more loops or meshes. Other topics included in this chapter are the voltage-to-current source transformations and vice versa, Thevenin's and Norton's theorems, the maximum power transfer theorem, linearity, superposition, efficiency, and regulation.

3.1 Nodal, Mesh, and Loop Equations

Network Topology is a branch of network theory concerned with the equations required to completely describe an electric circuit. In this text, we will only be concerned with the following two theorems.

Theorem 3.1

Let $N = $ *number of nodes in a circuit*; then $N - 1$ independent nodal equations are required to completely describe that circuit. These equations are obtained by setting the algebraic sum of the currents leaving each of the $N - 1$ nodes equal to zero.

Theorem 3.2

Let $L = M = $ *number of loops or meshes*, $B = $ *number of branches*, $N = $ *number of nodes* in a circuit; then $L = M = B - N + 1$ independent loop or mesh equations are required to completely describe that circuit. These equations are obtained by setting the algebraic sum of the voltage drops around each of the $L = M = B - N + 1$ loops or meshes equal to zero.

3.2 Analysis with Nodal Equations

In writing nodal equations, we perform the following steps:

1. For a circuit containing N nodes, we choose one of these as a reference node assumed to be zero volts or ground.

2. At each non-reference node we assign node voltages $v_1, v_2, ..., v_{n-1}$ where each of these voltages is measured with respect to the chosen reference node, i.e., ground.

3. If the circuit does not contain any voltage sources between nodes, we apply KCL and write a nodal equation for each of the node voltages $v_1, v_2, ..., v_{n-1}$.

4. If the circuit contains a voltage source between two nodes, say nodes j and k denoted as node variables v_j and v_k, we replace the voltage source with a short circuit thus forming a combined node and we write a nodal equation for this common node in terms of both v_j and v_k; then we relate the voltage source to the node variables v_j and v_k.

Example 3.1

Write nodal equations for the circuit shown in Figure 3.1, and solve for the unknowns of these equations using matrix theory, Cramer's rule, or the substitution method. Verify your answers with Excel® or MATLAB®. Please refer to Appendix A for discussion and examples.

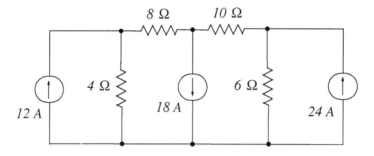

Figure 3.1. Circuit for Example 3.1

Solution:

We observe that there are 4 nodes and we denote these as ①, ②, ③, and G (for ground) as shown in Figure 3.2.

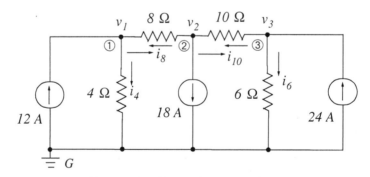

Figure 3.2. Circuit for Example 3.1

For convenience, we have denoted the currents with a subscript that corresponds to the resistor value through which it flows through; thus, the current that flows through the $4\ \Omega$ resistor is denoted as i_4, the current through the $8\ \Omega$ resistor is denoted as i_8, and so on. We will follow this practice in the subsequent examples.

For the circuit of Figure 3.2, we need $N - 1 = 4 - 1 = 3$ nodal equations. Let us choose node G (ground) as our reference node, and we assign voltages v_1, v_2, and v_3 at nodes ①, ②, and ③ respectively; these are to be measured with respect to the ground node G. Now, application of KCL at node ① yields

$$i_4 + i_8 - 12 = 0$$

or

$$i_4 + i_8 = 12 \tag{3.1}$$

where i_8 is the current flowing from left to right. Expressing (3.1) in terms of the node voltages, we get

$$\frac{v_1}{4} + \frac{v_1 - v_2}{8} = 12$$

or

$$\left(\frac{1}{4} + \frac{1}{8}\right)v_1 - \frac{1}{8}v_2 = 12$$

or

$$\boxed{3v_1 - v_2 = 96} \tag{3.2}$$

Next, application of KCL at node ② yields

$$i_8 + i_{10} + 18 = 0$$

or

$$i_8 + i_{10} = -18 \tag{3.3}$$

where i_8 is the current flowing from right to left [*] and i_{10} is the current that flows from left to right.

---

[*] *The direction of the current through the 8 Ω resistor from left to right in writing the nodal equation at Node 1, and from right to left in writing the nodal equation at Node 2, should not be confusing. Remember that we wrote independent node equations at independent nodes and, therefore, any assumptions made in writing the first equation need not be held in writing the second since the latter is independent of the first. Of course, we could have assumed that the current through the 8 Ω resistor flows in the same direction in both nodal equations. It is advantageous, however, to assign a (+) sign to all currents leaving the node in which we apply KCL. The advantage is that we can check, or even write the node equations by inspection. With reference to the above circuit and equation (3.1) for example, since $G = 1/R$, we denote the coefficients of v_1 (1/4 and 1/8 siemens) as **self conductances** and the coefficient of v_2 (−1/8) as **mutual conductance**. Likewise, in equation (3.3) the coefficients of v_2 (1/8 and 1/10 siemens) are the self conductances and the coefficients of v_1 (−1/8) and v_3 (−1/10) are the mutual conductances. Therefore, we can write a nodal equation at a particular node by inspection, that is, we assign plus (+) values to self conductances and minus (−) to mutual conductances.*

Expressing (3.3) in terms of node voltages, we get

$$\frac{v_2 - v_1}{8} + \frac{v_2 - v_3}{10} = -18$$

or

$$-\frac{1}{8}v_1 + \left(\frac{1}{8} + \frac{1}{10}\right)v_2 - \frac{1}{10}v_3 = -18$$

or

$$\boxed{5v_1 - 9v_2 + 4v_3 = 720} \qquad (3.4)$$

Similarly, application of KCL at node ③ yields

$$i_{10} + i_6 - 24 = 0$$

or

$$i_{10} + i_6 = 24$$

where i_{10} is the current flowing from right to left. Then, in terms of node voltages,

$$\frac{v_3 - v_2}{10} + \frac{v_3}{6} = 24 \qquad (3.5)$$

or

$$-\frac{1}{10}v_2 + \left(\frac{1}{10} + \frac{1}{6}\right)v_3 = 24$$

or

$$-3v_2 + 18v_3 = 720$$

or

$$\boxed{v_2 - 6v_3 = -240} \qquad (3.6)$$

Equations (3.2), (3.4), and (3.6) constitute a set of three simultaneous equations with three unknowns. We write them in matrix form as follows:

$$\underbrace{\begin{bmatrix} 3 & -1 & 0 \\ 5 & -9 & 4 \\ 0 & 1 & -6 \end{bmatrix}}_{G} \underbrace{\begin{bmatrix} v_1 \\ v_2 \\ v_3 \end{bmatrix}}_{V} = \underbrace{\begin{bmatrix} 96 \\ 720 \\ -240 \end{bmatrix}}_{I} \qquad (3.7)$$

We can use Cramer's rule or Gauss's elimination method as discussed in Appendix A, to solve (3.7) for the unknowns. Simultaneous solution yields $v_1 = 12\ V$, $v_2 = -60\ V$, and $v_3 = 30\ V$. With these values we can determine the current in each resistor, and the power absorbed or delivered by each device.

Check with MATLAB®:

```
G=[3 –1 0; 5 –9 4; 0 1 –6]; I=[96 720 -240]'; V=G\I;...
fprintf(' \n'); fprintf('v1 = %5.2f volts \t', V(1)); ...
fprintf('v2 = %5.2f volts \t', V(2)); fprintf('v3 = %5.2f volts', V(3)); fprintf(' \n')

v1 = 12.00 volts    v2 = -60.00 volts    v3 = 30.00 volts
```

Check with Excel®:

The spreadsheet of Figure 3.3 shows the solution of the equations of (3.7). The procedure is discussed in Appendix A.

	A	B	C	D	E	F	G	H
1	Spreadsheet for Matrix Inversion and Matrix Multiplication							
2								
3			3	-1	0			96
4	G=		5	-9	4		I=	720
5			0	1	-6			-240
6								
7			0.417	-0.050	-0.033			12
8	G⁻¹=		0.250	-0.150	-0.100		V=	-60
9			0.042	-0.025	-0.183			30

Figure 3.3. Spreadsheet for the solution of (3.7)

Example 3.2

For the circuit of Figure 3.4, write nodal equations in matrix form and solve for the unknowns using matrix theory, Cramer's rule, or Gauss's elimination method. Verify your answers with Excel or MATLAB. Please refer to Appendix A for procedures and examples. Then construct a table showing the voltages across, the currents through and the power absorbed or delivered by each device.

Solution:

We observe that there are 4 nodes and we denote these as ①, ②, ③, and G (for ground) as shown in Figure 3.5.

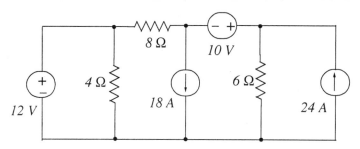

Figure 3.4. Circuit for Example 3.2

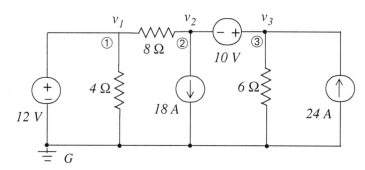

Figure 3.5. Circuit for Example 3.2 with assigned nodes and voltages

We assign voltages v_1, v_2, and v_3 at nodes ①, ②, and ③ respectively; these are to be measured with respect to the ground node G. We observe that v_1 is a known voltage, that is, $v_1 = 12\ V$ and thus our first equation is

$$v_1 = 12 \qquad\qquad (3.8)$$

Next, we move to node ② where we observe that there are three currents flowing out of this node, one to the left, one to the right, and one down. Therefore, our next nodal equation will contain three terms. We have no difficulty writing the term for the current flowing from node ② to node ①, and for the 18 A source; however, we encounter a problem with the third term because we cannot express it as term representing the current flowing from node ② to node ③. To work around this problem, we temporarily remove the 10 V voltage source and we replace it with a "short" thereby creating a *combined node* (or *generalized node* or *supernode* as some textbooks call it), and the circuit now looks as shown in Figure 3.6.

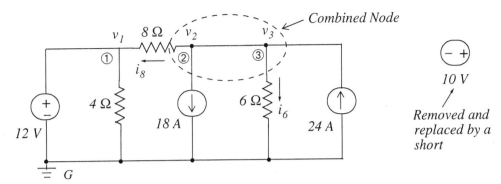

Figure 3.6. Circuit for Example 3.2 with a combined node

Now, application of KCL at this combined node yields the equation

$$i_8 + 18 + i_6 - 24 = 0$$

or

$$i_8 + i_6 = 6$$

or

$$\frac{v_2 - v_1}{8} + \frac{v_3}{6} = 6^* \tag{3.9}$$

or

$$-\frac{1}{8}v_1 + \frac{1}{8}v_2 + \frac{1}{6}v_3 = 6$$

or

$$\boxed{-3v_1 + 3v_2 + 4v_3 = 144} \tag{3.10}$$

To obtain the third equation, we reinsert the 10 V source between nodes ② and ③. Then,

$$\boxed{v_3 - v_2 = 10} \tag{3.11}$$

In matrix form, equations (3.8), (3.10), and (3.11) are

---

* The combined node technique allows us to combine two nodal equations into one but requires that we use the proper node designations. In this example, to retain the designation of node 2, we express the current i_8 as $\frac{v_2 - v_1}{8}$. Likewise, at node 3, we express the current i_6 as $\frac{v_3}{6}$.

$$
\underbrace{\begin{bmatrix} 1 & 0 & 0 \\ -3 & 3 & 4 \\ 0 & -1 & 1 \end{bmatrix}}_{G} \underbrace{\begin{bmatrix} v_1 \\ v_2 \\ v_3 \end{bmatrix}}_{V} = \underbrace{\begin{bmatrix} 12 \\ 144 \\ 10 \end{bmatrix}}_{I}
\qquad (3.12)
$$

Simultaneous solution yields $v_1 = 12\ V$, $v_2 = 20\ V$, and $v_3 = 30\ V$. From these we can find the current through each device and the power absorbed or delivered by each device.

Check with MATLAB:

```
G=[1 0 0; -3 3 4; 0 -1 1]; I=[12 144 10]'; V=G\I;...
fprintf(' \n'); fprintf('v1 = %5.2f volts \t', V(1)); ...
fprintf('v2 = %5.2f volts \t', V(2)); fprintf('v3 = %5.2f volts', V(3)); fprintf(' \n')

v1 = 12.00 volts    v2 = 20.00 volts    v3 = 30.00 volts
```

Check with Excel:

	A	B	C	D	E	F	G	H
1	Spreadsheet for Matrix Inversion and Matrix Multiplication							
2								
3		1	0	0			12	
4	G=	-3	3	4		I=	144	
5		0	-1	1			10	
6								
7		1.000	0.000	0.000			12	
8	G^{-1}=	0.429	0.143	-0.571		V=	20	
9		0.429	0.143	0.429			30	

Figure 3.7. Spreadsheet for the solution of (3.12)

Table 3.1 shows that the power delivered is equal to the power absorbed.

3.3 Analysis with Mesh or Loop Equations

In writing mesh or loop equations, we follow these steps:

1. For a circuit containing $M = L = B - N + 1$ meshes (or loops), we assign a mesh or loop current $i_1, i_2, ..., i_{n-1}$ for each mesh or loop.

2. If the circuit does not contain any current sources, we apply KVL around each mesh or loop.

TABLE 3.1 Table for Example 3.2

Device	Voltage (volts)	Current (amps)	Power (watts) Delivered	Absorbed
12 V Source	12	2	24	
10 V Source	10	19		190
18 A Source	20	18		360
24 A Source	30	24	720	
4 Ω Resistor	12	3		36
6 Ω Resistor	30	5		150
8 Ω Resistor	8	1		8
Total			744	744

3. If the circuit contains a current source between two meshes or loops, say meshes or loops j and k denoted as mesh variables i_j and i_k, we replace the current source with an open circuit thus forming a common mesh or loop, and we write a mesh or loop equation for this common mesh or loop in terms of both i_j and i_k. Then, we relate the current source to the mesh or loop variables i_j and i_k.

Example 3.3

For the circuit of Figure 3.8, write mesh equations in matrix form and solve for the unknowns using matrix theory, Cramer's rule, or Gauss's elimination method. Verify your answers with Excel or MATLAB. Please refer to Appendix A for procedures and examples. Then construct a table showing the voltages across, the currents through, and the power absorbed or delivered by each device.

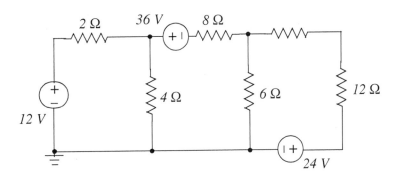

Figure 3.8. Circuit for Example 3.3

Solution:

For this circuit we need $M = L = B - N + 1 = 9 - 7 + 1 = 3$ mesh or loop equations and we arbitrarily assign currents i_1, i_2, and i_3 all in a clockwise direction as shown in Figure 3.9.

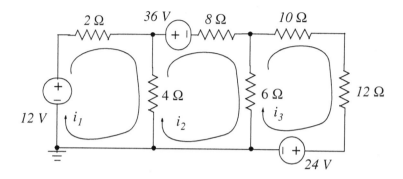

Figure 3.9. Circuit for Example 3.3

Applying KVL around each mesh we get:

Mesh #1: Starting with the left side of the $2 \ \Omega$ resistor, going clockwise, and observing the passive sign convention, we get the equation for this mesh as

$$2i_1 + 4(i_1 - i_2) - 12 = 0$$

or

$$\boxed{6i_1 - 4i_2 = 12} \tag{3.13}$$

Mesh #2: Starting with the lower end of the $4 \ \Omega$ resistor, going clockwise, and observing the passive sign convention, we get the equation

$$4(i_2 - i_1) + 36 + 8i_2 + 6(i_2 - i_3) = 0$$

or

$$\boxed{-4i_1 + 18i_2 - 6i_3 = -36} \tag{3.14}$$

Mesh #3: Starting with the lower end of the $6 \ \Omega$ resistor, going clockwise, and observing the passive sign convention, we get:

$$6(i_3 - i_2) + 10i_3 + 12i_3 + 24 = 0$$

or

$$\boxed{-6i_2 + 28i_3 = -24} \tag{3.15}$$

Note: For this example, we assigned all three currents with the same direction, i.e., clockwise. This, of course, was not mandatory; we could have assigned any direction in any mesh. It is advantageous, however, to assign the same direction to all currents. The advantage here is that we can check, or even write the mesh equations by inspection. This is best explained with the following observations:

1. With reference to the circuit of Figure 3.9 and equation (3.13), we see that current i_1 flows through the $2 \ \Omega$ and $4 \ \Omega$ resistors. We call these the *self resistances* of the first mesh. Their sum, i.e., $2 + 4 = 6$ is the coefficient of current i_1 in that equation. We observe that current i_2 also flows through the $4 \ \Omega$ resistor. We call this resistance the *mutual resistance* between the first and the second mesh. Since i_2 enters the lower end of the $4 \ \Omega$ resistor, and in writing equation (3.13) we have assumed that the upper end of this resistor has the plus (+) polarity, then in accordance with the passive sign convention, the voltage drop due to current i_2 is $-4i_2$ and this is the second term on the left side of (3.13).

2. In Mesh 2, the self resistances are the $4 \ \Omega$, $8 \ \Omega$, and $6 \ \Omega$ resistors whose sum, 18, is the coefficient of i_2 in equation (3.14). The $4 \ \Omega$ and $6 \ \Omega$ resistors are also the mutual resistances between the first and second, and the second and the third meshes respectively. Accordingly, the voltage drops due to the mutual resistances in the second equation have a minus (-) sign, i.e, $-4i_1$ and $-6i_3$.

3. The signs of the coefficients of i_2 and i_3 in (3.15) are similarly related to the self and mutual resistances in the third mesh.

Simplifying and rearranging (3.13), (3.14) and (3.15) we get:

$$3i_1 - 2i_2 = 6 \qquad (3.16)$$

$$2i_1 - 9i_2 + 3i_3 = 18 \qquad (3.17)$$

$$3i_2 - 14i_3 = 12 \qquad (3.18)$$

and in matrix form

$$\underbrace{\begin{bmatrix} 3 & -2 & 0 \\ 2 & -9 & 3 \\ 0 & 3 & -14 \end{bmatrix}}_{R} \underbrace{\begin{bmatrix} i_1 \\ i_2 \\ i_3 \end{bmatrix}}_{I} = \underbrace{\begin{bmatrix} 6 \\ 18 \\ 12 \end{bmatrix}}_{V} \qquad (3.19)$$

Simultaneous solution yields $i_1 = 0.4271$, $i_2 = -2.3593$, and $i_3 = -1.3627$ where the negative values for i_2 and i_3 indicate that the actual direction for these currents is counterclockwise.

Check with MATLAB:

R=[3 -2 0; 2 -9 3; 0 3 -14]; V=[6 18 12]'; I=R\V;...
fprintf(' \n'); fprintf('i1 = %5.2f amps \t', I(1)); ...
fprintf('i2 = %5.2f amps \t', I(2)); fprintf('i3 = %5.2f amps', I(3)); fprintf(' \n')

i1 = 0.43 amps   i2 = -2.36 amps   i3 = -1.36 amps

Excel produces the same answers as shown in Figure 3.10.

	A	B	C	D	E	F	G	H
1	Spreadsheet for Matrix Inversion and Matrix Multiplication							
2								
3		3	-2	0			6	
4	R=	2	-9	3		V=	18	
5		0	3	-14			12	
6								
7		0.397	-0.095	-0.020			0.4271	
8	R⁻¹=	0.095	-0.142	-0.031		I=	-2.3593	
9		0.020	-0.031	-0.078			-1.3627	

Figure 3.10. Spreadsheet for the solution of (3.19)

Table 3.2 shows that the power delivered by the voltage sources is equal to the power absorbed by the resistors.

TABLE 3.2 Table for Example 3.3

Device	Voltage (volts)	Current (amps)	Power (watts)	
			Delivered	Absorbed
12 V Source	12.000	0.427	5.124	
36 V Source	36.000	2.359	84.924	
24 V Source	24.000	1.363	32.712	
2 Ω Resistor	0.854	0.427		0.365
4 Ω Resistor	11.144	2.786		30.964
8 Ω Resistor	18.874	2.359		44.530
6 Ω Resistor	5.976	0.996		5.952
10 Ω Resistor	13.627	1.363		18.570
12 Ω Resistor	16.352	1.363		22.288
Total			122.760	122.669

Example 3.4

For the circuit of Figure 3.11, write loop equations in matrix form, and solve for the unknowns using matrix theory, Cramer's rule, or Gauss's elimination method. Verify your answers with Excel or MATLAB. Please refer to Appendix A for procedures and examples. Then, construct a table showing the voltages across, the currents through and the power absorbed or delivered by each device.

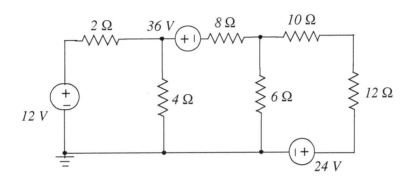

Figure 3.11. Circuit for Example 3.4

Solution:

This is the same circuit as that of the previous example where we found that we need 3 mesh or loop equations. We choose our loops as shown in Figure 3.12, and we assign currents i_1, i_2, and i_3, all in a clockwise direction.

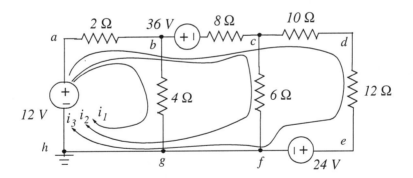

Figure 3.12. Circuit for Example 3.4 with assigned loops

Applying of KVL around each loop, we get:

Loop 1 (abgh): Starting with the left side of the $2\ \Omega$ resistor and complying with the passive sign convention, we get:

$$2(i_1 + i_2 + i_3) + 4i_1 - 12 = 0$$

or

$$6i_1 + 2i_2 + 2i_3 = 12$$

or

$$\boxed{3i_1 + i_2 + i_3 = 6} \tag{3.20}$$

Loop 2 (abcfgh): As before, starting with the left side of the $2 \ \Omega$ resistor and complying with the passive sign convention, we get:

$$2(i_1 + i_2 + i_3) + 36 + 8(i_2 + i_3) + 6i_2 - 12 = 0$$

or

$$2i_1 + 16i_2 + 10i_3 = -24$$

or

$$\boxed{i_1 + 8i_2 + 5i_3 = -12} \tag{3.21}$$

Loop 3 (abcdefgh): Likewise, starting with the left side of the $2 \ \Omega$ resistor and complying with the passive sign convention, we get:

$$2(i_1 + i_2 + i_3) + 36 + 8(i_2 + i_3) + 10i_3 + 12i_3 + 24 - 12 = 0$$

or

$$2i_1 + 10i_2 + 32i_3 = -48$$

or

$$\boxed{i_1 + 5i_2 + 16i_3 = -24} \tag{3.22}$$

and in matrix form

$$\underbrace{\begin{bmatrix} 3 & 1 & 1 \\ 1 & 8 & 5 \\ 1 & 5 & 16 \end{bmatrix}}_{R} \underbrace{\begin{bmatrix} i_1 \\ i_2 \\ i_3 \end{bmatrix}}_{I} = \underbrace{\begin{bmatrix} 6 \\ -12 \\ -24 \end{bmatrix}}_{V} \tag{3.23}$$

Solving with MATLAB we get:

```
R=[3 1 1; 1 8 5; 1 5 16]; V=[6 -12 -24]'; I=R\V;...
fprintf(' \n'); fprintf('i1 = %5.2f amps \t', I(1)); ...
fprintf('i2 = %5.2f amps \t', I(2)); fprintf('i3 = %5.2f amps', I(3)); fprintf(' \n')

i1 = 2.79 amps    i2 = -1.00 amps    i3 = -1.36 amps
```

Excel produces the same answers.

Table 3.3 shows that the power delivered by the voltage sources is equal to the power absorbed by the resistors and the values are approximately the same as those of the previous example.

TABLE 3.3 Table for Example 3.4

Device	Voltage (volts)	Current (amps)	Power (watts) Delivered	Absorbed
12 V Source	12.000	0.427	5.124	
36 V Source	36.000	2.359	84.924	
24 V Source	24.000	1.363	32.712	
2 Ω Resistor	0.854	0.427		0.365
4 Ω Resistor	11.146	2.786		31.053
8 Ω Resistor	18.872	2.359		44.519
6 Ω Resistor	5.982	0.997		5.964
10 Ω Resistor	13.627	1.363		18.574
12 Ω Resistor	16.352	1.363		22.283
Total			122.760	122.758

Example 3.5

For the circuit of figure 3.13, write mesh equations in matrix form and solve for the unknowns using matrix theory, Cramer's rule, or the substitution method. Verify your answers with Excel or MATLAB. Please refer to Appendix A for procedures and examples.

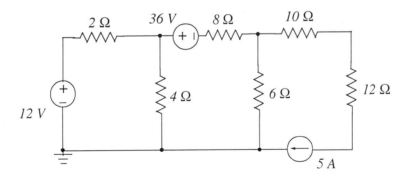

Figure 3.13. Circuit for Example 3.5

Solution:

This is the same circuit as those of the two previous examples except that the 24 V voltage source has been replaced by a 5 A current source. As before, we need $M = L = B - N + 1 = 9 - 7 + 1 = 3$ mesh or loop equations, and we assign currents i_1, i_2, and i_3 all in a clockwise direction as shown in Figure 3.14.

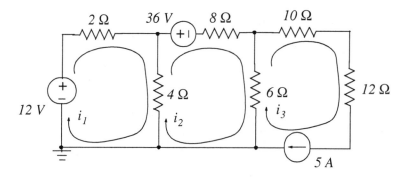

Figure 3.14. Circuit for Example 3.5 with assigned currents

For Meshes 1 and 2, the equations are the same as in Example 3.3 where we found them to be

$$6i_1 - 4i_2 = 12$$

or

$$3i_1 - 2i_2 = 6 \qquad (3.24)$$

and

$$-4i_1 + 18i_2 - 6i_3 = -36$$

or

$$2i_1 - 9i_2 + 3i_3 = 18 \qquad (3.25)$$

For Mesh 3, we observe that the current i_3 is just the current of the 5 A current source and thus our third equation is simply

$$i_3 = 5 \qquad (3.26)$$

and in matrix form,

$$\underbrace{\begin{bmatrix} 3 & -2 & 0 \\ 2 & -9 & 3 \\ 0 & 0 & 1 \end{bmatrix}}_{R} \underbrace{\begin{bmatrix} i_1 \\ i_2 \\ i_3 \end{bmatrix}}_{I} = \underbrace{\begin{bmatrix} 6 \\ 18 \\ 5 \end{bmatrix}}_{V}$$

(3.27)

Solving with MATLAB we get:

```
R=[3 -2 0; 2 -9 3; 0 0 1]; V=[6 18 5]'; I=R\V;...
fprintf(' \n'); fprintf('i1 = %5.2f amps \t', I(1)); ...
fprintf('i2 = %5.2f amps \t', I(2)); fprintf('i3 = %5.2f amps', I(3)); fprintf(' \n')

i1 = 2.09 amps   i2 = 0.13 amps   i3 =  5.00 amps
```

Example 3.6

Write mesh equations for the circuit of Figure 3.15 and solve for the unknowns using MATLAB or Excel. Then, compute the voltage drop across the $5\ A$ source.

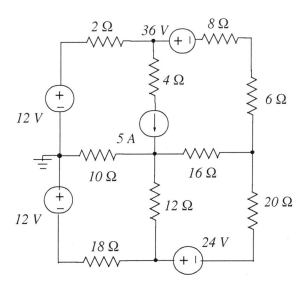

Figure 3.15. Circuit for Example 3.6

Solution:

Here, we would be tempted to assign mesh currents as shown in Figure 3.16. However, we will encounter a problem as explained below.

The currents i_3 and i_4 for Meshes 3 and 4 respectively present no problem; but for Meshes 1 and 2 we cannot write mesh equations for the currents i_1 and i_2 as shown because we cannot write a term which represents the voltage across the *5 A* current source. To work around this problem we temporarily remove (open) the *5 A* current source and we form a *"combined mesh"* (or *generalized mesh* or *supermesh* as some textbooks call it) and the current that flows around this combined mesh is as shown in Figure 3.17.

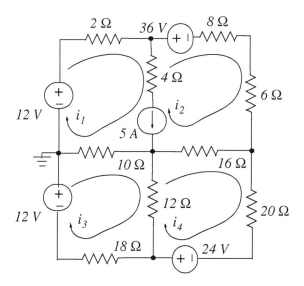

Figure 3.16. Circuit for Example 3.6 with erroneous current assignments

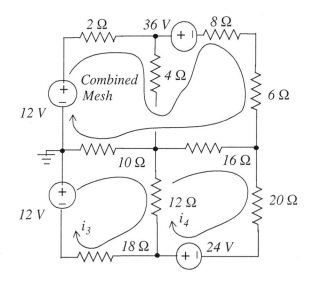

Figure 3.17. Circuit for Example 3.6 with correct current assignments

Now, we apply KVL around this combined mesh. We start at the left end of the *2 Ω* resistor, and we express the voltage drop across this resistor as $2i_1$ since in Mesh 1 the current is essentially i_1.

Continuing, we observe that there is no voltage drop across the $4 \ \Omega$ resistor since no current flows through it. The current now enters Mesh 2 where we encounter the $36 \ V$ drop due to the voltage source there, and the voltage drops across the $8 \ \Omega$ and $6 \ \Omega$ resistors are $8i_2$ and $6i_2$ respectively since in Mesh 2 the current now is really i_2. The voltage drops across the $16 \ \Omega$ and $10 \ \Omega$ resistors are expressed as in the previous examples and thus our first mesh equation is

$$2i_1 + 36 + 8i_2 + 6i_2 + 16(i_2 - i_4) + 10(i_1 - i_3) - 12 = 0$$

or

$$12i_1 + 30i_2 - 10i_3 - 16i_4 = -24$$

or

$$\boxed{6i_1 + 15i_2 - 5i_3 - 8i_4 = -12} \tag{3.28}$$

Now, we reinsert the 5 A current source between Meshes 1 and 2 and we obtain our second equation as

$$\boxed{i_1 - i_2 = 5} \tag{3.29}$$

For meshes 3 and 4, the equations are

$$10(i_3 - i_1) + 12(i_3 - i_4) + 18i_3 - 12 = 0$$

or

$$\boxed{5i_1 - 20i_3 + 6i_4 = -6} \tag{3.30}$$

and

$$16(i_4 - i_2) + 20i_4 - 24 + 12(i_4 - i_3) = 0$$

or

$$\boxed{4i_2 + 3i_3 - 12i_4 = -6} \tag{3.31}$$

and in matrix form

$$\underbrace{\begin{bmatrix} 6 & 15 & -5 & -8 \\ 1 & -1 & 0 & 0 \\ 5 & 0 & -20 & 6 \\ 0 & 4 & 3 & -12 \end{bmatrix}}_{R} \underbrace{\begin{bmatrix} i_1 \\ i_2 \\ i_3 \\ i_4 \end{bmatrix}}_{I} = \underbrace{\begin{bmatrix} -12 \\ 5 \\ -6 \\ -6 \end{bmatrix}}_{V} \tag{3.32}$$

We find the solution of (3.32) with the following MATLAB code.

```
R=[6 15 -5 -8; 1 -1 0 0; 5 0 -20 6; 0 4 3 -12]; V=[-12 5 -6 -6]'; I=R\V;...
fprintf(' \n');...
fprintf('i1 = %5.4f amps \t',I(1)); fprintf('i2 = %5.4f amps \t',I(2));...
fprintf('i3 = %5.4f amps \t',I(3)); fprintf('i4 = %5.4f amps',I(4)); fprintf(' \n')
```

```
i1=3.3975 amps  i2=-1.6025 amps  i3=1.2315 amps  i4=0.2737 amps
```

Now, we can find the voltage drop across the *5 A* current source by application of KVL around Mesh 1 using the following relation:

$$2 \times 3.3975 + 4 \times (3.3975 + 1.6025) + v_{5A} + 10 \times (3.3975 - 1.2315) - 12 = 0$$

This yields

$$v_{5A} = -36.455$$

We can verify this value by application of KVL around Mesh 2 where starting with the lower end of the *6 w* resistor and going counterclockwise we get

$$(6 + 8) \times 1.6025 - 36 + 4 \times (3.3975 + 1.6025) - 36.455 + 16 \times (1.6025 + 0.2737) = 0$$

With these values, we can also compute the power delivered or absorbed by each of the voltage sources and the current source.

3.4 Transformation between Voltage and Current Sources

In the previous chapter we stated that a voltage source maintains a constant voltage between its terminals regardless of the current that flows through it. This statement applies to an ideal voltage source which, of course, does not exist; for instance, no voltage source can supply infinite current to a short circuit. We also stated that a current source maintains a constant current regardless of the terminal voltage. This statement applies to an ideal current source which also does not exist; for instance, no current source can supply infinite voltage when its terminals are open-circuited.

A *practical voltage source* has an internal resistance which, to be accounted for, it is represented with an external resistance R_S in series with the voltage source v_S as shown in Figure 3.18 (a). Likewise a *practical current source* has an internal conductance which is represented as a resistance R_p (or conductance G_p) in parallel with the current source i_S as shown in Figure 3.18 (b).

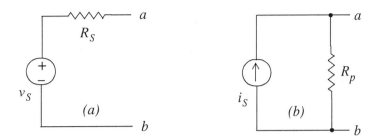

Figure 3.18. Practical voltage and current sources

In Figure 3.18 (a), the voltage of the source will always be v_S but the terminal voltage v_{ab} will be $v_{ab} = v_S - v_{R_s}$ if a load is connected at points a and b. Likewise, in Figure 3.18 (b) the current of the source will always be i_S but the terminal current i_{ab} will be $i_{ab} = i_S - i_{R_P}$ if a load is connected at points a and b.

Now, we will show that the networks of Figures 3.18 (a) and 3.18 (b) can be made equivalent to each other.

In the networks of Figures 3.19 (a) and 3.19 (b), the load resistor R_L is the same in both.

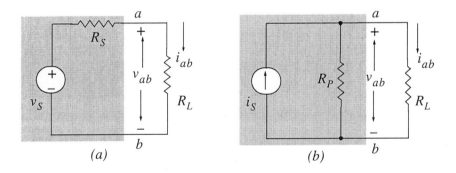

Figure 3.19. Equivalent sources

From the circuit of Figure 3.19 (a),

$$v_{ab} = \frac{R_L}{R_S + R_L} v_S \tag{3.33}$$

and

$$i_{ab} = \frac{v_S}{R_S + R_L} \tag{3.34}$$

From the circuit of Figure 3.19 (b),

$$v_{ab} = \frac{R_P R_L}{R_p + R_L} i_S \qquad (3.35)$$

and

$$i_{ab} = \frac{R_P}{R_p + R_L} i_S \qquad (3.36)$$

Since we want v_{ab} to be the same in both circuits 3.19 (a) and 3.19 (b), from (3.33) and (3.35) we get:

$$v_{ab} = \frac{R_L}{R_S + R_L} v_S = \frac{R_P R_L}{R_p + R_L} i_S \qquad (3.37)$$

Likewise, we want i_{ab} to be the same in both circuits 3.19 (a) and 3.19 (b). Then, from (3.34) and (3.36) we get:

$$i_{ab} = \frac{v_S}{R_S + R_L} = \frac{R_p}{R_p + R_L} i_S \qquad (3.38)$$

and for any R_L, from (3.37) and (3.38)

$$\boxed{v_S = R_p i_S} \qquad (3.39)$$

and

$$\boxed{R_p = R_S} \qquad (3.40)$$

Therefore, a voltage source v_S in series with a resistance R_S can be transformed to a current source i_S whose value is equal to v_S / R_S, in parallel with a resistance R_p whose value is the same as R_S.

Likewise, a current source i_S in parallel with a resistance R_p can be transformed to a voltage source v_S whose value is equal to $i_S \times R_S$, in series with a resistance whose value is the same as R_p.

The voltage-to-current source or current-to-voltage source transformation is not limited to a single resistance load; it applies to any load no matter how complex.

Example 3.7

Find the current i_{10} through the *10* Ω resistor in the circuit of Figure 3.20.

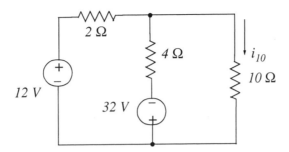

Figure 3.20. Circuit for Example 3.7

Solution:

This problem can be solved either by nodal or by mesh analysis; however, we will transform the voltage sources to current sources and we will replace the resistances with conductances except the $10 \ \Omega$ resistor. We will treat the $10 \ \Omega$ resistor as the load of this circuit so that we can compute the current i_{10} through it. Then, the circuit becomes as shown in Figure 3.21.

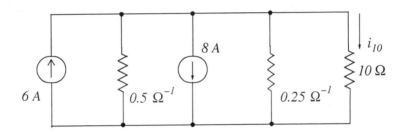

Figure 3.21. Circuit for Example 3.7 where voltage sources have been transformed to current sources

Combination of the two current sources and their conductances yields the circuit shown in Figure 3.22.

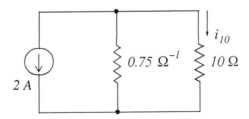

Figure 3.22. Circuit for Example 3.7 after combinations of current sources and conductances

Converting the $0.75 \ \Omega^{-1}$ conductance to a resistance and performing current-to-voltage source transformation, we get the circuit of Figure 3.23.

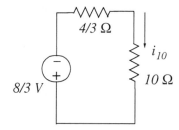

Figure 3.23. Circuit for Example 3.7 in its simplest form

Thus, the current through the $10 \, \Omega$ resistor is

$$i_{10} = \frac{-8/3}{10 + 4/3} = -4/17 \, A$$

3.5 Thevenin's Theorem

This theorem is perhaps the greatest time saver in circuit analysis, especially in electronic circuits. It states that we can replace a two terminal network by a voltage source v_{TH} in series with a resistance R_{TH} as shown in Figure 3.24.

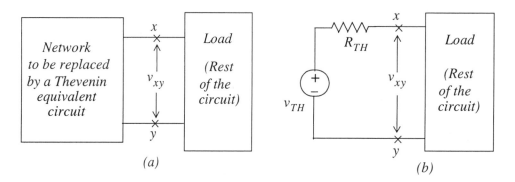

Figure 3.24. Replacement of a network by its Thevenin's equivalent

The network of Figure 3.24 (b) will be equivalent to the network of Figure 3.24 (a) if the load is removed in which case both networks will have the same open circuit voltages v_{xy} and consequently,

$$v_{TH} = v_{xy}$$

Therefore,

$$\boxed{v_{TH} = v_{xy \ open}}$$

(3.41)

The Thevenin resistance R_{TH} represents the equivalent resistance of the network being replaced by the Thevenin equivalent, and it is found from the relation

$$R_{TH} = \frac{v_{xy\ open}}{i_{xy\ short}} = \frac{v_{TH}}{i_{SC}} \qquad (3.42)$$

where i_{SC} stands for short-circuit current.

If the network to be replaced by a Thevenin equivalent contains independent sources only, we can find the Thevenin resistance R_{TH} by first shorting all (independent) voltage sources, opening all (independent) current sources, and calculating the *resistance looking into the direction that is opposite to the load when it has been disconnected from the rest of the circuit at terminals x and y*.

Example 3.8

Use Thevenin's theorem to find i_{LOAD} and v_{LOAD} for the circuit of Figure 3.25.

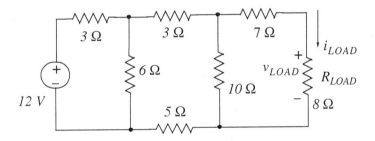

Figure 3.25. Circuit for Example 3.8

Solution:

We will apply Thevenin's theorem twice; first at terminals x and y and then at x' and y' as shown in Figure 3.26.

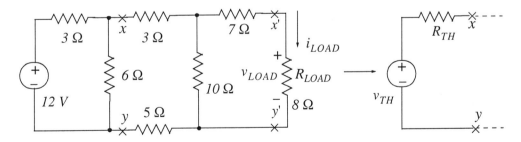

Figure 3.26. First step in finding the Thevenin equivalent of the circuit of Example 3.8

Breaking the circuit at $x - y$, we are left with the circuit shown in Figure 3.27.

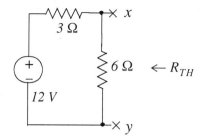

Figure 3.27. Second step in finding the Thevenin equivalent of the circuit of Example 3.8

Applying Thevenin's theorem at x and y and using the voltage division expression, we get

$$v_{TH} = v_{xy} = \frac{6}{3+6} \times 12 = 8 \ V$$

$$R_{TH}\Big|_{v_s = 0} = \frac{3 \times 6}{3+6} = 2 \ \Omega$$

(3.43)

and thus the equivalent circuit to the left of points x and x is as shown in Figure 3.28.

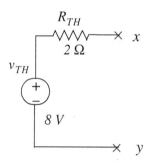

Figure 3.28. First Thevenin equivalent for the circuit of Example 3.8

Next, we attach the remaining part of the given circuit to the Thevenin equivalent of Figure 3.28, and the new circuit now is as shown in Figure 3.29.

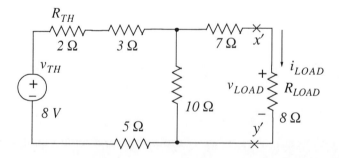

Figure 3.29. Circuit for Example 3.8 with first Thevenin equivalent

Now, we apply Thevenin's theorem at points x' and y' and we get the circuit of Figure 3.30.

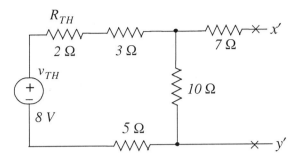

Figure 3.30. Applying Thevenin's theorem at points x' and y' for the circuit for Example 3.8

Using the voltage division expression, we get

$$v'_{TH} = v_{x'y'} = \frac{10}{2 + 3 + 10 + 5} \times 8 = 4 \ V$$

$$R'_{TH}\big|_{v_{TH} = 0} = [(2 + 3 + 5)||10] + 7 = 12 \ \Omega$$

This Thevenin equivalent with the load resistor attached to it, is shown in Figure 3.31.

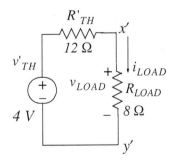

Figure 3.31. Entire circuit of Example 3.8 simplified by Thevenin's theorem

The voltage v_{LOAD} is found by application of the voltage division expression, and the current i_{LOAD} by Ohm's law as shown below.

$$v_{LOAD} = \frac{8}{12 + 8} \times 4 = 1.6 \ V$$

$$i_{LOAD} = \frac{4}{12 + 8} = 0.2 \ A$$

It is imperative to remember that when we compute the Thevenin equivalent resistance, we must always look towards the network portion which remains after disconnectinf the load at the x and y terminals. This is illustrated with the two examples that follow.

Let us consider the network of Figure 3.32 (a).

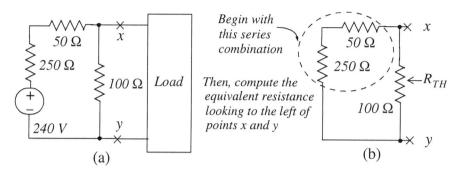

Figure 3.32. Computation of the Thevenin equivalent resistance when the load is to the right

This network contains no dependent sources; therefore, we can find the Thevenin equivalent by shorting the *240 V* voltage source, and computing the equivalent resistance looking to the left of points *x* and *y* as indicated in Figure 3.32 (b). Thus,

$$R_{TH} = (250 + 50)||100 = \frac{300 \times 100}{300 + 100} = 75 \ \Omega$$

Now, let us consider the network of Figure 3.33 (a).

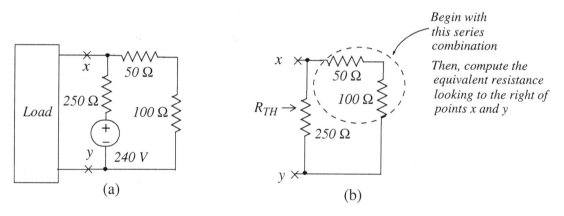

Figure 3.33. Computation of the Thevenin equivalent resistance when the load is to the left

This network contains no dependent sources; therefore, we can find the Thevenin equivalent by shorting the *240 V* voltage source, and computing the equivalent resistance looking to the right of points *x* and *y* as indicated in Figure 3.33 (b). Thus,

$$R_{TH} = (50 + 100)||250 = \frac{150 \times 250}{150 + 250} = 93.75 \ \Omega$$

We observe that, although the resistors in the networks of Figures 3.32 (a) and 3.33 (b) have the same values, the Thevenin resistance is different since it depends on the direction in which we look into (left or right).

Example 3.9

Use Thevenin's theorem to find i_{LOAD} and v_{LOAD} for the circuit of Figure 3.34.

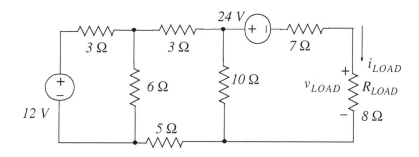

Figure 3.34. Circuit for Example 3.9

Solution:

This is the same circuit as the previous example except that a voltage source of *24 V* has been placed in series with the *7 Ω* resistor. By application of Thevenin's theorem at points *x* and *y* as before, and connecting the rest of the circuit, we get the circuit of Figure 3.35.

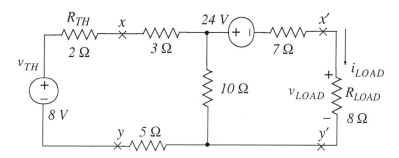

Figure 3.35. Circuit for Example 3.9 with first Thevenin equivalent

Next, disconnecting the load resistor and applying Thevenin's theorem at points *x'* and *y'* we get the circuit of Figure 3.36.

There is no current flow in the *7 Ω* resistor; therefore, the Thevenin voltage across the *x'* and *y'* points is the algebraic sum of the voltage drop across the *10 Ω* resistor and the *24 V* source, that is,

$$v'_{TH} = v_{x'y'} = \frac{10}{2 + 3 + 10 + 5} \times 8 - 24 = -20 \ V$$

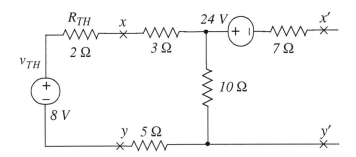

Figure 3.36. Applying Thevenin's theorem at points x' and y' for the circuit for Example 3.9

and the Thevenin resistance is the same as in the previous example, that is,

$$R'_{TH}\big|_{V_{TH} = 0} = [(2 + 3 + 5)||10] + 7 = 12 \ \Omega$$

Finally, connecting the load R_{LOAD} as shown in Figure 3.37, we compute v_{LOAD} and i_{LOAD} as follows:

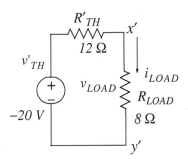

Figure 3.37. Final form of Thevenin equivalent with load connected for circuit of Example 3.9

$$v_{LOAD} = \frac{8}{12 + 8} \times (-20) = -8 \ V$$

$$i_{LOAD} = \frac{-20}{12 + 8} = -1 \ A$$

Example 3.10

For the circuit of Figure 3.38, use Thevenin's theorem to find i_{LOAD} and v_{LOAD}.

Solution:

This circuit contains a dependent voltage source whose value is twenty times the current through the 6 Ω resistor. We will apply Thevenin's theorem at points a and b as shown in Figure 3.39.

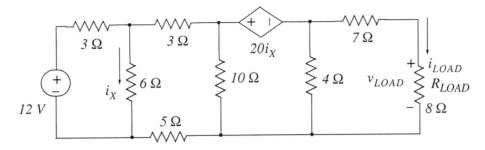

Figure 3.38. Circuit for Example 3.10

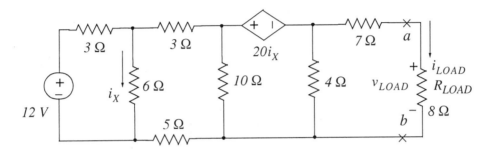

Figure 3.39. Application of Thevenin's theorem for Example 3.10

For the circuit of Figure 3.39, we cannot short the dependent source; therefore, we will find the Thevenin resistance from the relation

$$R_{TH} = \frac{v_{OC}}{i_{SC}} = \frac{v_{LOAD}\big|_{R_L \to \infty}}{i_{LOAD}\big|_{R_L \to 0}} \tag{3.44}$$

To find the open circuit voltage $v_{OC} = v_{ab}$, we disconnect the load resistor and our circuit now is as shown in Figure 3.40.

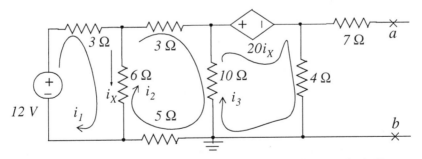

Figure 3.40. Circuit for finding $v_{OC} = v_{ab}$ of Example 3.10

We will use mesh analysis to find v_{OC} which is the voltage across the 4 Ω resistor. We chose mesh analysis since we only need three mesh equations whereas we would need five equations had we chosen nodal analysis. Please refer to Exercise 16 at the end of this chapter for a solution requiring nodal analysis.

Observing that $i_X = i_1 - i_2$, we write the three mesh equations for this network as

$$9i_1 - 6i_2 = 12$$
$$-6i_1 + 24i_2 - 10i_3 = 0 \qquad (3.45)$$
$$20(i_1 - i_2) + 4i_3 + 10(i_3 - i_2) = 0$$

and after simplification and combination of like terms, we write them in matrix form as

$$\underbrace{\begin{bmatrix} 3 & -2 & 0 \\ 3 & -12 & 5 \\ 10 & -15 & 7 \end{bmatrix}}_{R} \underbrace{\begin{bmatrix} I_1 \\ I_2 \\ I_3 \end{bmatrix}}_{I} = \underbrace{\begin{bmatrix} 4 \\ 0 \\ 0 \end{bmatrix}}_{V} \qquad (3.46)$$

Using the spreadsheet of Figure 3.41, we find that $i_3 = -3.53 \ A$

	A	B	C	D	E	F	G	H
1	Spreadsheet for Matrix Inversion and Matrix Multiplication							
3		3	-2	0			4	
4	R=	3	-12	5		V=	0	
5		10	-15	7			0	
7		0.106	-0.165	0.118			0.42	
8	R⁻¹=	-0.341	-0.247	0.176		I=	-1.36	
9		-0.882	-0.294	0.353			-3.53	

Figure 3.41. Spreadsheet for Example 3.10

Thus, the Thevenin voltage at points a and b is

$$v_{TH} = (-3.53) \times 4 = -14.18 \ V$$

Next, to find the Thevenin resistance R_{TH}, we must first compute the short circuit current I_{SC}. Accordingly, we place a short across points a and b and the circuit now is as shown in Figure 3.42 and we can find the short circuit current i_{SC} from the circuit of Figure 3.43 where $i_{SC} = i_4$

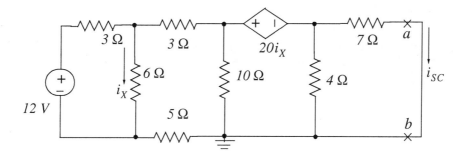

Figure 3.42. Circuit for finding $i_{SC} = i_{ab}$ *in Example 3.10*

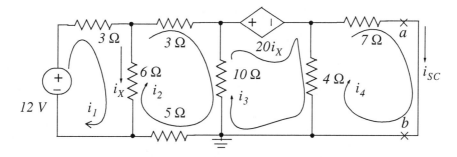

Figure 3.43. Mesh equations for finding $i_{SC} = i_{ab}$ *in Example 3.10*

The mesh equations for the circuit of Figure 3.43 are

$$9i_1 - 6i_2 = 12$$
$$-6i_1 + 24i_2 - 10i_3 = 0$$
$$20(i_1 - i_2) + 4(i_3 - i_4) + 10(i_3 - i_2) = 0 \qquad (3.47)$$
$$-4i_3 + 11i_4 = 0$$

and after simplification and combination of like terms, we write them in matrix form as

$$\underbrace{\begin{bmatrix} 3 & -2 & 0 & 0 \\ 3 & -12 & 5 & 0 \\ 10 & -15 & 7 & -2 \\ 0 & 0 & -4 & 11 \end{bmatrix}}_{R} \underbrace{\begin{bmatrix} i_1 \\ i_2 \\ i_3 \\ i_4 \end{bmatrix}}_{I} = \underbrace{\begin{bmatrix} 4 \\ 0 \\ 0 \\ 0 \end{bmatrix}}_{V} \qquad (3.48)$$

We will solve these using MATLAB as follows:

```
R=[3 −2 0 0; 3 −12 5 0; 10 −15 7 −2; 0 0 −4 11]; V=[4 0 0 0]'; I=R\V;
fprintf(' \n');...
fprintf('i1 = %3.4f A \t',I(1,1)); fprintf('i2 = %3.4f A \t',I(2,1));...
fprintf('i3 = %3.4f A \t',I(3,1)); fprintf('i4 = %3.4f A \t',I(4,1));...
fprintf(' \n');...fprintf(' \n')
```

```
i1 = 0.0173 A    i2 = -1.9741 A    i3 = -4.7482 A    i4 = -1.7266 A
```

Therefore,

$$i_{SC} = i_4 = -1.727$$

and

$$R_{TH} = \frac{v_{OC}}{i_{SC}} = \frac{-14.18}{-1.727} = 8.2 \ \Omega$$

The Thevenin equivalent is as shown in Figure 3.44.

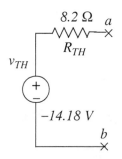

Figure 3.44. Final form of Thevenin's equivalent for the circuit of Example 3.10

Finally, with the load R_{LOAD} attached to points a and b, the circuit is as shown in Figure 3.45.

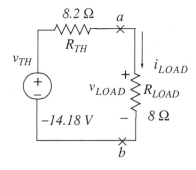

Figure 3.45. Circuit for finding v_{LOAD} and i_{LOAD} in Example 3.10

Therefore, using the voltage division expression and Ohm's law we get

$$v_{LOAD} = \frac{8}{8.2 + 8} \times (-14.18) = -7.00 \ V$$

$$i_{LOAD} = \frac{-14.18}{8.2 + 8} = -0.875 \ A$$

3.6 Norton's Theorem

This theorem is analogous to Thevenin's theorem and states that we can replace everything, except the load, in a circuit by an equivalent circuit containing only an independent current source which we will denote as i_N in parallel with a resistance which we will denote as R_N, as shown in Figure 3.46.

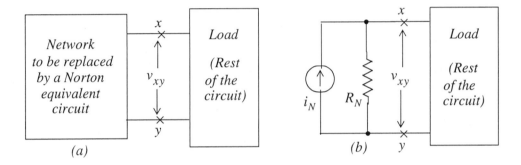

Figure 3.46. Replacement of a network by its Norton equivalent

The current source i_N has the value of the short circuit current which would flow if a short were connected between the terminals x and y, where the Norton equivalent is inserted, and the resistance R_N is found from the relation

$$R_N = \frac{v_{OC}}{i_{SC}} \tag{3.49}$$

where v_{OC} is the open circuit voltage which appears across the open terminals x and y.

Like Thevenin's, Norton's theorem is most useful when a series of computations involves changing the load of a network while the rest of the circuit remains unchanged.

Comparing the Thevenin's and Norton's equivalent circuits, we see that one can be derived from the other by replacing the Thevenin voltage and its series resistance with the Norton current source and its parallel resistance. Therefore, there is no need to perform separate computations for each of these equivalents; once we know Thevenin's equivalent we can easily draw the Norton equivalent and vice versa.

Example 3.11

Replace the network shown in Figure 3.47 by its Thevenin and Norton equivalents.

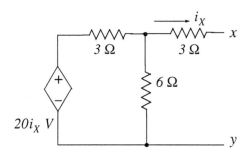

Figure 3.47. Network for Example 3.11

Solution:

We observe that no current flows through the *3 Ω* resistor; Therefore, $i_X = 0$ and the dependent current source is zero, i.e., a short circuit. Thus,

$$v_{TH} = v_{OC} = v_{xy} = 0$$

and also

$$i_{SC} = 0$$

This means that the given network is some mathematical model representing a resistance, but we cannot find this resistance from the expression

$$R_{TH} = R_N = \frac{v_{OC}}{i_{SC}}$$

since this results in the indeterminate form *0/0*. In this type of situations, we connect an external source (voltage or current) across the terminals *x* and *y*. For this example, we arbitrarily choose to connect a 1 volt source as shown in Figure 3.48.

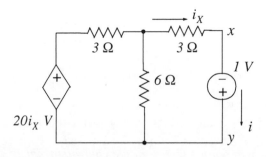

Figure 3.48. Network for Example 3.11 with an external voltage source connected to it.

In the circuit of Figure 3.48, the $1\ V$ source represents the open circuit voltage v_{OC} and the current i represents the short circuit current i_{SC}. Therefore, the Thevenin (or Norton) resistance will be found from the expression

$$R_{TH} = R_N = \frac{v_{OC}}{i_{SC}} = \frac{1\ V}{i} = \frac{1\ V}{i_X} \tag{3.50}$$

Now, we can find i from the circuit of Figure 3.49 by application of KCL at Node ①.

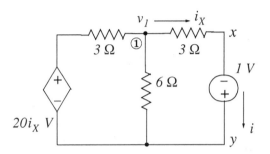

Figure 3.49. Circuit for finding i_X in Example 3.11

$$\frac{v_1 - 20i_X}{3} + \frac{v_1}{6} + i_X = 0 \tag{3.51}$$

where

$$i_X = \frac{v_1 - 1}{3} \tag{3.52}$$

Simultaneous solution of (3.51) and (3.52) yields $v_1 = 34/25$ and $i_X = 3/25$. Then, from (3.50),

$$R_{TH} = R_N = \frac{1}{3/25} = \frac{25}{3}$$

and the Thevenin and Norton equivalents are shown in Figure 3.50.

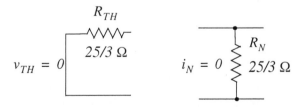

Figure 3.50. Thevenin's and Norton's equivalents for Example 3.11

3.7 Maximum Power Transfer Theorem

Consider the circuit shown in Figure 3.51. We want to find the value of R_{LOAD} which will absorb maximum power from the voltage source v_S whose internal resistance is R_S.

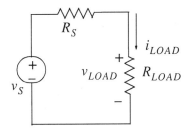

Figure 3.51. Circuit for computation of maximum power delivered to the load R_{LOAD}

The power p_{LOAD} delivered to the load is found from

$$p_{LOAD} = v_{LOAD} \times i_{LOAD} = \left(\frac{R_{LOAD}}{R_S + R_{LOAD}} v_S \right) \left(\frac{v_S}{R_S + R_{LOAD}} \right)$$

or

$$p_{LOAD} = \frac{R_{LOAD}}{(R_S + R_{LOAD})^2} v_S^2 \qquad (3.53)$$

To find the value of R_{LOAD} which will make p_{LOAD} maximum, we differentiate (3.53) with respect to R_{LOAD}. Recalling that

$$\frac{d}{dx}\left(\frac{u}{v}\right) = \frac{v\dfrac{d}{dx}(u) - u\dfrac{d}{dx}(v)}{v^2}$$

and differentiating (3.53), we get

$$\frac{dp_{LOAD}}{dR_{LOAD}} = \frac{(R_S + R_{LOAD})^2 v_S^2 - v_S^2 R_{LOAD}(2)(R_S + R_{LOAD})}{(R_S + R_{LOAD})^4} \qquad (3.54)$$

and (3.54) will be zero if the numerator is set equal to zero, that is,

$$(R_S + R_{LOAD})^2 v_S^2 - v_S^2 R_{LOAD}(2)(R_S + R_{LOAD}) = 0$$

or

$$R_S + R_{LOAD} = 2R_{LOAD}$$

or

$$\boxed{R_{LOAD} = R_S} \tag{3.55}$$

Therefore, we conclude that a voltage source with internal series resistance R_S or a current source with internal parallel resistance R_P delivers maximum power to a load R_{LOAD} when $R_{LOAD} = R_S$ or $R_{LOAD} = R_P$. For example, in the circuits of 3.52, the voltage source v_S and current source i_N deliver maximum power to the adjustable[*] load when $R_{LOAD} = R_S = R_P = 5 \ \Omega$

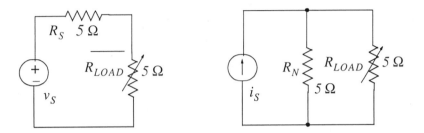

Figure 3.52. Circuits where R_{LOAD} is set to receive maximum power

We can use Excel or MATLAB to see that the load receives maximum power when it is set to the same value as that of the resistance of the source. Figure 3.53 shows a spreadsheet with various values of an adjustable resistive load. We observe that the power is maximum when $R_{LOAD} = 5 \ \Omega$.

The condition of maximum power transfer is also referred to as *resistance matching* or *impedance matching*. We will define the term "impedance" in Chapter 6.

The maximum power transfer theorem is of great importance in electronics and communications applications where it is desirable to receive maximum power from a given circuit and efficiency is not an important consideration. On the other hand, in power systems, this application is of no use since the intent is to supply a large amount of power to a given load by making the internal resistance R_S as small as possible.

3.8 Linearity

A *linear passive element* is one in which there is a linear voltage-current relationship such as

$$v_R = Ri_R \qquad v_L = L\frac{d}{dt}i_L \qquad i_C = C\frac{d}{dt}v_C \tag{3.56}$$

---

* *An adjustable resistor is usually denoted with an arrow as shown in Figure 3.52.*

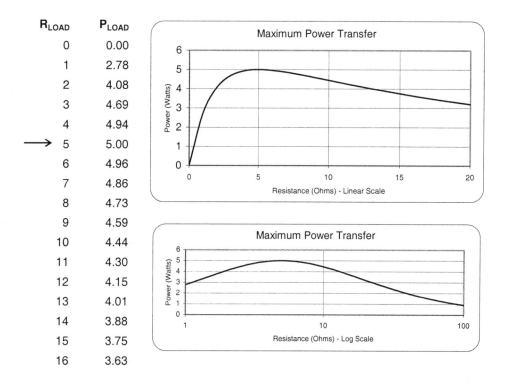

R_{LOAD}	P_{LOAD}
0	0.00
1	2.78
2	4.08
3	4.69
4	4.94
→ 5	5.00
6	4.96
7	4.86
8	4.73
9	4.59
10	4.44
11	4.30
12	4.15
13	4.01
14	3.88
15	3.75
16	3.63

Figure 3.53. Spreadsheet to illustrate maximum power transfer to a resistive load

Definition 3.1

A *linear dependent source* is a dependent voltage or current source whose output voltage or current is proportional only to the first power of some voltage or current variable in the circuit or a *linear* combination (the sum or difference of such variables). For example, $v_{xy} = 2v_1 - 3i_2$ is a linear relationship but $p = vi = Ri^2 = v^2/R$ and $i = I_S e^{v/nV_T}$ are non-linear.

Definition 3.2

A *linear circuit* is a circuit which is composed entirely of independent sources, linear dependent sources and linear passive elements or a combination of these.

Circuit Analysis I with MATLAB Applications
Orchard Publications

3.9 Superposition Principle

The *principle of superposition* states that the response (a desired voltage or current) in any branch of a linear circuit having more than one independent source can be obtained as the sum of the responses caused by each independent source acting alone with all other independent voltage sources replaced by short circuits and all other independent current sources replaced by open circuits.

Note: *Dependent sources (voltage or current) must not be superimposed* since their values depend on the voltage across or the current through some other branch of the circuit. Therefore, all dependent sources must always be left intact in the circuit while superposition is applied.

Example 3.12

For the circuit of Figure 3.54, compute i_6 by application of the superposition principle.

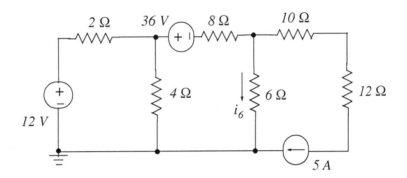

Figure 3.54. Circuit for Example 3.12

Solution:

Let i'_6 represent the current due to the *12 V* source acting alone, i''_6 the current due to the *36 V* source acting alone, and i'''_6 the current due to the *5 A* source acting alone. Then, by the principle of superposition,

$$i_6 = i'_6 + i''_6 + i'''_6$$

First, to find i'_6 we short the *36 V* voltage source and open the *5 A* current source. The circuit then reduces to that shown in Figure 3.55.

Applying Thevenin's theorem at points x and y of Figure 3.55, we obtain the circuit of Figure 3.56 and from it we get

$$v_{xy} = v_{TH} = \frac{4 \times 12}{2 + 4} = 8 \ V$$

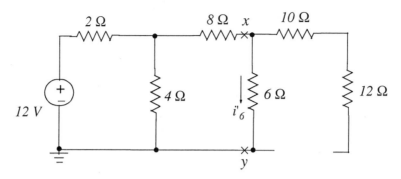

Figure 3.55. Circuit for finding i'_6 in Example 3.12

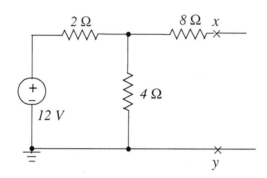

Figure 3.56. Circuit for computing the Thevenin voltage to find i'_6 in Example 3.12

Next, we will use the circuit of Figure 3.57 to find the Thevenin resistance.

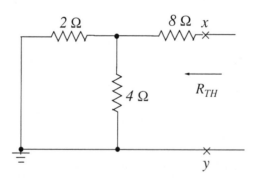

Figure 3.57. Circuit for computing the Thevenin resistance to find i'_6 in Example 3.12

$$R_{TH} = 8 + \frac{4 \times 2}{4 + 2} = \frac{28}{3} \ \Omega$$

We find the current i'_6 from Figure 3.58.

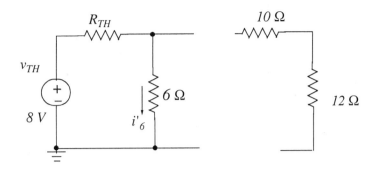

Figure 3.58. Circuit for computing i'_6 in Example 3.12

$$i'_6 = \frac{8}{28/3 + 6} = \frac{12}{23} \ A \qquad\qquad (3.57)$$

Next, the current i''_6 due to the 36 V source acting alone is found from the circuit of Figure 3.59.

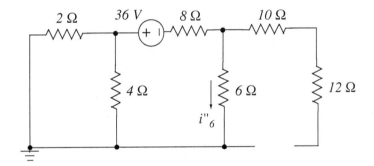

Figure 3.59. Circuit for finding i''_6 in Example 3.12

and after combination of the 2 Ω and 4 Ω parallel resistors to a single resistor, the circuit simplifies to that shown in Figure 3.60.

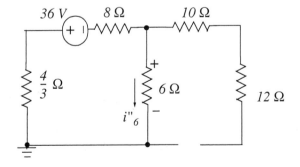

Figure 3.60. Simplification of the circuit of Figure 3.59 to compute i''_6 for Example 3.12

From the circuit of Figure 3.60, we get

$$i''_6 = -\frac{36}{4/3 + 8 + 6} = -\frac{54}{23} \; A \tag{3.58}$$

Finally, to find i'''_6, we short the voltage sources, and with the $5 \; A$ current source acting alone the circuit reduces to that shown in Figure 3.61.

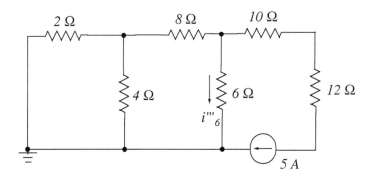

Figure 3.61. Circuit for finding i'''_6 in Example 3.12

Replacing the $2 \; \Omega$, $4 \; \Omega$, and $8 \; \Omega$ resistors by a single resistor, we get

$$\frac{2 \times 4}{2 + 4} + 8 = \frac{28}{3} \; \Omega$$

and the circuit of Figure 3.61 reduces to that shown in Figure 3.62.

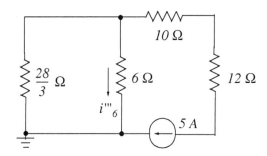

Figure 3.62. Simplification of the circuit of Figure 3.59 to compute i'''_6 for Example 3.12

We will use the current division expression in the circuit of Figure 3.62 to find i'''_6. Thus,

$$i'''_6 = \frac{28/3}{28/3 + 6} \times (-5) = -\frac{70}{23} \tag{3.59}$$

Therefore, from (3.57), (3.58), and (3.59) we get

$$i_6 = i'_6 + i''_6 + i'''_6 = \frac{12}{23} - \frac{54}{23} - \frac{70}{23} = -\frac{112}{23}$$

or

$$i_6 = -4.87 \ A \qquad\qquad (3.60)$$

and this is the same value as that of Example 3.5.

3.10 Circuits with Non-Linear Devices

Most electronic circuits contain non-linear devices such as diodes and transistors whose i - v (current-voltage) relationships are non-linear. However, for small signals (voltages or currents) these circuits can be represented by linear equivalent circuit models. A detailed discussion of these is beyond the scope of this text; however we will see how operational amplifiers can be represented by equivalent linear circuits in the next chapter.

If a circuit contains only one non-linear device, such as a diode, and all the other devices are linear, we can apply Thevenin's theorem to reduce the circuit to a Thevenin equivalent in series with the non-linear element. Then, we can analyze the circuit using a graphical solution. The procedure is illustrated with the following example.

Example 3.13

For the circuit of Figure 3.63, the $i - v$ characteristics of the diode D are shown in figure 3.64. We wish to find the voltage v_D across the diode and the current i_D through this diode using a graphical solution.

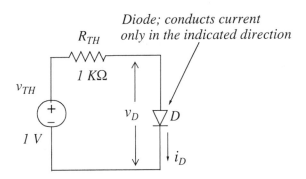

Figure 3.63. Circuit for Example 3.13

Solution:

The current i_D through the diode is also the current through the resistor. Then, by KVL

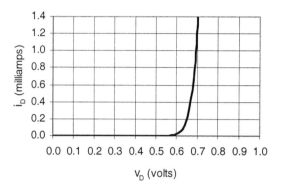

Figure 3.64. Diode i-v characteristics

$$v_R + v_D = 1 \ V$$

or

$$Ri_D = -v_D + 1$$

or

$$i_D = -\frac{1}{R}v_D + \frac{1}{R} \tag{3.61}$$

We observe that (3.61) is an equation of a straight line and the two points are obtained from it by first letting $v_D = 0$, then, $i_D = 0$. We obtain the straight line shown in Figure 3.65 which is plotted on the same graph as the given diode $i - v$ characteristics.

The intersection of the non-linear curve and the straight line yields the voltage and the current of the diode where we find that $v_D = 0.665V$ and $i_D = 0.335 \ mA$.

Check:

Since this is a series circuit, $i_R = 0.335 \ mA$ also. Therefore, the voltage drop v_R across the resistor is $v_R = 1 \ k\Omega \times 0.335 \ mA = 0.335 \ V$. Then, by KVL

$$v_R + v_D = 0.335 + 0.665 = 1 \ V$$

Diode Voltage (Volts)	Diode Current (milliamps)
0.00	0.000
0.02	0.000
0.04	0.000
0.06	0.000
0.08	0.000
0.10	0.000
0.12	0.000
0.14	0.000
0.16	0.000
0.18	0.000
0.20	0.000
0.22	0.000
0.24	0.000
0.26	0.000
0.28	0.000
0.30	0.000

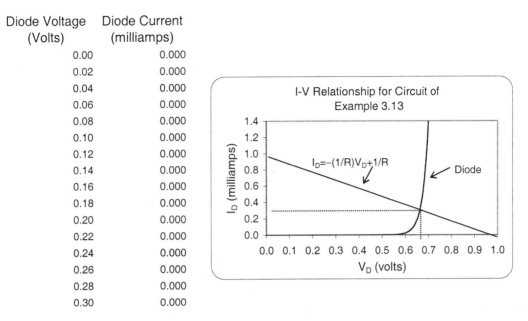

Figure 3.65. Curves for determining voltage and current in a diode

3.11 Efficiency

We have learned that the power absorbed by a resistor can be found from $p_R = i^2 R$ and this power is transformed into heat. In a long length of a conductive material, such as copper, this lost power is known as $i^2 R$ loss and thus the energy received by the load is equal to the energy transmitted minus the $i^2 R$ loss. Accordingly, we define *efficiency* η as

$$Efficiency = \eta = \frac{Output}{Input} = \frac{Output}{Output + Loss}$$

The efficiency η is normally expressed as a percentage. Thus,

$$\boxed{\% \ Efficiency = \% \ \eta = \frac{Output}{Input} \times 100 = \frac{Output}{Output + Loss} \times 100} \tag{3.62}$$

Obviously, a good efficiency should be close to *100%*

Example 3.14

In a two-story industrial building, the total load on the first floor draws an average of 60 amperes during peak activity, while the total load of the second floor draws 40 amperes at the same time. The building receives its electric power from a *480 V* source. Assuming that the total resistance of the cables (copper conductors) on the first floor is $1\ \Omega$ and on the second floor is $1.6\ \Omega$, compute the efficiency of transmission.

Solution:

First, we draw a circuit that represents the electrical system of this building. This is shown in Figure 3.66.

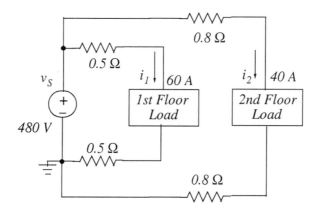

Figure 3.66. Circuit for Example 3.14

Power p_S supplied by the source:

$$p_S = v_S(i_1 + i_2) = 480 \times (60 + 40) = 48 \ kilowatts \tag{3.63}$$

Power loss between source and 1st floor load:

$$p_{loss1} = i_1^2(0.5\ \Omega + 0.5\ \Omega) = 60^2 \times 1 = 3.6 \ kilowatts \tag{3.64}$$

Power loss between source and 2nd floor load:

$$p_{loss2} = i_2^2(0.8\ \Omega + 0.8\ \Omega) = 40^2 \times 1.6 = 2.56 \ kilowatts \tag{3.65}$$

Total power loss:

$$p_{loss} = p_{loss1} + p_{loss2} = 3.60 + 2.56 = 6.16 \ kilowatts \tag{3.66}$$

Total power p_L received by 1st and 2nd floor loads:

$$p_L = p_S - p_{loss} = 48.00 - 6.16 = 41.84 \ kilowatts \tag{3.67}$$

$$\% \ Efficiency = \% \ \eta = \frac{Output}{Input} \times 100 = \frac{41.84}{48.00} \times 100 = 87.17 \ \% \tag{3.68}$$

3.12 Regulation

The *regulation* is defined as the ratio of the change in load voltage when the load changes from no load (NL) to full load (FL) divided by the full load. Thus, denoting the no-load voltage as v_{NL} and the full-load voltage as v_{FL}, the regulation is defined as In other words,

$$Regulation = \frac{v_{NL} - v_{FL}}{v_{FL}}$$

The regulation is also expressed as a percentage. Thus,

$$\%Regulation = \frac{v_{NL} - v_{FL}}{v_{FL}} \times 100 \qquad (3.69)$$

Example 3.15

Compute the regulation for the 1st floor load of the previous example.

Solution:

The current drawn by 1st floor load is given as 60 A and the total resistance from the source to the load as $1\ \Omega$. Then, the total voltage drop in the conductors is $60 \times 1 = 60\ V$. Therefore, the full-load voltage of the load is $v_{FL} = (480 - 60 = 420\ V)$ and the percent regulation is

$$\% \ Regulation = \frac{v_{NL} - v_{FL}}{v_{FL}} \times 100 = \frac{480 - 420}{420} \times 100 = 14.3\ \%$$

3.13 Summary

- When using nodal analysis, for a circuit that contains N nodes, we must write $N - 1$ independent nodal equations in order to completely describe that circuit. When the presence of voltage sources in a circuit seem to complicate the nodal analysis because we do not know the current through those voltage sources, we create combined nodes as illustrated in Example 3.2.

- When using nodal analysis, for a circuit that contains M meshes or L loops, B branches, and N nodes, we must write $L = M = B - N + 1$ independent loop or mesh equations in order to completely describe that circuit. When the presence of current sources in a circuit seem to complicate the mesh or loop analysis because we do not know the voltage across those current sources, we create combined meshes as illustrated in Example 3.6.

- A practical voltage source has an internal resistance and it is represented by a voltage source whose value is the value of the ideal voltage source in series with a resistance whose value is the value of the internal resistance.

- A practical current source has an internal conductance and it is represented by a current source whose value is the value of the ideal current source in parallel with a conductance whose value is the value of the internal conductance.

- A practical voltage source v_S in series with a resistance R_S can be replaced by a current source i_S whose value is v_S/i_S in parallel with a resistance R_P whose value is the same as R_S

- A practical current source i_S in parallel with a resistance R_P can be replaced by a voltage source v_S whose value is equal to $i_S \times R_S$ in series with a resistance R_S whose value is the same as R_P

- Thevenin's theorem states that in a two terminal network we can be replace everything except the load, by a voltage source denoted as v_{TH} in series with a resistance denoted as R_{TH}. The value of v_{TH} represents the open circuit voltage where the circuit is isolated from the load and R_{TH} is the equivalent resistance of that part of the isolated circuit. If a given circuit contains independent voltage and independent current sources only, the value of R_{TH} can be found by first shorting all independent voltage sources, opening all independent current sources, and calculating the resistance looking into the direction which is opposite to the disconnected load. If the circuit contains dependent sources, the value of R_{TH} must be computed from the relation $R_{TH} = v_{OC}/i_{SC}$

- Norton's theorem states that in a two terminal network we can be replace everything except the load, by a current source denoted as i_N in parallel with a resistance denoted as R_N. The value of i_N represents the short circuit current where the circuit is isolated from the load and R_N is the equivalent resistance of that part of the isolated circuit. If the circuit contains independent voltage and independent current sources only, the value of R_N can be found by first shorting all independent voltage sources, opening all independent current sources, and calculating the resistance looking into the direction which is opposite to the disconnected load. If the circuit contains dependent sources, the value of R_N must be computed from the relation $R_N = v_{OC}/i_{SC}$

- The maximum power transfer theorem states that a voltage source with a series resistance R_S or a current source with parallel resistance R_S delivers maximum power to a load R_{LOAD} when $R_{LOAD} = R_S$ or $R_{LOAD} = R_N$

- Linearity implies that there is a linear voltage–current relationship.

- A linear circuit is composed entirely of independent voltage sources, independent current sources, linear dependent sources, and linear passive devices such as resistors, inductors, and capacitors.

- The principle of superposition states that the response (a desired voltage or current) in any branch of a linear circuit having more than one independent source can be obtained as the sum of the responses caused by each independent source acting alone with all other independent voltage sources replaced by short circuits and all other independent current sources replaced by open circuits.

- Efficiency is defined as the ratio of output to input and thus it is never greater than unity. It is normally expressed as a percentage.

- Regulation is defined as the ratio of $v_{NL} - v_{FL}$ to v_{FL} and ideally should be close to zero. It is normally expressed as a percentage.

3.14 Exercises

Multiple Choice

1. The voltage across the 2 Ω resistor in the circuit of Figure 3.67 is

   A. *6 V*

   B. *16 V*

   C. *−8 V*

   D. *32 V*

   E. *none of the above*

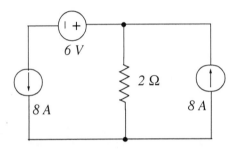

Figure 3.67. Circuit for Question 1

2. The current *i* in the circuit of Figure 3.68 is

   A. *−2 A*

   B. *5 A*

   C. *3 A*

   D. *4 A*

   E. *none of the above*

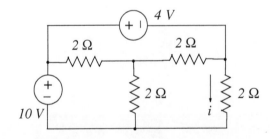

Figure 3.68. Circuit for Question 2

3. The node voltages shown in the partial network of Figure 3.69 are relative to some reference node which is not shown. The current i is

A. $-4 A$

B. $8/3 A$

C. $-5 A$

D. $-6 A$

E. *none of the above*

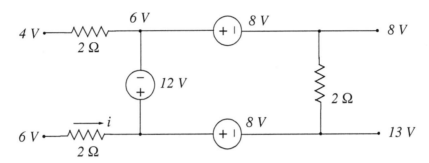

Figure 3.69. Circuit for Question 3

4. The value of the current i for the circuit of Figure 3.70 is

A. $-3 A$

B. $-8 A$

C. $-9 A$

D. $6 A$

E. *none of the above*

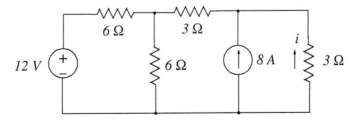

Figure 3.70. Circuit for Question 4

5. The value of the voltage v for the circuit of Figure 3.71 is

A. *4 V*

B. *6 V*

C. *8 V*

D. *12 V*

E. *none of the above*

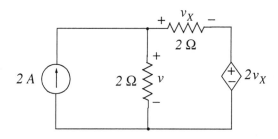

Figure 3.71. Circuit for Question 5

6. For the circuit of Figure 3.72, the value of k is dimensionless. For that circuit, no solution is possible if the value of k is

A. *2*

B. *1*

C. *∞*

D. *0*

E. *none of the above*

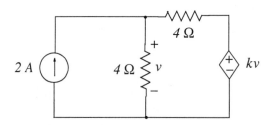

Figure 3.72. Circuit for Question 6

7. For the network of Figure 3.73, the Thevenin equivalent resistance R_{TH} to the right of terminals a and b is

A. *1*

B. *2*

C. *5*

D. *10*

E. *none of the above*

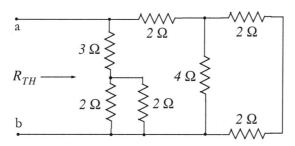

Figure 3.73. Network for Question 7

8. For the network of Figure 3.74, the Thevenin equivalent voltage V_{TH} across terminals a and b is

A. *−3 V*

B. *−2 V*

C. *1 V*

D. *5 V*

E. *none of the above*

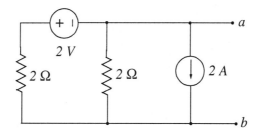

Figure 3.74. Network for Question 8

9. For the network of Figure 3.75, the Norton equivalent current source I_N and equivalent parallel resistance R_N across terminals a and b are

A. *1 A, 2 Ω*

B. *1.5 A, 25 Ω*

C. *4 A, 2.5 Ω*

D. *0 A, 5Ω*

E. *none of the above*

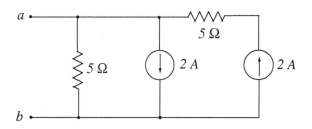

Figure 3.75. Network for Question 9

10. In applying the superposition principle to the circuit of Figure 3.76, the current i due to the *4 V* source acting alone is

A. *8 A*

B. *−1 A*

C. *4 A*

D. *−2 A*

E. *none of the above*

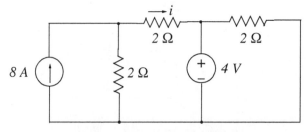

Figure 3.76. Network for Question 10

Problems

1. Use nodal analysis to compute the voltage across the 18 A current source in the circuit of Figure 3.77. Answer: *1.12 V*

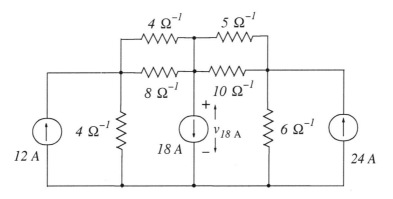

Figure 3.77. Circuit for Problem 1

2. Use nodal analysis to compute the voltage $v_{6\,\Omega}$ in the circuit of Figure 3.78. Answer: *21.6 V*

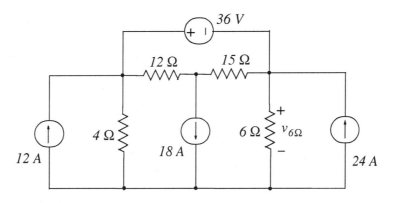

Figure 3.78. Circuit for Problem 2

3. Use nodal analysis to compute the current through the $6\,\Omega$ resistor and the power supplied (or absorbed) by the dependent source shown in Figure 3.79. Answers: *–3.9 A, –499.17 w*

4. Use mesh analysis to compute the voltage v_{36A} in Figure 3.80. Answer: *86.34 V*

5. Use mesh analysis to compute the current through the $i_{6\Omega}$ resistor, and the power supplied (or absorbed) by the dependent source shown in Figure 3.81. Answers: *–3.9 A, –499.33 w*

6. Use mesh analysis to compute the voltage $v_{10\Omega}$ in Figure 3.82. Answer: *0.5 V*

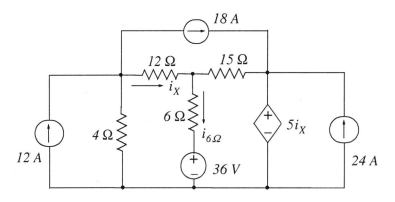

Figure 3.79. Circuit for Problem 3

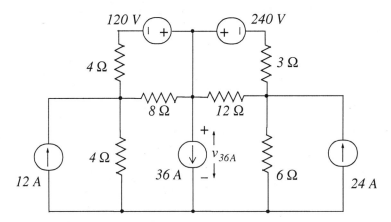

Figure 3.80. Circuit for Problem 4

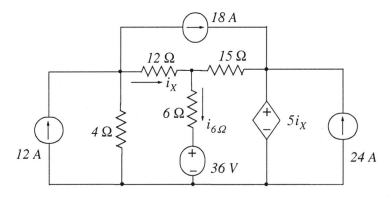

Figure 3.81. Circuit for Problem 5

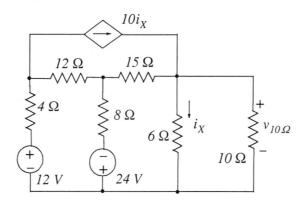

Figure 3.82. Circuit for Problem 6

7. Compute the power absorbed by the *10 Ω* resistor in the circuit of Figure 3.83 using any method.
   Answer: *1.32 w*

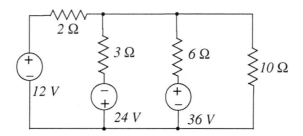

Figure 3.83. Circuit for Problem 7

8. Compute the power absorbed by the *20 Ω* resistor in the circuit of Figure 3.84 using any method. Answer: *73.73 w*

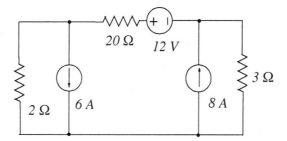

Figure 3.84. Circuit for Problem 8

9. In the circuit of Figure 3.85:

   a. To what value should the load resistor R_{LOAD} should be adjusted to so that it will absorb maximum power? Answer: *2.4 Ω*

b. What would then the power absorbed by R_{LOAD} be? Answer: *135 w*

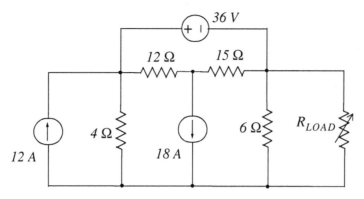

Figure 3.85. Circuit for Problem 9

10. Replace the network shown in Figure 3.86 by its Norton equivalent.
    Answers: $i_N = 0, R_N = 23.75 \; \Omega$

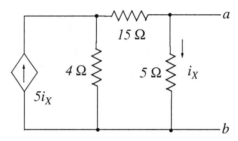

Figure 3.86. Circuit for Problem 10

11. Use the superposition principle to compute the voltage v_{18A} in the circuit of Figure 3.87.
    Answer: *1.12 V*

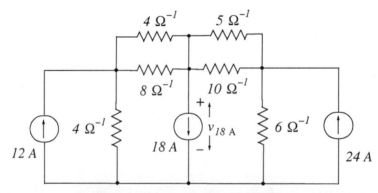

Figure 3.87. Circuit for Problem 11

12. Use the superposition principle to compute voltage $v_{6\,\Omega}$ in the circuit of Figure 3.88.

    Answer: *21.6 V*

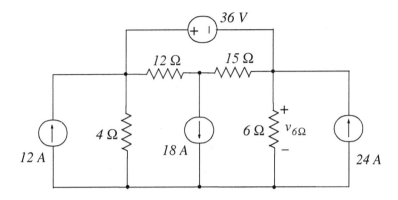

Figure 3.88. Circuit for Problem 12

13. In the circuit of Figure 3.89, v_{S1} and v_{S2} are adjustable voltage sources in the range $-50 \le V \le 50$ V, and R_{S1} and R_{S2} represent their internal resistances. Table 3.4 shows the results of several measurements. In Measurement 3 the load resistance is adjusted to the same value as Measurement 1, and in Measurement 4 the load resistance is adjusted to the same value as Measurement 2. For Measurements 5 and 6 the load resistance is adjusted to *1* Ω. Make the necessary computations to fill-in the blank cells of this table.

TABLE 3.4 Table for Problem 13

Measurement	Switch S_1	Switch S_2	v_{S1} (V)	v_{S2} (V)	i_{LOAD} (A)
1	Closed	Open	48	0	16
2	Open	Closed	0	36	6
3	Closed	Open		0	−5
4	Open	Closed	0	−42	
5	Closed	Closed	15	18	
6	Closed	Closed		24	0

Answers: *−15 V , −7 A , 11 A , −24 V*

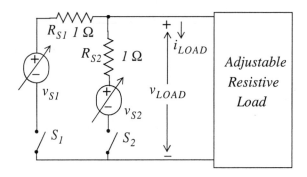

Figure 3.89. Network for Problem 13

14. Compute the efficiency of the electrical system of Figure 3.90. Answer: *76.6%*

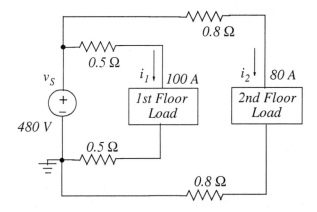

Figure 3.90. Electrical system for Problem 14

15. Compute the regulation for the 2st floor load of the electrical system of Figure 3.91.
Answer: *36.4%*

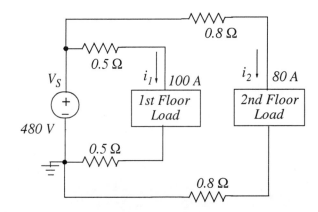

Figure 3.91. Circuit for Problem 15

16. Write a set of nodal equations and then use MATLAB to compute i_{LOAD} and v_{LOAD} for the circuit of Example 3.10 which is repeated as Figure 3.92 for convenience.

    Answers: -0.96 A, -7.68 V

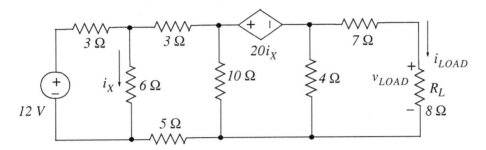

Figure 3.92. Circuit for Problem 16

3.15 Answers to Exercises

Multiple Choice

1. E The current entering Node A is equal to the current leaving that node. Therefore, there is no current through the $2\ \Omega$ resistor and the voltage across it is zero.

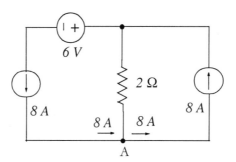

2. C From the figure below, $V_{AC} = 4\ V$. Also, $V_{AB} = V_{BC} = 2\ V$ and $V_{AD} = 10\ V$. Then, $V_{BD} = V_{AD} - V_{AB} = 10 - 2 = 8\ V$ and $V_{CD} = V_{BD} - V_{BC} = 8 - 2 = 6\ V$. Therefore, $i = 6/2 = 3\ A$.

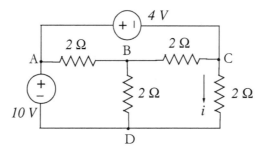

3. A From the figure below we observe that the node voltage at A is $6\ V$ relative to the reference node which is not shown. Therefore, the node voltage at B is $6 + 12 = 18\ V$ relative to the same reference node. The voltage across the resistor is $V_{BC} = 18 - 6 = 12\ V$ and the direction of current through the $3\ \Omega$ resistor is opposite to that shown since Node B is at a higher potential than Node C. Thus $i = -12/3 = -4\ A$

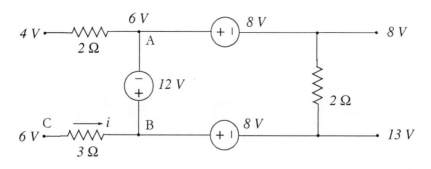

4. E We assign node voltages at Nodes A and B as shown below.

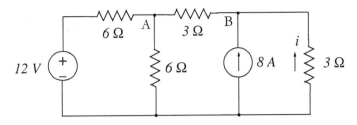

At Node A

$$\frac{V_A - 12}{6} + \frac{V_A}{6} + \frac{V_A - V_B}{3} = 0$$

and at Node B

$$\frac{V_B - V_A}{3} + \frac{V_B}{3} = 8$$

These simplify to

$$\frac{2}{3}V_A - \frac{1}{3}V_B = 2$$

and

$$-\frac{1}{3}V_A + \frac{2}{3}V_B = 8$$

Multiplication of the last equation by 2 and addition with the first yields $V_B = 18$ and thus $i = -18/3 = -6\,A$.

5. E Application of KCL at Node A of the circuit below yields

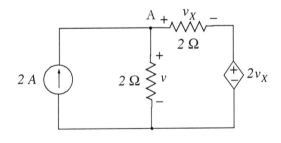

$$\frac{v}{2} + \frac{v - 2v_X}{2} = 2$$

or

$$v - v_X = 2$$

Also by KVL

$$v = v_X + 2v_X$$

and by substitution

$$v_X + 2v_X - v_X = 2$$

or

$$v_X = 1$$

and thus

$$v = v_X + 2v_X = 1 + 2 \times 1 = 3 \ V$$

6. A Application of KCL at Node A of the circuit below yields

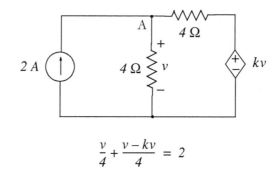

$$\frac{v}{4} + \frac{v - kv}{4} = 2$$

or

$$\frac{1}{4}(2v - kv) = 2$$

and this relation is meaningless if $k = 2$. Thus, this circuit has solutions only if $k \neq 2$.

7. B The two $2 \ \Omega$ resistors on the right are in series and the two $2 \ \Omega$ resistors on the left shown in the figure below are in parallel.

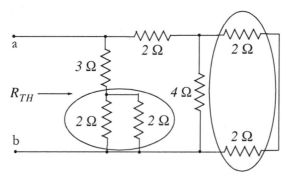

Starting on the right side and proceeding to the left we get $2 + 2 = 4$, $4 \| 4 = 2$, $2 + 2 = 4$, $4 \| (3 + 2 \| 2) = 4 \| (3 + 1) = 4 \| 4 = 2 \ \Omega$.

8. A Replacing the current source and its $2 \ \Omega$ parallel resistance with an equivalent voltage source in series with a $2 \ \Omega$ resistance we get the network shown below.

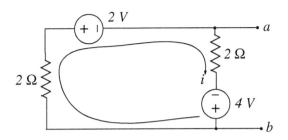

By Ohm's law,

$$i = \frac{4-2}{2+2} = 0.5 \ A$$

and thus

$$v_{TH} = v_{ab} = 2 \times 0.5 + (-4) = -3 \ V$$

9. D The Norton equivalent current source I_N is found by placing a short across the terminals a and b. This short shorts out the $5 \ \Omega$ resistor and thus the circuit reduces to the one shown below.

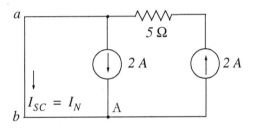

By KCL at Node A,

$$I_N + 2 = 2$$

and thus $I_N = 0$

The Norton equivalent resistance R_N is found by opening the current sources and looking to the right of terminals a and b. When this is done, the circuit reduces to the one shown below.

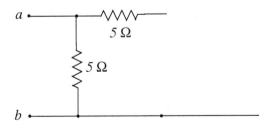

Therefore, $R_N = 5 \ \Omega$ and the Norton equivalent circuit consists of just a $5 \ \Omega$ resistor.

10. B With the *4 V* source acting alone, the circuit is as shown below.

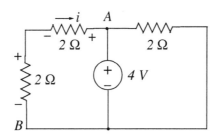

We observe that $v_{AB} = 4\ V$ and thus the voltage drop across each of the 2 Ω resistors to the left of the *4 V* source is *2 V* with the indicated polarities. Therefore,

$$i = -2/2 = -1\ A$$

Problems

1. We first replace the parallel conductances with their equivalents and the circuit simplifies to that shown below.

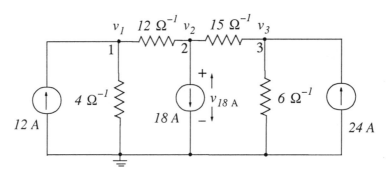

Applying nodal analysis at Nodes 1, 2, and 3 we get:

Node 1:

$$16v_1 - 12v_2 = 12$$

Node 2:

$$-12v_1 + 27v_2 - 15v_3 = -18$$

Node 3:

$$-15v_2 + 21v_3 = 24$$

Simplifying the above equations, we get:

$$4v_1 - 3v_2 \qquad\quad = 3$$
$$-4v_1 + 9v_2 - 5v_3 = -6$$
$$-5v_2 + 7v_3 = 8$$

Addition of the first two equations above and grouping with the third yields

$$6v_2 - 5v_3 = -3$$

$$-5v_2 + 7v_3 = 8$$

For this problem we are only interested in $v_2 = v_{18\,A}$. Therefore, we will use Cramer's rule to solve for v_2. Thus,

$$v_2 = \frac{D_2}{\Delta} \qquad D_2 = \begin{bmatrix} -3 & -5 \\ 8 & 7 \end{bmatrix} = -21 + 40 = 19 \qquad \Delta = \begin{bmatrix} 6 & -5 \\ -5 & 7 \end{bmatrix} = 42 - 25 = 17$$

and

$$v_2 = v_{18\,A} = 19/17 = 1.12\ V$$

2. Since we cannot write an expression for the current through the *36 V* source, we form a combined node as shown on the circuit below.

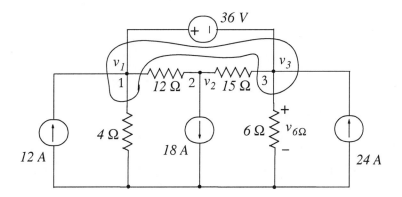

At Node 1 (combined node):

$$\frac{v_1}{4} + \frac{v_1 - v_2}{12} + \frac{v_3 - v_2}{15} + \frac{v_3}{6} - 12 - 24 = 0$$

and at Node 2,

$$\frac{v_2 - v_1}{12} + \frac{v_2 - v_3}{15} = -18$$

Also,

$$v_1 - v_3 = 36$$

Simplifying the above equations, we get:

$$\frac{1}{3}v_1 - \frac{3}{20}v_2 + \frac{7}{30}v_3 = 36$$

$$-\frac{1}{12}v_1 + \frac{3}{20}v_2 - \frac{1}{15}v_3 = -18$$

$$v_1 \qquad\qquad -v_3 = 36$$

Addition of the first two equations above and multiplication of the third by $-1/4$ yields

$$\frac{1}{4}v_1 + \frac{1}{6}v_3 = 18$$

$$-\frac{1}{4}v_1 + \frac{1}{4}v_3 = -9$$

and by adding the last two equations we get

$$\frac{5}{12}v_3 = 9$$

or

$$v_3 = v_{6\,\Omega} = \frac{108}{5} = 21.6V$$

Check with MATLAB:

```
format rat
R=[1/3 –3/20 7/30; –1/12 3/20 –1/15; 1 0 –1];
I=[36 –18 36]';
V=R\I;
fprintf('\n'); disp('v1='); disp(V(1)); disp('v2='); disp(V(2)); disp('v3='); disp(V(3))

v1=
    288/5
v2=
    -392/5
v3=
    108/5
```

3. We assign node voltages v_1, v_2, v_3, v_4 and current i_Y as shown in the circuit below. Then,

$$\frac{v_1}{4} + \frac{v_1 - v_2}{12} + 18 - 12 = 0$$

and

$$\frac{v_2 - v_1}{12} + \frac{v_2 - v_3}{12} + \frac{v_2 - v_4}{6} = 0$$

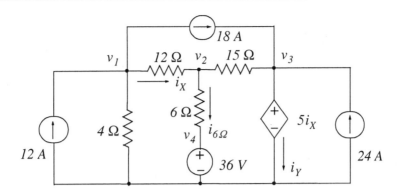

Simplifying the last two equations above, we get

$$\frac{1}{3}v_1 - \frac{1}{12}v_2 = -6$$

and

$$-\frac{1}{12}v_1 + \frac{19}{60}v_2 - \frac{1}{15}v_3 - \frac{1}{6}v_4 = 0$$

Next, we observe that $i_X = \frac{v_1 - v_2}{12}$, $v_3 = 5i_X$ and $v_4 = 36\ V$. Then $v_3 = \frac{5}{12}(v_1 - v_2)$ and by substitution into the last equation above, we get

$$-\frac{1}{12}v_1 + \frac{19}{60}v_2 - \frac{1}{15} \times \frac{5}{12}(v_1 - v_2) - \frac{1}{6}36 = 0$$

or

$$-\frac{1}{9}v_1 + \frac{31}{90}v_2 = 6$$

Thus, we have two equations with two unknowns, that is,

$$\frac{1}{3}v_1 - \frac{1}{12}v_2 = -6$$

$$-\frac{1}{9}v_1 + \frac{31}{90}v_2 = 6$$

Multiplication of the first equation above by $1/3$ and addition with the second yields

$$\frac{19}{60}v_2 = 4$$

or

$$v_2 = 240/19$$

We find v_1 from

$$\frac{1}{3}v_1 - \frac{1}{12}v_2 = -6$$

Thus,

$$\frac{1}{3}v_1 - \frac{1}{12} \times \frac{240}{19} = -6$$

or

$$v_1 = -282/19$$

Now, we find v_3 from

$$v_3 = \frac{5}{12}(v_1 - v_2) = \frac{5}{12}\left(\frac{-282}{19} - \frac{240}{19}\right) = -\frac{435}{38}$$

Therefore, the node voltages of interest are:

$$v_1 = -282/19 \ V$$

$$v_2 = 240/19 \ V$$

$$v_3 = -435/38 \ V$$

$$v_4 = 36 \ V$$

The current through the $6 \ \Omega$ resistor is

$$i_{6 \ \Omega} = \frac{v_2 - v_4}{6} = \frac{240/19 - 36}{6} = -\frac{74}{19} = -3.9 \ A$$

To compute the power supplied (or absorbed) by the dependent source, we must first find the current i_Y. It is found by application of KCL at node voltage v_3. Thus,

$$i_Y - 24 - 18 + \frac{v_3 - v_2}{15} = 0$$

or

$$i_Y = 42 - \frac{-435/38 - 240/19}{15}$$

$$= 42 + \frac{915/38}{15} = \frac{1657}{38}$$

and

$$p = v_3 i_Y = -\frac{435}{38} \times \frac{1657}{38} = -\frac{72379}{145} = -499.17 \ w$$

that is, the dependent source supplies power to the circuit.

4. Since we cannot write an expression for the *36 A* current source, we temporarily remove it and we form a combined mesh for Meshes 2 and 3 as shown below.

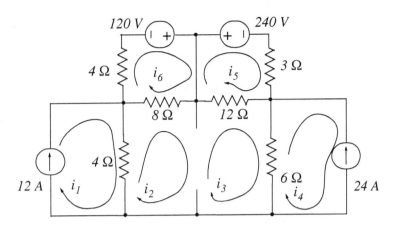

Mesh 1:

$$i_1 = 12$$

Combined mesh (2 and 3):

$$-4i_1 + 12i_2 + 18i_3 - 6i_4 - 8i_5 - 12i_6 = 0$$

or

$$-2i_1 + 6i_2 + 9i_3 - 3i_4 - 4i_5 - 6i_6 = 0$$

We now re-insert the *36 A* current source and we write the third equation as

$$i_2 - i_3 = 36$$

Mesh 4:

$$i_4 = -24$$

Mesh 5:

$$-8i_2 + 12i_5 = 120$$

or

$$-2i_2 + 3i_5 = 30$$

Mesh 6:

$$-12i_3 + 15i_6 = -240$$

or

$$-4i_3 + 5i_6 = -80$$

Thus, we have the following system of equations:

$$i_1 = 12$$
$$-2i_1 + 6i_2 + 9i_3 - 3i_4 - 4i_5 - 6i_6 = 0$$
$$i_2 - i_3 = 36$$
$$i_4 = -24$$
$$-2i_2 + 3i_5 = 30$$
$$-4i_3 + 5i_6 = -80$$

and in matrix form

$$
\underbrace{\begin{bmatrix} 1 & 0 & 0 & 0 & 0 & 0 \\ -2 & 6 & 9 & -3 & -4 & -6 \\ 0 & 1 & -1 & 0 & 0 & 0 \\ 0 & 0 & 0 & 1 & 0 & 0 \\ 0 & -2 & 0 & 0 & 3 & 0 \\ 0 & 0 & -4 & 0 & 0 & 5 \end{bmatrix}}_{R} \cdot \underbrace{\begin{bmatrix} i_1 \\ i_2 \\ i_3 \\ i_4 \\ i_5 \\ i_6 \end{bmatrix}}_{I} = \underbrace{\begin{bmatrix} 12 \\ 0 \\ 36 \\ -24 \\ 30 \\ -80 \end{bmatrix}}_{V}
$$

We find the currents i_1 through i_6 with the following MATLAB code:

```
R=[1 0 0 0 0 0; -2 6 9 -3 -4 -6;...
   0 1 -1 0 0 0;0 0 0 1 0 0;...
   0 -2 0 0 3 0;0 0 -4 0 0 5];
V=[12 0 36 -24 30 -80]';
I=R\V;
fprintf('\n');...
   fprintf('i1=%7.2f A \t', I(1));...
   fprintf('i2=%7.2f A \t', I(2));...
   fprintf('i3=%7.2f A \t', I(3));...
   fprintf('\n');...
   fprintf('i4=%7.2f A \t', I(4));...
   fprintf('i5=%7.2f A \t', I(5));...
   fprintf('i6=%7.2f A \t', I(6));...
   fprintf('\n')
```

```
i1= 12.00 A    i2=   6.27 A    i3= -29.73 A
i4= -24.00 A   i5=  14.18 A    i6= -39.79 A
```

Now, we can find the voltage $v_{36\,A}$ by application of KVL around Mesh 3. Thus,

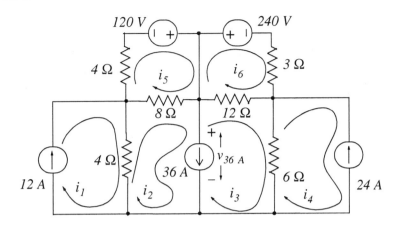

$$v_{36\,A} = v_{12\,\Omega} + v_{6\,\Omega} = 12 \times [(-29.73) - (-39.79)] + 6 \times [(-29.73) - (24.00)]$$

or

$$v_{36\,A} = 86.34 \ V$$

To verify that this value is correct, we apply KVL around Mesh 2. Thus, we must show that

$$v_{4\,\Omega} + v_{8\,\Omega} + v_{36\,A} = 0$$

By substitution of numerical values, we find that

$$4 \times [6.27 - 12] + 8 \times [6.27 - 14.18] + 86.34 = 0.14$$

5. This is the same circuit as that of Problem 3. We will show that we obtain the same answers using mesh analysis.

We assign mesh currents as shown below.

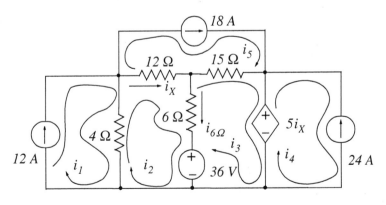

Mesh 1:

$$i_1 = 12$$

Mesh 2:

$$-4i_1 + 22i_2 - 6i_3 - 12i_5 = -36$$

or

$$-2i_1 + 11i_2 - 3i_3 - 6i_5 = -18$$

Mesh 3:

$$-6i_2 + 21i_3 - 15i_5 + 5i_X = 36$$

and since $i_X = i_2 - i_5$, the above reduces to

$$-6i_2 + 21i_3 - 15i_5 + 5i_2 - 5i_5 = 36$$

or

$$-i_2 + 21i_3 - 20i_5 = 36$$

Mesh 4:

$$i_4 = -24$$

Mesh 5:

$$i_5 = 18$$

Grouping these five independent equations we get:

$$
\begin{aligned}
i_1 &= 12 \\
-2i_1 + 11i_2 - 3i_3 \quad -6i_5 &= -18 \\
-i_2 + 21i_3 \quad -20i_5 &= 36 \\
i_4 \quad &= -24 \\
i_5 &= 18
\end{aligned}
$$

and in matrix form,

$$
\underbrace{\begin{bmatrix} 1 & 0 & 0 & 0 & 0 \\ -2 & 11 & -3 & 0 & -6 \\ 0 & -1 & 21 & 0 & -20 \\ 0 & 0 & 0 & 1 & 0 \\ 0 & 0 & 0 & 0 & 1 \end{bmatrix}}_{R} \cdot \underbrace{\begin{bmatrix} i_1 \\ i_2 \\ i_3 \\ i_4 \\ i_5 \end{bmatrix}}_{I} = \underbrace{\begin{bmatrix} 12 \\ -18 \\ 36 \\ -24 \\ 18 \end{bmatrix}}_{V}
$$

We find the currents i_1 through i_5 with the following MATLAB code:

```
R=[1 0 0 0 0; -2 11 -3 0 -6; 0 -1 21 0 -20; ...
    0 0 0 1 0;0 0 0 0 1];
V=[12 -18 36 -24 18]';
I=R\V;
```

```
fprintf('\n');...
  fprintf('i1=%7.2f A \t', I(1));...
  fprintf('i2=%7.2f A \t', I(2));...
  fprintf('i3=%7.2f A \t', I(3));...
  fprintf('\n');...
  fprintf('i4=%7.2f A \t', I(4));...
  fprintf('i5=%7.2f A \t', I(5));...
  fprintf('\n')
```

```
i1=   12.00 A    i2=   15.71 A    i3=   19.61 A
i4=  -24.00 A    i5=   18.00 A
```

By inspection,

$$i_{6\,\Omega} = i_2 - i_3 = 15.71 - 19.61 = -3.9\ A$$

Next,

$$p_{5i_x} = 5i_x(i_3 - i_4) = 5(i_2 - i_5)(i_3 - i_4)$$

$$= 5(15.71 - 18.00)(19.61 + 24.00) = -499.33\ w$$

These are the same answers as those we found in Problem 3.

6. We assign mesh currents as shown below and we write mesh equations.

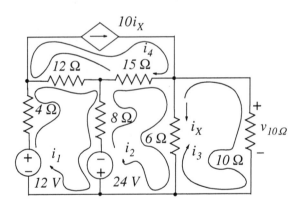

Mesh 1:

$$24i_1 - 8i_2 - 12i_4 - 24 - 12 = 0$$

or

$$6i_1 - 2i_2 - 3i_4 = 9$$

Mesh 2:

$$-8i_1 + 29i_2 - 6i_3 - 15i_4 = -24$$

Mesh 3:

$$-6i_2 + 16i_3 = 0$$

or

$$-3i_2 + 8i_3 = 0$$

Mesh 4:

$$i_4 = 10i_X = 10(i_2 - i_3)$$

or

$$10i_2 - 10i_3 - i_4 = 0$$

Grouping these four independent equations we get:

$$6i_1 - 2i_2 \qquad - 3i_4 = 9$$
$$-8i_1 + 29i_2 - 6i_3 - 15i_4 = -24$$
$$-3i_2 + 8i_3 \qquad = 0$$
$$10i_2 - 10i_3 - i_4 = 0$$

and in matrix form,

$$\underbrace{\begin{bmatrix} 6 & -2 & 0 & -3 \\ -8 & 29 & -6 & -15 \\ 0 & -3 & 8 & 0 \\ 0 & 10 & -10 & -1 \end{bmatrix}}_{R} \cdot \underbrace{\begin{bmatrix} i_1 \\ i_2 \\ i_3 \\ i_4 \end{bmatrix}}_{I} = \underbrace{\begin{bmatrix} 9 \\ -24 \\ 0 \\ 0 \end{bmatrix}}_{V}$$

We find the currents i_1 through i_4 with the following MATLAB code:

```
R=[6 -2 0 -3; -8 29 -6 -15; 0 -3 8 0 ; 0 10 -10 -1];
V=[9 -24 0 0]';
I=R\V;
fprintf('\n');...
   fprintf('i1=%7.2f A \t', I(1));...
   fprintf('i2=%7.2f A \t', I(2));...
   fprintf('i3=%7.2f A \t', I(3));...
   fprintf('i4=%7.2f A \t', I(4));...
   fprintf('\n')

   i1= 1.94 A    i2= 0.13 A    i3= 0.05 A    i4= 0.79 A
```

Now, we find $v_{10\Omega}$ by Ohm's law, that is,

$$v_{10\Omega} = 10i_3 = 10 \times 0.05 = 0.5 \text{ V}$$

The same value is obtained by computing the voltage across the $6 \ \Omega$ resistor, that is,

$$v_{6\Omega} = 6(i_2 - i_3) = 6(0.13 - 0.05) = 0.48 \text{ V}$$

7. Voltage-to-current source transformation yields the circuit below.

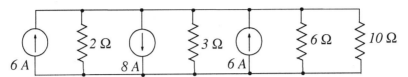

By combining all current sources and all parallel resistors except the 10 Ω resistor, we obtain the simplified circuit below.

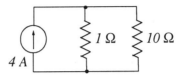

Applying the current division expression, we get

$$i_{10\,\Omega} = \frac{1}{1+10} \times 4 = \frac{4}{11}\ A$$

and thus

$$P_{10\,\Omega} = i_{10\,\Omega}^2(10) = \left(\frac{4}{11}\right)^2 \times 10 = \frac{16}{121} \times 10 = \frac{160}{121} = 1.32\ w$$

8. Current-to-voltage source transformation yields the circuit below.

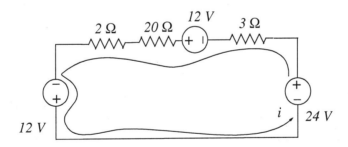

From this series circuit,

$$i = \frac{\Sigma v}{\Sigma R} = \frac{48}{25}\ A$$

and thus

$$P_{20\,\Omega} = i^2(20) = \left(\frac{48}{25}\right)^2 \times 20 = \frac{2304}{625} \times 20 = 73.73\ w$$

9. We remove R_{LOAD} from the rest of the rest of the circuit and we assign node voltages v_1, v_2, and v_3. We also form the combined node as shown on the circuit below.

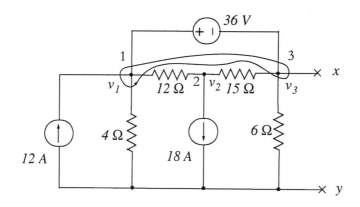

Node 1:

$$\frac{v_1}{4} + \frac{v_1 - v_2}{12} - 12 + \frac{v_3 - v_2}{15} + \frac{v_3}{6} = 0$$

or

$$\frac{1}{3}v_1 - \frac{3}{20}v_2 + \frac{7}{30}v_3 = 12$$

Node 2:

$$\frac{v_2 - v_1}{12} + \frac{v_2 - v_3}{15} = -18$$

or

$$-\frac{1}{12}v_1 + \frac{3}{20}v_2 - \frac{1}{15}v_3 = -18$$

Also,

$$v_1 - v_3 = 36$$

For this problem, we are interested only in the value of v_3 which is the Thevenin voltage v_{TH}, and we could find it by Gauss's elimination method. However, for convenience, we will group these three independent equations, express these in matrix form, and use MATLAB for their solution.

$$\frac{1}{3}v_1 - \frac{3}{20}v_2 + \frac{7}{30}v_3 = 12$$

$$-\frac{1}{12}v_1 + \frac{3}{20}v_2 - \frac{1}{15}v_3 = -18$$

$$v_1 \qquad\quad - v_3 = 36$$

and in matrix form,

$$\underbrace{\begin{bmatrix} \dfrac{1}{3} & -\dfrac{3}{20} & \dfrac{7}{30} \\[2ex] -\dfrac{1}{12} & \dfrac{3}{20} & -\dfrac{1}{15} \\[2ex] 1 & 0 & -1 \end{bmatrix}}_{G} \cdot \underbrace{\begin{bmatrix} v_1 \\ v_2 \\ v_3 \end{bmatrix}}_{V} = \underbrace{\begin{bmatrix} 12 \\ -18 \\ 36 \end{bmatrix}}_{I}$$

We find the voltages v_1 through v_3 with the following MATLAB code:

```
G=[1/3 −3/20 7/30; −1/12 3/20 −1/15; 1 0 −1];
I=[12 −18 36]'; V=G\I;
fprintf('\n');...
   fprintf('v1=%7.2f V \t', V(1)); fprintf('v2=%7.2f V \t', V(2)); fprintf('v3=%7.2f V \t', V(3));
   fprintf('\n')

   v1= 0.00 V    v2= -136.00 V    v3= -36.00 V
```

Thus,

$$v_{TH} = v_3 = -36 \ V$$

To find R_{TH} we short circuit the voltage source and we open the current sources. The circuit then reduces to the resistive network below.

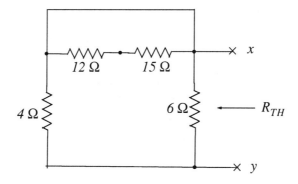

We observe that the resistors in series are shorted out and thus the Thevenin resistance is the parallel combination of the $4 \ \Omega$ and $6 \ \Omega$ resistors, that is,

$$4 \ \Omega \parallel 6 \ \Omega = 2.4 \ \Omega$$

and the Thevenin equivalent circuit is as shown below.

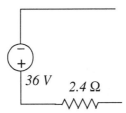

Now, we connect the load resistor R_{LOAD} at the open terminals and we get the simple series circuit shown below.

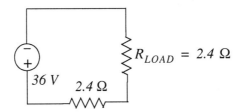

a. For maximum power transfer,

$$R_{LOAD} = 2.4 \ \Omega$$

b. Power under maximum power transfer condition is

$$p_{MAX} = i^2 R_{LOAD} = \left(\frac{36}{2.4 + 2.4}\right)^2 \times 2.4 = 7.5^2 \times 2.4 = 135 \ w$$

10. We assign a node voltage Node 1 and a mesh current for the mesh on the right as shown below.

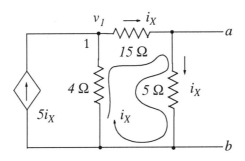

At Node 1:

$$\frac{v_1}{4} + i_X = 5i_X$$

Mesh on the right:

$$(15 + 5)i_X = v_1$$

and by substitution into the node equation above,

$$\frac{20i_X}{4} + i_X = 5i_X$$

or

$$6i_X = 5i_X$$

but this can only be true if $i_X = 0$.

Then,

$$i_N = \frac{v_{OC}}{R_N} = \frac{v_{ab}}{R_N} = \frac{5 \times i_X}{R_N} = \frac{5 \times 0}{R_N} = 0$$

Thus, the Norton current source is open as shown below.

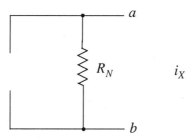

To find R_N we insert a *1 A* current source as shown below.

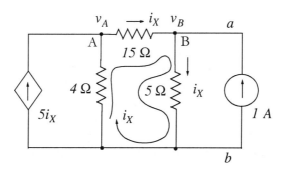

At Node A:

$$\frac{v_A}{4} + \frac{v_A - v_B}{15} = 5i_X$$

But

$$v_B = (5\ \Omega) \times i_X = 5i_X$$

and by substitution into the above relation

$$\frac{v_A}{4} + \frac{v_A - v_B}{15} = v_B$$

or

$$\frac{19}{60}v_A - \frac{16}{15}v_B = 0$$

At Node B:

$$\frac{v_B - v_A}{15} + \frac{v_B}{5} = 1$$

or

$$-\frac{1}{15}v_A + \frac{4}{15}v_B = 1$$

For this problem, we are interested only in the value of v_B which we could find by Gauss's elimination method. However, for convenience, we will use MATLAB for their solution.

$$\frac{19}{60}v_A - \frac{16}{15}v_B = 0$$

$$-\frac{1}{15}v_A + \frac{4}{15}v_B = 1$$

and in matrix form,

$$\underbrace{\begin{bmatrix} \dfrac{19}{60} & -\dfrac{16}{15} \\ -\dfrac{1}{15} & \dfrac{4}{15} \end{bmatrix}}_{G} \cdot \underbrace{\begin{bmatrix} v_A \\ v_B \end{bmatrix}}_{V} = \underbrace{\begin{bmatrix} 0 \\ 1 \end{bmatrix}}_{I}$$

We find the voltages v_1 and v_2 with the following MATLAB code:

```
G=[19/60 -16/15; -1/15 4/15];
I=[0 1]'; V=G\I;
fprintf('\n');...
   fprintf('vA=%7.2f V \t', V(1)); fprintf('vB=%7.2f V \t', V(2));
   fprintf('\n')
```

      vA= 80.00 V   vB= 23.75 V

Now, we can find the Norton equivalent resistance from the relation

$$R_N = \frac{V_{ab}}{I_{SC}} = \frac{V_B}{1} = 23.75 \ \Omega$$

11. This is the same circuit as that of Problem 1. Let v'_{18A} be the voltage due to the $12\ A$ current source acting alone. The simplified circuit with assigned node voltages is shown below where the parallel conductances have been replaced by their equivalents.

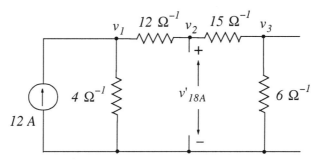

The nodal equations at the three nodes are

$$16v_1 - 12v_2 \qquad = 12$$
$$-12v_1 + 27v_2 - 15v_3 = 0$$
$$-15v_2 + 21v_3 = 0$$

or

$$4v_1 - 3v_2 \qquad = 3$$
$$-4v_1 + 9v_2 - 5v_3 = 0$$
$$-5v_2 + 7v_3 = 0$$

Since $v_2 = v'_{18A}$, we only need to solve for v_2. Adding the first 2 equations above and grouping with the third we obtain

$$6v_2 - 5v_3 = 3$$
$$-5v_2 + 7v_3 = 0$$

Multiplying the first by 7 and the second by 5 we get

$$42v_2 - 35v_3 = 21$$
$$-25v_2 + 35v_3 = 0$$

and by addition of these we get

$$v_2 = v'_{18A} = \frac{21}{17}\ V$$

Next, we let v''_{18A} be the voltage due to the $18\ A$ current source acting alone. The simplified circuit with assigned node voltages is shown below where the parallel conductances have been replaced by their equivalents.

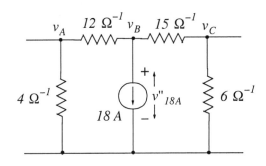

The nodal equations at the three nodes are

$$16v_A - 12v_B \qquad = 0$$
$$-12v_A + 27v_B - 15v_C = -18$$
$$-15v_B + 21v_C = 0$$

or

$$4v_A - 3v_B \qquad = 0$$
$$-4v_A + 9v_B - 5v_C = -6$$
$$-5v_B + 7v_C = 0$$

Since $v_B = v''_{18A}$, we only need to solve for v_B. Adding the first 2 equations above and grouping with the third we obtain

$$6v_B - 5v_C = -6$$
$$-5v_B + 7v_C = 0$$

Multiplying the first by *7* and the second by *5* we get

$$42v_B - 35v_C = -42$$
$$-25v_B + 35v_C = 0$$

and by addition of these we get

$$v_B = v''_{18A} = \frac{-42}{17} \ V$$

Finally, we let v'''_{18A} be the voltage due to the *24 A* current source acting alone. The simplified circuit with assigned node voltages is shown below where the parallel conductances have been replaced by their equivalents.

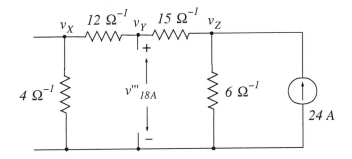

The nodal equations at the three nodes are

$$16v_X - 12v_Y \qquad = 0$$
$$-12v_A + 27v_Y - 15v_Z = 0$$
$$-15v_B + 21v_Z = 24$$

or

$$4v_X - 3v_Y \qquad = 0$$
$$-4v_X + 9v_Y - 5v_Z = 0$$
$$-5v_Y + 7v_Z = 8$$

Since $v_Y = v'''_{18A}$, we only need to solve for v_Y. Adding the first 2 equations above and grouping with the third we obtain

$$6v_Y - 5v_Z = 0$$
$$-5v_Y + 7v_Z = 0$$

Multiplying the first by 7 and the second by 5 we get

$$42v_Y - 35v_Z = 0$$
$$-25v_Y + 35v_Z = 40$$

and by addition of these we get

$$v_Y = v'''_{18A} = \frac{40}{17} \ V$$

and thus

$$v_{18A} = v'_{18A} + v''_{18A} + v'''_{18A} = \frac{21}{17} + \frac{-42}{17} + \frac{40}{17} = \frac{19}{17} = 1.12 \ V$$

This is the same answer as in Problem 1.

12. This is the same circuit as that of Problem 2. Let $v'_{6\,\Omega}$ be the voltage due to the *12 A* current source acting alone. The simplified circuit is shown below.

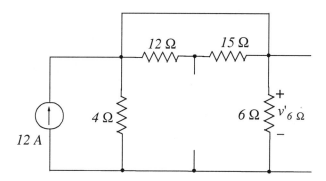

The *12 Ω* and *15 Ω* resistors are shorted out and the circuit is further simplified to the one shown below.

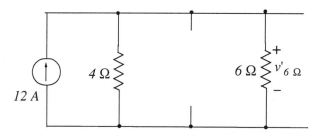

The voltage $v'_{6\,\Omega}$ is computed easily by application of the current division expression and multiplication by the *6 Ω* resistor. Thus,

$$v'_{6\,\Omega} = \left(\frac{4}{4+6} \times 12\right) \times 6 = \frac{144}{5} \ V$$

Next, we let $v''_{6\,\Omega}$ be the voltage due to the *18 A* current source acting alone. The simplified circuit is shown below. The letters A, B, and C are shown to visualize the circuit simplification process.

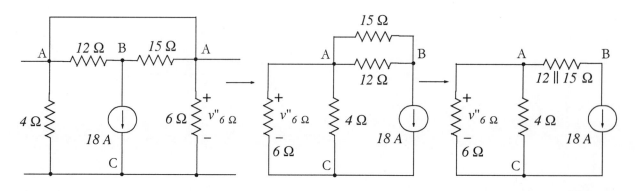

The voltage $v''_{6\,\Omega}$ is computed easily by application of the current division expression and multiplication by the $6\ \Omega$ resistor. Thus,

$$v''_{6\,\Omega} = \left[\frac{4}{4+6} \times (-18)\right] \times 6 = \frac{-216}{5}\ V$$

Now, we let $v'''_{6\,\Omega}$ be the voltage due to the $24\ A$ current source acting alone. The simplified circuit is shown below.

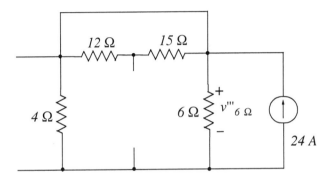

The $12\ \Omega$ and $15\ \Omega$ resistors are shorted out and voltage $v'''_{6\,\Omega}$ is computed by application of the current division expression and multiplication by the $6\ \Omega$ resistor. Thus,

$$v'''_{6\,\Omega} = \left(\frac{4}{4+6} \times 24\right) \times 6 = \frac{288}{5}\ V$$

Finally, we let $v^{iv}_{6\,\Omega}$ be the voltage due to the $36\ V$ voltage source acting alone. The simplified circuit is shown below.

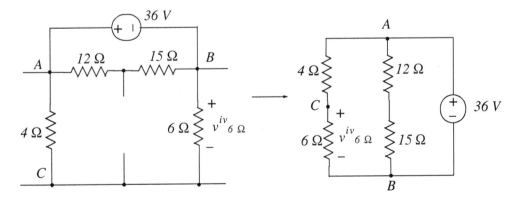

By application of the voltage division expression we find that

$$v^{iv}_{6\,\Omega} = \frac{6}{4+6} \times (-36) = -\frac{108}{5}$$

Therefore,

$$v_{6\,\Omega} = v'_{6\,\Omega} + v''_{6\,\Omega} + v'''_{6\,\Omega} + v^{iv}_{6\,\Omega} = \frac{144}{5} - \frac{216}{5} + \frac{288}{5} - \frac{108}{5} = \frac{108}{5} = 21.6\ V$$

This is the same answer as that of Problem 2.

13. The circuit for Measurement 1 is shown below.

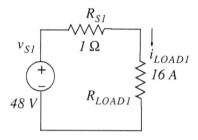

Let $R_{eq1} = R_{S1} + R_{LOAD1}$. Then,

$$R_{eq1} = \frac{v_{S1}}{i_{LOAD1}} = \frac{48}{16} = 3\ \Omega$$

For Measurement 3 the load resistance is the same as for Measurement 1 and the load current is given as $-5\ A$. Therefore, for Measurement 3 we find that

$$v_{S1} = R_{eq1}(-5) = 3 \times (-5) = -15\ V$$

and we enter this value in the table below.

The circuit for Measurement 2 is shown below.

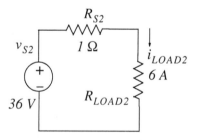

Let $R_{eq2} = R_{S1} + R_{LOAD2}$. Then,

$$R_{eq2} = \frac{v_{S2}}{i_{LOAD2}} = \frac{36}{6} = 6\ \Omega$$

For Measurement 4 the load resistance is the same as for Measurement 2 and v_{S2} is given as $-42\ V$. Therefore, for Measurement 4 we find that

$$i_{LOAD2} = \frac{v_{S2}}{R_{eq2}} = -\frac{42}{6} = -7 \text{ A}$$

and we enter this value in the table below.

The circuit for Measurement 5 is shown below.

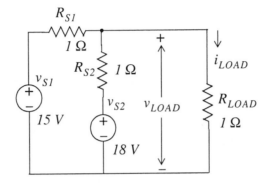

Replacing the voltage sources with their series resistances to their equivalent current sources with their parallel resistances and simplifying, we get the circuit below.

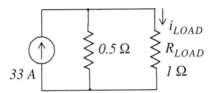

Application of the current division expression yields

$$i_{LOAD} = \frac{0.5}{0.5 + 1} \times 33 = 11 \text{ A}$$

and we enter this value in the table below.

The circuit for Measurement 6 is shown below.

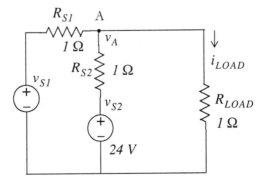

We observe that i_{LOAD} will be zero if $v_A = 0$ and this will occur when $v_{S1} = -24$. This can be shown to be true by writing a nodal equation at Node A. Thus,

$$\frac{v_A - (-24)}{1} + \frac{v_A - 24}{1} + 0 = 0$$

or $v_A = 0$

Measurement	Switch S_1	Switch S_2	v_{S1} (V)	v_{S2} (V)	i_L (A)
1	Closed	Open	48	0	16
2	Open	Closed	0	36	6
3	Closed	Open	-15	0	-5
4	Open	Closed	0	-42	-7
5	Closed	Closed	15	18	11
6	Closed	Closed	-24	24	0

14. The power supplied by the voltage source is

$$p_S = v_S(i_1 + i_2) = 480(100 + 80) = 86,400\ w = 86.4\ Kw$$

The power loss on the 1st floor is

$$p_{LOSS1} = i_1^2(0.5 + 0.5) = 100^2 \times 1 = 10,000\ w = 10\ Kw$$

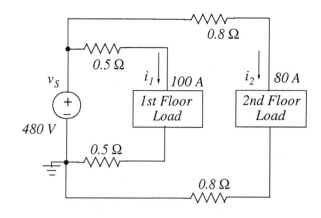

The power loss on the 2nd floor is

$$p_{LOSS2} = i_2^2(0.8 + 0.8) = 80^2 \times 1.6 = 10,240\ w = 10.24\ Kw$$

and thus the total loss is

$$Total\ loss\ =\ 10 + 10.24\ =\ 20.24\ Kw$$

Then,

$$Output\ power\ =\ Input\ power - power\ losses\ =\ 86.4 - 20.24\ =\ 66.16Kw$$

and

$$\%\ Efficiency\ =\ \eta\ =\ \frac{Output}{Input} \times 100\ =\ \frac{66.16}{86.4} \times 100\ =\ 76.6\%$$

This is indeed a low efficiency.

15. The voltage drop on the second floor conductor is

$$v_{cond}\ =\ R_T i_2\ =\ 1.6 \times 80\ =\ 128\ V$$

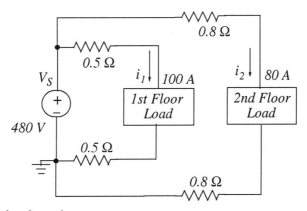

and thus the full-load voltage is

$$v_{FL}\ =\ 480 - 128\ =\ 352\ V$$

Then,

$$\%\ Regulation\ =\ \frac{v_{NL} - v_{FL}}{v_{FL}} \times 100\ =\ \frac{480 - 352}{352} \times 100\ =\ 36.4\%$$

This is a very poor regulation.

16. We assign node voltages and we write nodal equations as shown below.

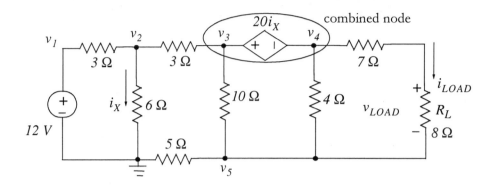

$$v_1 = 12$$

$$\frac{v_2 - v_1}{3} + \frac{v_2}{6} + \frac{v_2 - v_3}{3} = 0$$

$$\frac{v_3 - v_2}{3} + \frac{v_3 - v_5}{10} + \frac{v_4 - v_5}{4} + \frac{v_4 - v_5}{7+8} = 0$$

$$v_3 - v_4 = 20i_X$$

where $i_X = \dfrac{v_2}{6}$ and thus

$$v_5 = \frac{10}{3}v_2$$

$$\frac{v_5}{5} + \frac{v_5 - v_3}{10} + \frac{v_5 - v_4}{4} + \frac{v_5 - v_4}{7+8} = 0$$

Collecting like terms and rearranging we get

$$v_1 \qquad\qquad\qquad\qquad = 12$$

$$\frac{-1}{3}v_1 + \frac{5}{6}v_2 + \frac{-1}{3}v_3 \qquad\qquad = 0$$

$$\frac{-1}{3}v_2 + \frac{13}{30}v_3 + \frac{19}{60}v_4 - \frac{19}{60}v_5 = 0$$

$$-\frac{10}{3}v_2 \quad + v_3 \quad - v_4 \qquad\qquad = 0$$

$$-\frac{1}{10}v_3 - \frac{19}{60}v_4 + \frac{37}{60}v_5 = 0$$

and in matrix form

$$
\underbrace{
\begin{bmatrix}
1 & 0 & 0 & 0 & 0 \\
\dfrac{-1}{3} & \dfrac{5}{6} & \dfrac{-1}{3} & 0 & 0 \\
0 & \dfrac{-1}{3} & \dfrac{13}{30} & \dfrac{19}{60} & -\dfrac{19}{60} \\
0 & -\dfrac{10}{3} & 1 & -1 & 0 \\
0 & 0 & -\dfrac{1}{10} & -\dfrac{19}{60} & \dfrac{37}{60}
\end{bmatrix}}_{G}
\cdot
\underbrace{
\begin{bmatrix}
v_1 \\ v_2 \\ v_3 \\ v_4 \\ v_5
\end{bmatrix}}_{V}
=
\underbrace{
\begin{bmatrix}
12 \\ 0 \\ 0 \\ 0 \\ 0
\end{bmatrix}}_{I}
$$

We will use MATLAB to solve the above.

```
G=[1  0  0  0  0;...
-1/3 5/6 -1/3 0 0;...
0 -1/3 13/30 19/60 -19/60;...
0 -10/3 1 -1 0;...
0 0 -1/10 -19/60 37/60];
I=[12 0 0 0 0]'; V=G\I;
fprintf('\n');...
  fprintf('v1 = %7.2f V \n',V(1));...
  fprintf('v2 = %7.2f V \n',V(2));...
  fprintf('v3 = %7.2f V \n',V(3));...
  fprintf('v4 = %7.2f V \n',V(4));...
  fprintf('v5 = %7.2f V \n',V(5));...
  fprintf('\n'); fprintf('\n')
```

```
v1  =    12.00 V
v2  =    13.04 V
v3  =    20.60 V
v4  =   -22.87 V
v5  =    -8.40 V
```

Now,

$$
i_{LOAD} = \frac{v_4 - v_5}{8 + 7} = \frac{-22.87 - (-8.40)}{15} = -0.96 \, A
$$

and

$$
v_{LOAD} = 8i_{LOAD} = 8 \times (-0.96) = -7.68 \, V
$$

NOTES

Chapter 4

Introduction to Operational Amplifiers

This chapter is an introduction to amplifiers. It discusses amplifier gain in terms of decibels *(dB)* and provides an overview of operational amplifiers, their characteristics and applications. Numerous formulas for the computation of the gain are derived and several practical examples are provided.

4.1 Signals

A *signal* is any waveform that serves as a means of communication. It represents a fluctuating electric quantity, such as voltage, current, electric or magnetic field strength, sound, image, or any message transmitted or received in telegraphy, telephony, radio, television, or radar. A typical signal which varies with time is shown in figure 4.1 where $f(t)$ can be any physical quantity such as voltage, current, temperature, pressure, and so on.

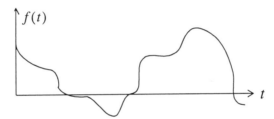

Figure 4.1. A signal that changes with time

4.2 Amplifiers

An *amplifier* is an electronic circuit which increases the magnitude of the input signal. The symbol of a typical amplifier is a triangle as shown in Figure 4.2.

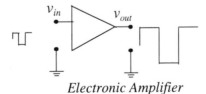

Electronic Amplifier

Figure 4.2. Symbol for electronic amplifier

An electronic (or electric) circuit which produces an output that is smaller than the input is called an *attenuator*. A resistive voltage divider is a typical attenuator.

Chapter 4 Introduction to Operational Amplifiers

An amplifier can be classified as a voltage amplifier, current amplifier, or power amplifier.

The *gain* of an amplifier is the ratio of the output to the input. Thus for a voltage amplifier,

$$Voltage\ Gain\ =\ \frac{Output\ Voltage}{Input\ Voltage}$$

or

$$G_v\ =\ \frac{v_{out}}{v_{in}} \tag{4.1}$$

The *current gain* G_i and *power gain* G_p are defined similarly.

Note 1: Throughout this text, the common (base 10) logarithm of a number x will be denoted as $log(x)$ while its natural (base e) logarithm will be denoted as $ln(x)$.

4.3 Decibels

The ratio of any two values of the same quantity (power, voltage or current) can be expressed in *decibels* (dB). For instance, we say that an amplifier has *10 dB* power gain or a transmission line has a power loss of *7 dB* (or gain *−7 dB*). If the gain (or loss) is *0 dB*, the output is equal to the input.

We must remember that a negative voltage or current gain G_v or G_i indicates that there is a *180°* phase difference between the input and the output waveforms. For instance, if an amplifier has a gain of −100 (dimensionless number), it means that the output is 180 degrees out-of-phase with the input. Therefore, to avoid misinterpretation of gain or loss, we use absolute values of power, voltage and current when these are expressed in dB.

By definition,

$$dB\ =\ 10log\left|\frac{p_{out}}{p_{in}}\right| \tag{4.2}$$

Therefore,

10 dB represents a power ratio of *10*

10n dB represents a power ratio of *10^n*

It is useful to remember that

20 dB represents a power ratio of *100*

30 dB represents a power ratio of *1000*

60 dB represents a power ratio of *1000000*

Also,

1 *dB* represents a power ratio of approximately *1.25*

3 *dB* represents a power ratio of approximately *2*

7 *dB* represents a power ratio of approximately *5*

From these, we can estimate other values. For instance, *4 dB* = *3 dB* + *1 dB* which is equivalent to a power ratio of approximately $2 \times 1.25 = 2.5$. Likewise, *27 dB* = *20 dB* + *7 dB* and this is equivalent to a power ratio of approximately $100 \times 5 = 500$.

Since $y = log x^2 = 2log x$ and $p = v^2/R = i^2 R$, if we let $R = 1$, the dB values for voltage and current ratios become:

$$dB_v = 10log \left| \frac{v_{out}}{v_{in}} \right|^2 = 20log \left| \frac{v_{out}}{v_{in}} \right| \tag{4.3}$$

and

$$dB_i = 10log \left| \frac{i_{out}}{i_{in}} \right|^2 = 20log \left| \frac{i_{out}}{i_{in}} \right| \tag{4.4}$$

Example 4.1

Compute the gain in dB_w for the amplifier shown in Figure 4.3.

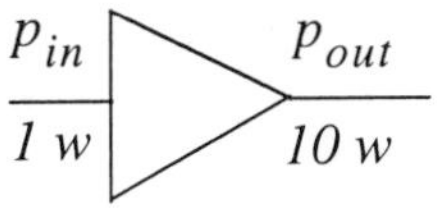

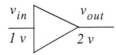

Figure 4.3. Amplifier for Example 4.1

Solution:

$$dB_w = 10log \frac{p_{out}}{p_{in}} = 10log \frac{10}{1} = 10log\,10 = 10 \times 1 = 10 \ \text{dBw}$$

Example 4.2

Compute the gain in dB_v for the amplifier shown in Figure 4.4, given that $log\,2 = 0.3$.

Figure 4.4. Amplifier for Example 4.2

Solution:

$$dB_v = 20log\frac{v_{out}}{v_{in}} = 20log\frac{2}{1} = 20log\,0.3 = 20 \times 0.3 = 6 \ \ dBv$$

4.4 Bandwidth and Frequency Response

Like electric filters, amplifiers exhibit a band of frequencies over which the output remains nearly constant. Consider, for example, the magnitude of the output voltage $|v_{out}|$ of an electric or electronic circuit as a function of radian frequency ω as shown in Figure 4.5.

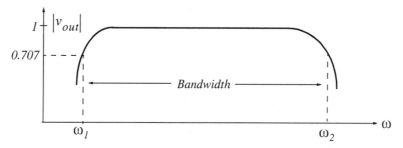

Figure 4.5. Typical bandwidth of an amplifier

As shown above, the *bandwidth* is $BW = \omega_2 - \omega_1$ where ω_1 and ω_2 are the *lower and upper cutoff frequencies* respectively. At these frequencies, $|v_{out}| = \sqrt{2}/2 = 0.707$ and these two points are known as the *3-dB down* or *half-power points*. They derive their name from the fact that power $p = v^2/R = i^2 R$, and for $R = 1$ and $v = \sqrt{2}/2 = 0.707$ or $i = \sqrt{2}/2 = 0.707$, the power is $1/2$, that is, the power is "halved". Alternately, we can define the bandwidth as the frequency band between half-power points.

Most amplifiers are used with a *feedback* path which returns (feeds) some or all its output to the input as shown in Figure 4.6.

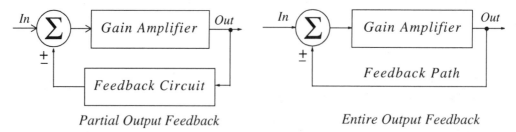

Partial Output Feedback   *Entire Output Feedback*

Figure 4.6. Gain amplifiers used with feedback

In Figure 4.6, the symbol Σ (Greek capital letter sigma) inside the circle denotes the summing point where the output signal, or portion of it, is combined with the input signal. This summing point may be also indicated with a large plus (+) symbol inside the circle. The positive (+) sign below the summing point implies *positive feedback* which means that the output, or portion of it, is added to the input. On the other hand, the negative (–) sign implies *negative feedback* which means that the output, or portion of it, is subtracted from the input. Practically, all amplifiers use used with negative feedback since positive feedback causes circuit instability.

4.5 The Operational Amplifier

The *operational amplifier* or simply *op amp* is the most versatile electronic amplifier. It derives it name from the fact that it is capable of performing many mathematical operations such as addition, multiplication, differentiation, integration, analog-to-digital conversion or vice versa. It can also be used as a comparator and electronic filter. It is also the basic block in analog computer design. Its symbol is shown in Figure 4.7.

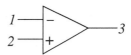

Figure 4.7. Symbol for operational amplifier

As shown above the op amp has two inputs but only one output. For this reason it is referred to as *differential input, single ended output* amplifier. Figure 4.8 shows the internal construction of a typical op amp. This figure also shows terminals V_{CC} and V_{EE}. These are the voltage sources required to power up the op amp. Typically, V_{CC} is +15 volts and V_{EE} is −15 volts. These terminals are not shown in op amp circuits since they just provide power, and do not reveal any other useful information for the op amp's circuit analysis.

4.6 An Overview of the Op Amp

The op amp has the following important characteristics:

1. Very high input impedance (resistance)

2. Very low output impedance (resistance)

3. Capable of producing a very large gain that can be set to any value by connection of external resistors of appropriate values

4. Frequency response from DC to frequencies in the MHz range

5. Very good stability

6. Operation to be performed, i.e., addition, integration etc. is done externally with proper selection of passive devices such as resistors, capacitors, diodes, and so on.

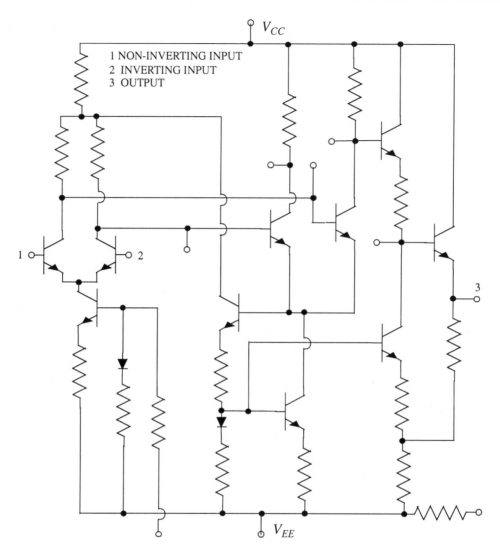

1 NON-INVERTING INPUT
2 INVERTING INPUT
3 OUTPUT

Figure 4.8. Internal Devices of a Typical Op Amp

An op amp is said to be connected in the *inverting mode* when an input signal is connected to the inverting (−) input through an external resistor R_{in} whose value along with the feedback resistor R_f determine the op amp's gain. The non-inverting (+) input is grounded through an external resistor R as shown in Figure 4.9.

For the circuit of Figure 4.9, the voltage gain G_v is

$$G_v = \frac{v_{out}}{v_{in}} = -\frac{R_f}{R_{in}} \tag{4.5}$$

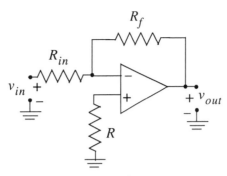

Figure 4.9. Circuit of Inverting op amp

Note 2: The resistor R connected between the non-inverting (+) input and ground serves only as a current limiting device, and thus it does not influence the op amp's gain. It will be omitted in our subsequent discussion.

Note 3: The input voltage v_{in} and the output voltage v_{out} as indicated in the circuit of Figure 4.9, should not be interpreted as open circuits; these designations imply that an input voltage of any waveform may be applied at the input terminals and the corresponding output voltage appears at the output terminals.

As shown in the formula of (4.5), the gain for this op amp configuration is the ratio $-R_f/R_{in}$ where R_f is the feedback resistor which allows portion of the output to be fed back to the input. The minus (–) sign in the gain ratio $-R_f/R_{in}$ implies that the output signal has opposite polarity from that of the input signal; hence the name inverting amplifier. Therefore, when the input signal is positive (+) the output will be negative (–) and vice versa. For example, if the input is +1 volt DC and the op amp gain is 100, the output will be −100 volts DC. For AC (sinusoidal) signals, the output will be 180 degrees out-of-phase with the input. Thus, if the input is 1 volt AC and the op amp gain is 5, the output will be −5 volts AC or 5 volts AC with 180 degrees out-of-phase with the input.

Example 4.3

Compute the voltage gain G_v and then the output voltage v_{out} for the inverting op amp circuit shown in Figure 4.10, given that $v_{in} = 1\ mV$. Plot v_{in} and v_{out} as mV versus time on the same set of axes.

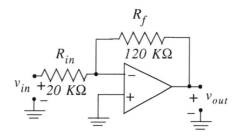

Figure 4.10. Circuit for Example 4.3

Solution:

This is an inverting amplifier and thus the voltage gain G_v is

$$G_v = -\frac{R_f}{R_{in}} = -\frac{120\ K\Omega}{20\ K\Omega}$$

or

$$G_v = -6$$

and since

$$G_v = \frac{v_{out}}{v_{in}}$$

the output voltage is

$$v_{out} = G_v v_{in} = -6 \times 1$$

or

$$v_{out} = -6\ mV$$

The voltages v_{in} and v_{out} are plotted as shown in Figure 4.11.

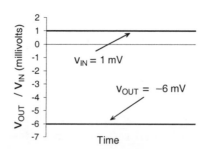

Figure 4.11. Input and output waveforms for the circuit of Example 4.3

Example 4.4

Compute the voltage gain G_v and then the output voltage v_{out} for the inverting op amp circuit shown in Figure 4.12, given that $v_{in} = sint \ mV$. Plot v_{in} and v_{out} as mV versus time on the same set of axes.

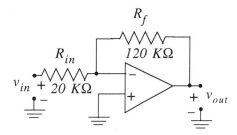

Figure 4.12. Circuit for Example 4.4

Solution:

This is the same circuit as that of the previous example except that the input is a sine wave with unity amplitude and the voltage gain G_v is the same as before, that is,

$$G_v = -\frac{R_f}{R_{in}} = -\frac{120 \ K\Omega}{20 \ K\Omega} = -6$$

and the output voltage is

$$v_{out} = G_v v_{in} = -6 \times sint = -6 sint \ mV$$

The voltages v_{in} and v_{out} are plotted as shown in Figure 4.13.

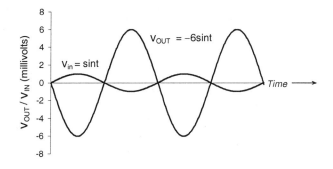

Figure 4.13. Input and output waveforms for the circuit of Example 4.4

An op amp is said to be connected in the *non-inverting mode* when an input signal is connected to the non-inverting (+) input through an external resistor R which serves as a current limiter, and the inverting (–) input is grounded through an external resistor R_{in} as shown in Figure 4.14. In our subsequent discussion, the resistor R will represent the internal resistance of the applied voltage v_{in}.

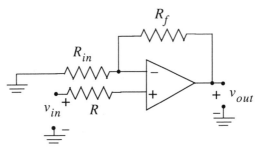

Figure 4.14. Circuit of non-inverting op amp

For the circuit of Figure 4.14, the voltage gain G_v is

$$G_v = \frac{v_{out}}{v_{in}} = 1 + \frac{R_f}{R_{in}} \tag{4.6}$$

As indicated by the relation of (4.6), the gain for this op amp configuration is $1 + R_f/R_{in}$ and therefore, in the non-inverting mode the op amp output signal has the same polarity as the input signal; hence, the name non-inverting amplifier. Thus, when the input signal is positive (+) the output will be also positive and if the input is negative, the output will be also negative. For example, if the input is $+1\ mV\ DC$ and the op amp gain is 75, the output will be $+75\ mV\ DC$. For AC signals the output will be in-phase with the input. For example, if the input is $0.5\ V\ AC$ and the op amp gain is $G_v = 1 + 19\ K\Omega/1\ K\Omega = 20$, the output will be $10\ V\ AC$ and in-phase with the input.

Example 4.5

Compute the voltage gain G_v and then the output voltage v_{out} for the non-inverting op amp circuit shown in Figure 4.15, given that $v_{in} = 1\ mV$. Plot v_{in} and v_{out} as mV versus time on the same set of axes.

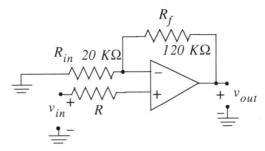

Figure 4.15. Circuit for Example 4.5

Solution:

The voltage gain G_v is

$$G_v = \frac{v_{out}}{v_{in}} = 1 + \frac{R_f}{R_{in}} = 1 + \frac{120 \; K\Omega}{20 \; K\Omega} = 1 + 6 = 7$$

and thus

$$v_{out} = G_v v_{in} = 7 \times 1 \; mV = 7 \; mV$$

The voltages v_{in} and v_{out} are plotted as shown in Figure 4.16.

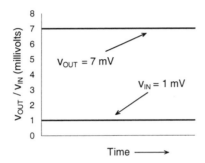

Figure 4.16. Input and output waveforms for the circuit of Example 4.5

Example 4.6

Compute the voltage gain G_v and then the output voltage v_{out} for the non-inverting op amp circuit shown in Figure 4.17, given that $v_{in} = sint \; mV$. Plot v_{in} and v_{out} as mV versus time on the same set of axes.

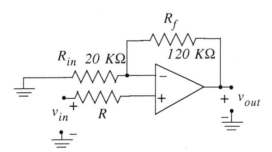

Figure 4.17. Circuit for Example 4.6

Solution:

This is the same circuit as in the previous example except that the input is a sinusoid. Therefore, the voltage gain G_v is the same as before, that is,

$$G_v = \frac{v_{out}}{v_{in}} = 1 + \frac{R_f}{R_{in}} = 1 + \frac{120 \; K\Omega}{20 \; K\Omega} = 1 + 6 = 7$$

and the output voltage is

$$v_{out} = G_v v_{in} = 7 \times sint = 7\,sint\;mV$$

The voltages v_{in} and v_{out} are plotted as shown in Figure 4.18.

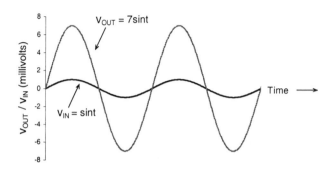

Figure 4.18. Input and output waveforms for the circuit of Example 4.6

Quite often an op amp is connected as shown in Figure 4.19.

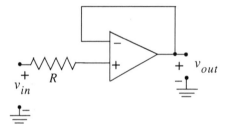

Figure 4.19. Circuit of unity gain op amp

For the circuit of Figure 4.19, the voltage gain G_v is

$$G_v = \frac{v_{out}}{v_{in}} = 1 \tag{4.7}$$

and thus

$$v_{out} = v_{in} \tag{4.8}$$

For this reason, the op amp circuit of Figure 4.19 it is called *unity gain amplifier*. For example, if the input voltage is *5 mV DC* the output will also be *5 mV DC*, and if the input voltage is *2 mV AC*, the output will also be *2 mV AC*. The unity gain op amp is used to provide a very high resistance between a voltage source and the load connected to it. An example will be given in Section 4.8.

4.7 Active Filters

An *active filter* is an electronic circuit consisting of an amplifier and other devices such as resistors and capacitors. In contrast, a *passive filter* is a circuit which consists of passive devices such as resistors, capacitors and inductors. Operational amplifiers are used extensively as active filters.

A *low-pass filter* transmits (passes) all frequencies below a *critical (cutoff)* frequency denoted as ω_C, and *attenuates* (blocks) all frequencies above this cutoff frequency. An op amp low-pass filter is shown in Figure 4.20 and its frequency response in Figure 4.21.

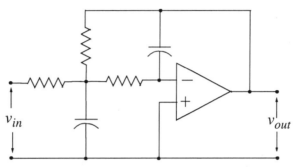

Figure 4.20. A low-pass active filter

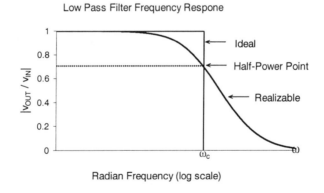

Figure 4.21. Frequency response for amplitude of a low-pass filter

In Figure 4.21, the straight vertical and horizontal lines represent the ideal (unrealizable) and the smooth curve represents the practical (realizable) low-pass filter characteristics. The vertical scale represents the magnitude of the ratio of output-to-input voltage v_{out}/v_{in}, that is, the gain G_v. The cutoff frequency ω_c is the frequency at which the maximum value of v_{out}/v_{in} which is unity, falls to $0.707 \times G_v$, and as mentioned before, this is the *half power* or the *−3 dB* point.

A *high-pass filter* transmits (passes) all frequencies above a critical (cutoff) frequency ω_c, and attenuates (blocks) all frequencies below the cutoff frequency. An op amp high-pass filter is shown in Figure 4.22 and its frequency response in Figure 4.23.

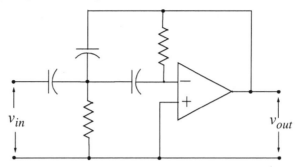

Figure 4.22. A high-pass active filter

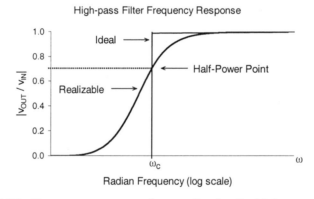

Figure 4.23. Frequency response for amplitude of a high-pass filter

In Figure 4.23, the straight vertical and horizontal lines represent the ideal (unrealizable) and the smooth curve represents the practical (realizable) high-pass filter characteristics. The vertical scale represents the magnitude of the ratio of output-to-input voltage v_{out}/v_{in}, that is, the gain G_v. The cutoff frequency ω_c is the frequency at which the maximum value of v_{out}/v_{in} which is unity, falls to $0.707 \times G_v$, i.e., the *half power* or the $-3\ dB$ point.

A *band-pass* filter transmits (passes) the band (range) of frequencies between the critical (cutoff) frequencies denoted as ω_1 and ω_2, where the maximum value of G_v which is unity, falls to $0.707 \times G_v$, while it attenuates (blocks) all frequencies outside this band. An op amp band-pass filter and its frequency response are shown below. An op amp band-pass filter is shown in Figure 4.24 and its frequency response in Figure 4.25.

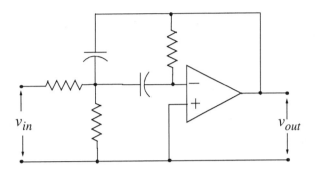

Figure 4.24. An active band-pass filter

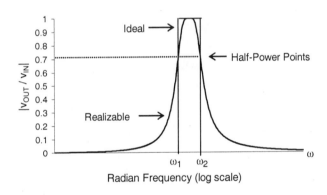

Figure 4.25. Frequency response for amplitude of a band-pass filter

A *band-elimination* or *band-stop* or *band-rejection* filter attenuates (rejects) the band (range) of frequencies between the critical (cutoff) frequencies denoted as ω_1 and ω_2, where the maximum value of G_v which is unity, falls to $0.707 \times G_v$, while it transmits (passes) all frequencies outside this band. An op amp band-stop filter is shown in Figure 4.26 and its frequency response in Figure 4.27.

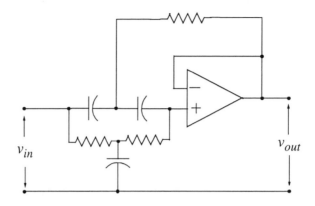

Band Elimination Active Filter Circuit

Figure 4.26.   An active band-elimination filter

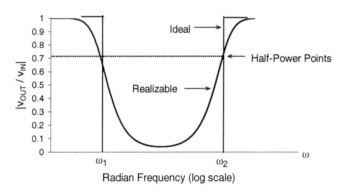

Figure 4.27.   Frequency response for amplitude of a band-elimination filter

4.8 Analysis of Op Amp Circuits

The procedure for analyzing an op amp circuit (finding voltages, currents and power) is the same as for the other circuits which we have studied thus far. That is, we can apply Ohm's law, KCL and KVL, superposition, Thevenin's and Norton's theorems. When analyzing an op amp circuit, we must remember that in any op-amp:

a. *The currents into both input terminals are zero*

b. *The voltage difference between the input terminals of an op amp is zero*

c. *For circuits containing op amps, we will assume that the reference (ground) is the common terminal of the two power supplies.* For simplicity, the power supplies will not be shown.

We will provide several examples to illustrate the analysis of op amp circuits without being concerned about its internal operation; this is discussed in electronic circuit analysis books.

Example 4.7

The op amp circuit shown in Figure 4.28 is called *inverting op amp*. Prove that the voltage gain G_v is as given in (4.9) below, and draw its equivalent circuit showing the output as a dependent source.

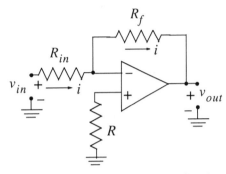

Figure 4.28. Circuit for deriving the gain of an inverting op amp

$$G_v = \frac{v_{out}}{v_{in}} = -\frac{R_f}{R_{in}} \tag{4.9}$$

Proof:

No current flows through the (–) input terminal of the op amp; therefore the current i which flows through resistor R_{in} flows also through resistor R_f. Also, since the (+) input terminal is grounded and there is no voltage drop between the (–) and (+) terminals, the (–) input is said to be at *virtual ground*. From the circuit of Figure 4.28,

$$v_{out} = -R_f i$$

where

$$i = \frac{v_{in}}{R_{in}}$$

and thus

$$v_{out} = -\frac{R_f}{R_{in}} v_{in}$$

or

$$G_v = \frac{v_{out}}{v_{in}} = -\frac{R_f}{R_{in}}$$

The input and output parts of the circuit are shown in Figure 4.29 with the virtual ground being the same as the circuit ground.

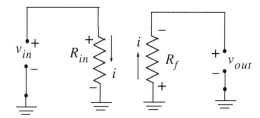

Figure 4.29. Input and output parts of the inverting op amp

These two circuits are normally drawn with the output as a dependent source as shown in Figure 4.30. This is the *equivalent circuit of the inverting op amp* and, as mentioned in Chapter 1, the dependent source is a Voltage Controlled Voltage Source (VCVS).

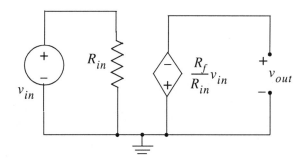

Figure 4.30. Equivalent circuit of the inverting op amp

Example 4.8

The op amp circuit shown in Figure 4.31 is called *non-inverting op amp*. Prove that the voltage gain G_v is as given in (4.10) below, and draw its equivalent circuit showing the output as a dependent source.

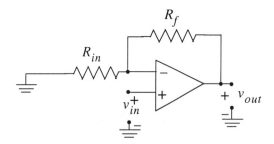

Figure 4.31. Circuit of non-inverting op amp

$$G_v = \frac{v_{out}}{v_{in}} = 1 + \frac{R_f}{R_{in}} \qquad (4.10)$$

Proof:

Let the voltages at the (−) and (+) terminals be denoted as v_1 and v_2 respectively as shown in Figure 4.32.

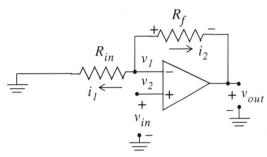

Figure 4.32. Non-inverting op amp circuit for derivation of (4.10)

By application of KCL at v_1

$$i_1 + i_2 = 0$$

or

$$\frac{v_1}{R_{in}} + \frac{v_1 - v_{out}}{R_f} = 0 \qquad (4.11)$$

There is no potential difference between the (−) and (+) terminals; therefore, $v_1 - v_2 = 0$ or $v_1 = v_2 = v_{in}$. Relation (4.11) then can be written as

$$\frac{v_{in}}{R_{in}} + \frac{v_{in} - v_{out}}{R_f} = 0$$

or

$$\left(\frac{1}{R_{in}} + \frac{1}{R_f}\right)v_{in} = \frac{v_{out}}{R_f}$$

Rearranging, we get

$$G_v = \frac{v_{out}}{v_{in}} = 1 + \frac{R_f}{R_{in}}$$

and its equivalent circuit is as shown in Figure 4.33. The dependent source of this equivalent circuit is also a VCVS.

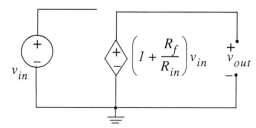

Figure 4.33. Equivalent circuit of the non-inverting op amp

Example 4.9

If, in the non-inverting op amp circuit of the previous example, we replace R_{in} with an open circuit $(R_{in} \rightarrow \infty)$ and R_f with a short circuit $(R_f \rightarrow 0)$, prove that the voltage gain G_v is

$$G_v = \frac{v_{out}}{v_{in}} = 1 \tag{4.12}$$

and thus

$$v_{out} = v_{in} \tag{4.13}$$

Proof:

With R_{in} open and R_f shorted, the non-inverting amplifier of the previous example reduces to the circuit of Figure 4.34.

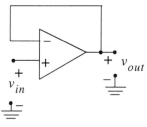

Figure 4.34. Circuit of Figure 4.32 with R_{in} open and R_f shorted

The voltage difference between the (+) and (−) terminals is zero; then $v_{out} = v_{in}$.

We will obtain the same result if we consider the non-inverting op amp gain $G_v = \frac{v_{out}}{v_{in}} = 1 + \frac{R_f}{R_{in}}$.
Then, letting $R_f \rightarrow 0$, the gain reduces to $G_v = 1$ and for this reason this circuit is called *unity gain*

amplifier or *voltage follower*. It is also called *buffer amplifier* because it can be used to "buffer" (isolate) one circuit from another when one "loads" the other as we will see on the next example.

Example 4.10

For the circuit of Figure4.35

a. With the load R_{LOAD} disconnected, compute the open circuit voltage v_{ab}

b. With the load connected, compute the voltage v_{LOAD} across the load R_{LOAD}

c. Insert a buffer amplifier between a and b and compute the new voltage v_{LOAD} across the same load R_{LOAD}

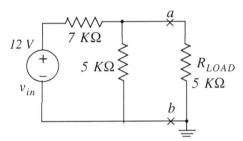

Figure 4.35. Circuit for Example 4.10

Solution:

a. With the load R_{LOAD} disconnected the circuit is as shown in Figure 4.36.

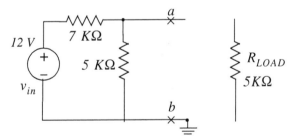

Figure 4.36. Circuit for Example 4.10 with the load disconnected

The voltage across terminals a and b is

$$v_{ab} = \frac{5 \ K\Omega}{7 \ K\Omega + 5 \ K\Omega} \times 12 = 5 \ V$$

b. With the load R_{LOAD} reconnected the circuit is as shown in Figure 4.37.

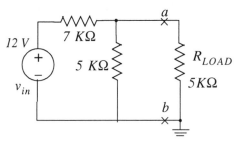

Figure 4.37. Circuit for Example 4.10 with the load reconnected

$$v_{LOAD} = \frac{5\ K\Omega \parallel 5\ K\Omega}{7\ K\Omega + 5\ K\Omega \parallel 5\ K\Omega} \times 12 = 3.16\ V$$

Here, we observe that the load R_{LOAD} "loads down" the load voltage from *5 V* to *3.16 V* and this voltage may not be sufficient for proper operation of the load.

c. With the insertion of the buffer amplifier between points *a* and *b* and the load, the circuit now is as shown in Figure 4.38.

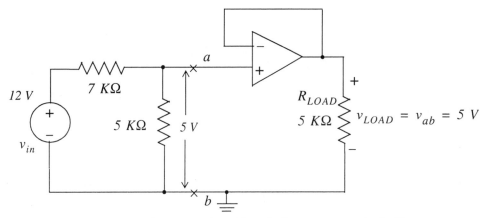

Figure 4.38. Circuit for Example 4.10 with the insertion of a buffer op amp

From the circuit of Figure 4.38, we observe that the voltage across the load is *5 V* as desired.

Example 4.11

The op amp circuit shown in Figure 4.39 is called *summing circuit* or *summer* because the output is the summation of the weighted inputs. Prove that for this circuit,

$$v_{out} = -R_f\left(\frac{v_{in1}}{R_{in1}} + \frac{v_{in2}}{R_{in2}}\right) \tag{4.14}$$

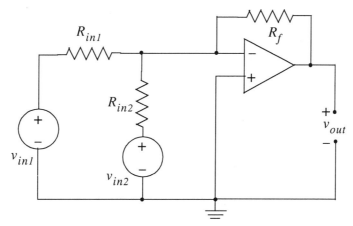

Figure 4.39. Two-input summing op amp circuit

Proof:

We recall that the voltage across the (−) and (+) terminals is zero. We also observe that the (+) input is grounded, and thus the voltage at the (−) terminal is at "virtual ground". Then, by application of KCL at the (−) terminal, we get

$$\frac{v_{in1}}{R_{in1}} + \frac{v_{in2}}{R_{in2}} + \frac{v_{out}}{R_f} = 0$$

and solving for v_{out} we get (4.14). Alternately, we can apply the principle of superposition to derive this relation.

Example 4.12

Compute the output voltage v_{out} for the amplifier circuit shown in Figure 4.40.

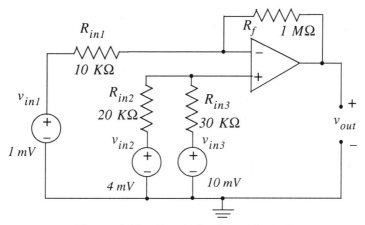

Figure 4.40. Circuit for Example 4.12

Solution:

Let v_{out1} be the output due to v_{in1} acting alone, v_{out2} be the output due to v_{in2} acting alone, and v_{out3} be the output due to v_{in3} acting alone. Then by superposition,

$$v_{out} = v_{out1} + v_{out2} + v_{out3}$$

First, with v_{in1} acting alone and v_{in2} and v_{in3} shorted, the circuit becomes as shown in Figure 4.41.

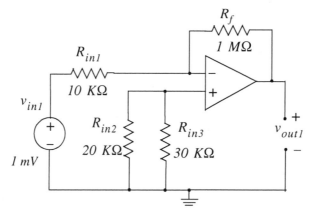

Figure 4.41. Circuit for Example 4.12 with v_{in1} acting alone

We recognize this as an inverting amplifier whose voltage gain G_v is

$$G_v = 1 \ M\Omega / 10 \ K\Omega = 100$$

and thus

$$v_{out1} = (100)(-1 \ mV) = -100 \ mV \qquad (4.15)$$

Next, with v_{in2} acting alone and v_{in1} and v_{in3} shorted, the circuit becomes as shown in Figure 4.42.

The circuit of Figure 4.42 as a non-inverting op amp whose voltage gain G_v is

$$G_v = 1 + 1 \ M\Omega / 10 \ K\Omega \doteq 101$$

and the voltage at the plus (+) input is computed from the voltage divider circuit shown in Figure 4.43.

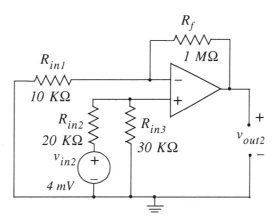

Figure 4.42. Circuit for Example 4.12 with v_{in2} acting alone

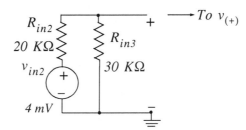

Figure 4.43. Voltage divider circuit for the computation of $v_{(+)}$ with v_{in2} acting alone

Then,

$$v_{(+)} = \frac{R_{in3}}{R_{in2} + R_{in3}} \times v_{in2} = \frac{30\ K\Omega}{50\ K\Omega} \times 4\ mV = 2.4\ mV$$

and thus

$$v_{out2} = 101 \times 2.4\ mV = 242.4\ mV \qquad (4.16)$$

Finally, with v_{in3} acting alone and v_{in1} and v_{in2} shorted, the circuit becomes as shown in Figure 4.44.

The circuit of Figure 4.44 is also a non-inverting op amp whose voltage gain G_v is

$$G_v = 1 + 1\ M\Omega/10\ K\Omega = 101$$

and the voltage at the plus (+) input is computed from the voltage divider circuit shown in Figure 4.45.

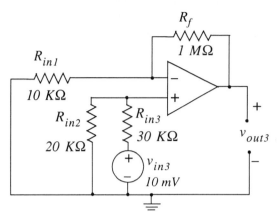

Figure 4.44. Circuit for Example 4.12 with v_{in3} acting alone

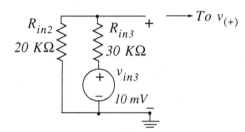

Figure 4.45. Voltage divider circuit for the computation of $v_{(+)}$ with v_{in3} acting alone

Then,

$$v_{(+)} = \frac{R_{in2}}{R_{in2} + R_{in3}} \times v_{in2} = \frac{20 \ K\Omega}{50 \ K\Omega} \times 10 \ mV = 4 \ mV$$

and thus

$$v_{out3} = 101 \times 4 \ mV = 404 \ mV \tag{4.17}$$

Therefore, from (4.15), (4.16) and (4.17),

$$v_{out} = v_{out1} + v_{out2} + v_{out3} = -100 + 242.4 + 404 = 546.4 \ mV$$

Example 4.13

For the circuit shown in Figure 4.46, derive an expression for the voltage gain G_v in terms of the external resistors R_1, R_2, R_3, and R_f.

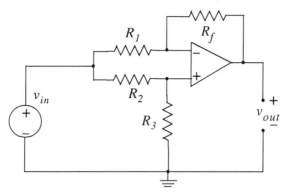

Figure 4.46. Circuit for Example 4.13

Solution:

We apply KCL at nodes v_1 and v_2 as shown in Figure 4.47.

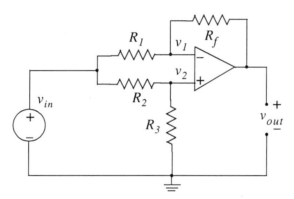

Figure 4.47. Application of KCL for the circuit of Example 4.13

At node v_1:

$$\frac{v_1 - v_{in}}{R_1} + \frac{v_1 - v_{out}}{R_f} = 0$$

or

$$\left(\frac{1}{R_1} + \frac{1}{R_f}\right)v_1 = \frac{v_{in}}{R_1} + \frac{v_{out}}{R_f}$$

or

$$\left(\frac{R_1 + R_f}{R_1 R_f}\right)v_1 = \frac{R_f v_{in} + R_1 v_{out}}{R_1 R_f}$$

or

$$v_1 = \frac{R_f v_{in} + R_1 v_{out}}{R_1 + R_f} \tag{4.18}$$

At node v_2:

$$\frac{v_2 - v_{in}}{R_2} + \frac{v_2}{R_3} = 0$$

or

$$v_2 = \frac{R_3 v_{in}}{R_2 + R_3} \tag{4.19}$$

and since $v_2 = v_1$, we rewrite (4.19) as

$$v_1 = \frac{R_3 v_{in}}{R_2 + R_3} \tag{4.20}$$

Equating the right sides of (4.18) and (4.20) we get

$$\frac{R_f v_{in} + R_1 v_{out}}{R_1 + R_f} = \frac{R_3 v_{in}}{R_2 + R_3}$$

or

$$R_f v_{in} + R_1 v_{out} = \frac{R_3 v_{in}}{R_2 + R_3}(R_1 + R_f)$$

Dividing both sides of the above relation by $R_1 v_{in}$ and rearranging, we get

$$\frac{v_{out}}{v_{in}} = \frac{R_3(R_1 + R_f)}{R_1(R_2 + R_3)} - \frac{R_f}{R_1}$$

and after simplification

$$G_v = \frac{v_{out}}{v_{in}} = \frac{R_1 R_3 - R_2 R_f}{R_1(R_2 + R_3)} \tag{4.21}$$

4.9 Input and Output Resistance

The input and output resistances are very important parameters in amplifier circuits.

The *input resistance* R_{in} of a circuit is defined as the ratio of the applied voltage v_S to the current i_S drawn by the circuit, that is,

$$\boxed{R_{in} = \frac{v_S}{i_S}} \tag{4.22}$$

Therefore, in an op amp circuit the input resistance provides a measure of the current i_S which the amplifier draws from the voltage source v_S. Of course, we want i_S to be as small as possible; accordingly, we must make the input resistance R_{in} as high as possible.

Example 4.14

Compute the input resistance R_{in} of the inverting op amp amplifier shown in Figure 4.47 in terms of R_I and R_f.

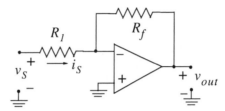

Figure 4.48. Circuit for Example 4.14

Solution:

By definition,

$$R_{in} = \frac{v_S}{i_S} \tag{4.23}$$

and since no current flows into the minus (−) terminal of the op amp and this terminal is at virtual ground, it follows that

$$i_S = \frac{v_S}{R_I} \tag{4.24}$$

From (4.23) and (4.24) we observe that

$$\boxed{R_{in} = R_I} \tag{4.25}$$

It is therefore, desirable to make R_I as high as possible. However, if we make R_I very high such as $10\ M\Omega$, for a large gain, say 100, the value of the feedback resistor R_f should be $1\ G\Omega$. Obviously, this is an impractical value. Fortunately, a large gain can be achieved with the circuit of Problem 8.

Example 4.15

Compute the input resistance R_{in} of the op amp shown in Figure 4.49.

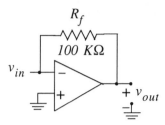

Figure 4.49. Circuit for Example 4.15

Solution:

In the circuit of Figure 4.49, v_{in} is the voltage at the minus (–) terminal; not the source voltage v_S. Therefore, there is no current i_S drawn by the op amp. In this case, we apply a test (hypothetical) current i_X as shown in Figure 4.49, and we treat v_{in} as the source voltage.

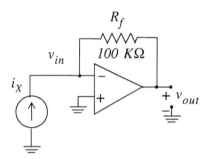

Figure 4.50. Circuit for Example 4.15 with a test current source

We observe that v_{in} is zero (virtual ground). Therefore,

$$R_{in} = \frac{v_{in}}{i_X} = \frac{0}{i_X} = 0$$

By definition, *the output resistance R_{out} is the ratio of the open circuit voltage to the short circuit current*, that is,

$$\boxed{R_{out} = \frac{v_{OC}}{i_{SC}}} \tag{4.26}$$

The output resistance R_{out} is not the same as the load resistance. The output resistance provides a measure of the change in output voltage when a load which is connected at the output terminals draws current from the circuit. It is desirable to have an op amp with very low output resistance as illustrated by the following example.

Example 4.16

The output voltage of an op amp decreases by *10%* when a *5 KΩ* load is connected at the output terminals. Compute the output resistance R_{out}.

Solution:

Consider the output portion of the op amp shown in Figure 4.51.

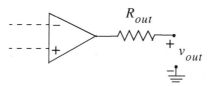

Figure 4.51. Partial circuit for Example 4.16

With no load connected at the output terminals,

$$v_{out} = v_{OC} = G_v v_{in} \tag{4.27}$$

With a load R_{LOAD} connected at the output terminals, the load voltage v_{LOAD} is

$$v_{LOAD} = \frac{R_{LOAD}}{R_{out} + R_{LOAD}} \times v_{out} \tag{4.28}$$

and from (4.27) and (4.28)

$$v_{LOAD} = \frac{R_{LOAD}}{R_{out} + R_{LOAD}} \times G_v v_{in} \tag{4.29}$$

Therefore,

$$\frac{v_{LOAD}}{v_{OC}} = 0.9 = \frac{5\ K\Omega}{R_{out} + 5\ K\Omega}$$

and solving for R_{out} we get

$$R_{out} = 555\ \Omega$$

We observe from (4.29) that as $R_{out} \to 0$, relation (4.29) reduces to $v_{LOAD} = G_v v_{in}$ and by comparison with (4.27), we see that $v_{LOAD} = v_{OC}$

4.10 Summary

- A signal is any waveform representing a fluctuating electric quantity, such as voltage, current, electric or magnetic field strength, sound, image, or any message transmitted or received in telegraphy, telephony, radio, television, or radar. al that changes with time.

- An amplifier is an electronic circuit which increases the magnitude of the input signal.

- The gain of an amplifier is the ratio of the output to the input. It is normally expressed in decibel (dB) units where by definition $dB = 10log|p_{out}/p_{in}|$

- Frequency response is the band of frequencies over which the output remains fairly constant.

- The lower and upper cutoff frequencies are those where the output is 0.707 of its maximum value. They are also known as half-power points.

- Most amplifiers are used with feedback where the output, or portion of it, is fed back to the input.

- The operational amplifier (op amp) is the most versatile amplifier and its main features are:

  1. Very high input impedance (resistance)

  2. Very low output impedance (resistance)

  3. Capable of producing a very large gain that can be set to any value by connection of external resistors of appropriate values

  4. Frequency response from DC to frequencies in the MHz range

  5. Very good stability

  6. Operation to be performed, i.e., addition, integration etc. is done externally with proper selection of passive devices such as resistors, capacitors, diodes, and so on.

- The gain of an inverting op amp is the ratio $-R_f/R_{in}$ where R_f is the feedback resistor which allows portion of the output to be fed back to the minus (–) input. The minus (–) sign implies that the output signal has opposite polarity from that of the input signal.

- The gain of an non-inverting op amp is $1 + R_f/R_{in}$ where R_f is the feedback resistor which allows portion of the output to be fed back to the minus (–) input which is grounded through the R_{in} resistor. The output signal has the same polarity from that of the input signal.

- In a unity gain op amp the output is the same as the input. A unity gain op amp is used to provide a very high resistance between a voltage source and the load connected to it.

- Op amps are also used as active filters.

- A low-pass filter transmits (passes) all frequencies below a critical (cutoff) frequency denoted as ω_C and attenuates (blocks) all frequencies above this cutoff frequency.

- A high-pass filter transmits (passes) all frequencies above a critical (cutoff) frequency ω_c, and attenuates (blocks) all frequencies below the cutoff frequency.

- A band-pass filter transmits (passes) the band (range) of frequencies between the critical (cutoff) frequencies denoted as ω_1 and ω_2, where the maximum value of G_v which is unity, falls to $0.707 \times G_v$, while it attenuates (blocks) all frequencies outside this band.

- A band-elimination or band-stop or band-rejection filter attenuates (rejects) the band (range) of frequencies between the critical (cutoff) frequencies denoted as ω_1 and ω_2, where the maximum value of G_v which is unity, falls to $0.707 \times G_v$, while it transmits (passes) all frequencies outside this band.

- A summing op amp is a circuit with two or more inputs.

- The input resistance is the ratio of the applied voltage v_S to the current i_S drawn by the circuit, that is, $R_{in} = v_S / i_S$

- The output resistance (not to be confused with the load resistance) is the ratio of the open circuit voltage when the load is removed from the circuit, to the short circuit current which is the current that flows through a short circuit connected at the output terminals, that is, $R_o = v_{OC} / i_{SC}$

4.11 Exercises

Multiple Choice

1. In the op amp circuit of Figure 4.52 $v_{in} = 2\ V$, $R_{in} = 1\ K\Omega$, and it is desired to have $v_{out} = 8\ V$. This will be obtained if the feedback resistor R_f has a value of

   A. *1 KΩ*

   B. *2 KΩ*

   C. *3 KΩ*

   D. *4 KΩ*

   E. *none of the above*

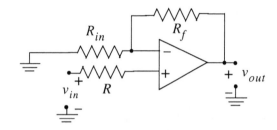

Figure 4.52. Circuit for Question 1

2. In the circuit of Figure 4.53 $v_{in} = 6\ V$, $R_{in} = 2\ K\Omega$, and $R_f = 3\ K\Omega$. Then v_{out} will be

   A. *–9 V*

   B. *9 V*

   C. *–4 V*

   D. *4 V*

   E. *none of the above*

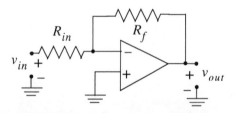

Figure 4.53. Circuit for Question 2

3. In the circuit of Figure 4.54 $i_S = 2\ mA$ and $R_f = 5\ K\Omega$. Then v_{out} will be

A. $\infty\ V$

B. $0\ V$

C. $10\ V$

D. $-10\ V$

E. *none of the above*

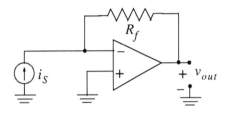

Figure 4.54. Circuit for Question 3

4. In the circuit of Figure 4.55 $i_S = 4\ mA$ and $R = 3\ K\Omega$. Then v_{out} will be

A. $\infty\ V$

B. $0\ V$

C. indeterminate

D. $-12\ V$

E. *none of the above*

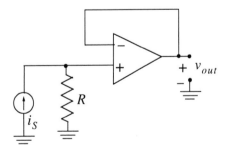

Figure 4.55. Circuit for Question 4

5. In the circuit of Figure 4.56 $v_{in} = 4\ V$, $R_{in} = 12\ K\Omega$, $R_f = 18\ K\Omega$, and $R_{LOAD} = 6\ K\Omega$. Then i will be

A. $-1\ mA$

B. $1\ mA$

C. $-4/3\ mA$

D. $4/3\ mA$

E. none of the above

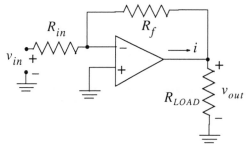

Figure 4.56. Circuit for Question 5

6. In the circuit of Figure 4.57 $v_{in} = 1\ V$ and all resistors have the same value. Then v_{out} will be

A. $-2\ V$

B. $2\ V$

C. $-4\ V$

D. $4\ V$

E. none of the above

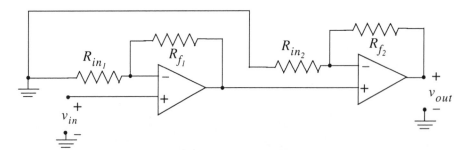

Figure 4.57. Circuit for Question 6

7. In the circuit of Figure 4.58 $v_{in_1} = 2\ V$, $v_{in_2} = 4\ V$, and $R_{in} = R_f = 1\ K\Omega$. Then v_{out} will be

   A. $-2\ V$

   B. $2\ V$

   C. $-8\ V$

   D. $8\ V$

   E. *none of the above*

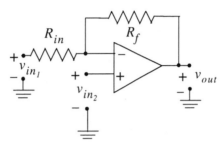

Figure 4.58. Circuit for Question 7

8. In the circuit of Figure 4.59 $v_{in} = 30\ V$. Then v_{out} will be

   A. $-5\ V$

   B. $-10\ V$

   C. $-15\ V$

   D. $-90\ V$

   E. *none of the above*

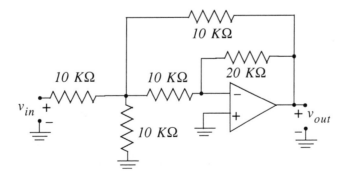

Figure 4.59. Circuit for Question 8

9. For the circuit of Figure 4.60, the input resistance R_{in} is

   A. *1 KΩ*

   B. *2 KΩ*

   C. *4 KΩ*

   D. *8 KΩ*

   E. *none of the above*

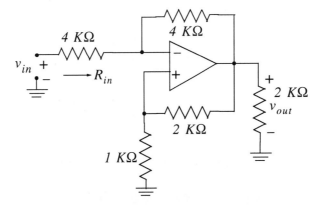

Figure 4.60. Network for Question 9

10. For the circuit of Figure 4.61, the current *i* is

   A. *–40 A*

   B. *40 A*

   C. *–400 A*

   D. *400 A*

   E. *none of the above*

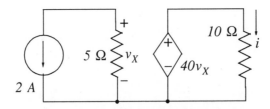

Figure 4.61. Network for Question 10

Problems

1. For the circuit of Figure 4.62, compute v_{out2}. Answer: *–0.9 V*

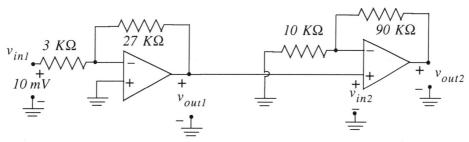

Figure 4.62. Circuit for Problem 1

2. For the circuit of Figure 4.63, compute $i_{5K\Omega}$. Answer: *4μA*

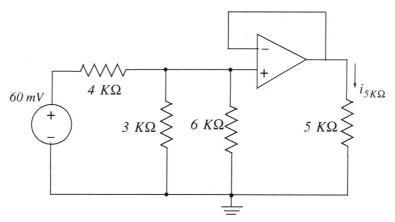

Figure 4.63. Circuit for Problem 2

3. For the circuit of Figure 4.64, R_{in1}, R_{in2}, and R_{in3} represent the internal resistances of the input voltages v_{in1}, v_{in2}, and v_{in3} respectively. Derive an expression for v_{out} in terms of the input voltage sources and their internal resistances, and the feedback resistance R_f.

Answer: $v_{out} = R_f \left(\dfrac{v_{in3}}{R_{in3}} - \dfrac{v_{in2}}{R_{in2}} - \dfrac{v_{in1}}{R_{in1}} \right)$

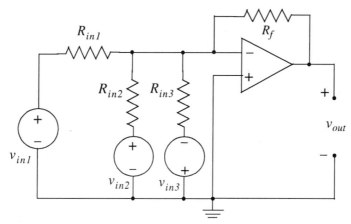

Figure 4.64. Circuit for Problem 3

4. For the circuit of Figure 4.65, compute v_{out}. Answer: *–40 mV*

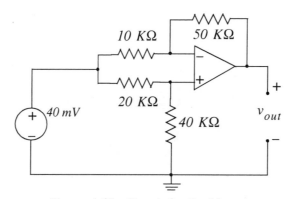

Figure 4.65. Circuit for Problem 4

5. The op-amp circuit of Figure 4.66 (a) can be represented by its equivalent circuit shown in Figure 4.66 (b). For the circuit of Figure 4.67 (c), compute the value of R_L so that it will receive maximum power. Answer: *3.75 KΩ*

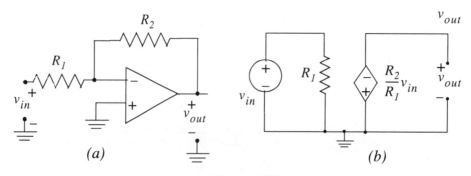

Figure 4.66.

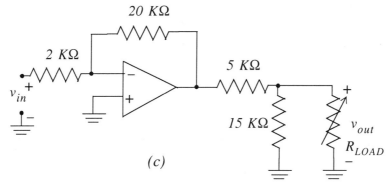

Figure 4.67. Circuits for Problem 5

6. For the circuit of Figure 4.68, compute $v_{5K\Omega}$ using Thevenin's theorem. Answer: *20 mV*

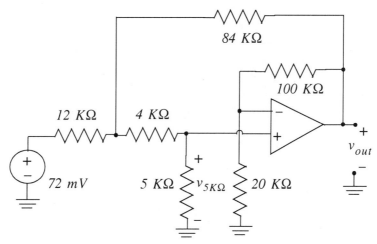

Figure 4.68. Circuit for Problem 6

7. For the circuit of Figure 4.69, compute the gain $G_v = v_{out}/v_{in}$. Answer: $-(2/37)$

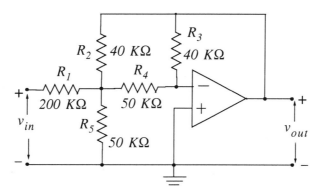

Figure 4.69. Circuit for Problem 7

8. For the circuit of Figure 4.70, show that the gain is given by

$$G_v = \frac{v_{out}}{v_{in}} = -\frac{1}{R_1}\left[R_5 + R_3\left(\frac{R_5}{R_4} + 1\right)\right]$$

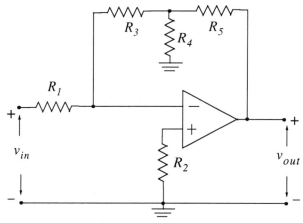

Figure 4.70. Circuit for Problem 8

4.12 Answers to Exercises

Multiple Choice

1. C For $v_{in} = 2$ and $v_{out} = 8$, the gain must be $G_v = 4$ or $1 + R_f/R_{in} = 4$. Therefore, $R_f = 3 \ K\Omega$

2. A $v_{out} = -R_f/R_{in} \times v_{in} = -9 \ V$

3. D All current flows through R_f and the voltage drop across it is $-(2 \ mA \times 5 \ K\Omega) = -10 \ V$

4. E All current flows through R and the voltage drop across it is $4 \ mA \times 3 \ K\Omega = 12 \ V$. Since this circuit is a unity gain amplifier, it follows that $v_{out} = 12 \ V$ also.

5. C $v_{out} = -(18/12) \times 4 = -6 \ V$. Therefore, $i_{LOAD} = v_{out}/R_{LOAD} = -6 \ V/6 \ K\Omega = -1 \ mA$. Applying KCL at the plus (+) terminal of v_{out} we get

$$ i = \frac{-6 \ V}{6 \ K\Omega} + \frac{-6 \ V - 4 \ V}{18 \ K\Omega + 12 \ K\Omega} = -1 - \frac{1}{3} = -\frac{4}{3} \ mA $$

6. D The gain of each of the non-inverting op amps is 2. Thus, the output of the first op amp is $2 \ V$ and the output of the second is $4 \ V$.

7. E By superposition, v_{out_1} due to v_{in_1} acting alone is $-2 \ V$ and v_{out_2} due to v_{in_2} acting alone is $8 \ V$. Therefore, $v_{out} = -2 + 8 = 6 \ V$

8. B We assign node voltage v_A as shown below and we replace the encircled port by its equivalent. $v_{in} = 30 \ V$. Then v_{out} will be

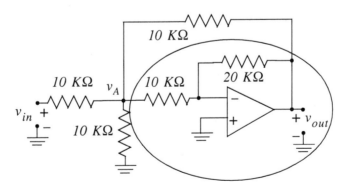

We now attach the remaining resistors and the entire equivalent circuit is shown below.

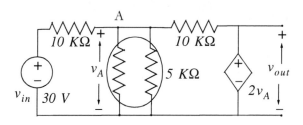

Application of KCL at Node A yields

$$\frac{v_A - 30}{10} + \frac{v_A}{5} + \frac{v_A - (-2v_A)}{10} = 0$$

and thus $v_A = 30/6 = 5\ V$

Therefore,

$$v_{out} = -2v_A = -10V$$

NOTE: For this circuit, the magnitude of the voltage is less than the magnitude of the input voltage. Therefore, this circuit is an attenuator, not an amplifier. Op amps are not configured for attenuation. This circuit is presented just for instructional purposes. A better and simpler attenuator is a voltage divider circuit.

9. C   The voltage gain for this circuit is $4\ K\Omega/4\ K\Omega = 1$ and thus $v_{out} = -v_{in}$. The voltage v at the minus (–) input of the op amp is zero as proved below.

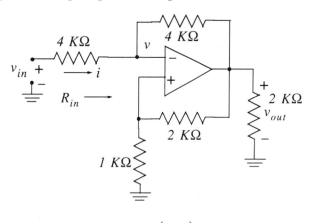

$$\frac{v - v_{in}}{4} + \frac{v - (-v_{in})}{4} = 0$$

or

$$v = 0$$

Then

$$i = \frac{v_{in}}{4\ K\Omega}$$

and

$$R_{in} = \frac{v_{in}}{v_{in}/4\ K\Omega} = 4\ K\Omega$$

10. **A** For this circuit, $v_X = -10\ V$ and thus $40v_X = -400\ V$. Then, $i = -400/10 = -40\ A$

Problems

1.
$$v_{out1} = -(27/3) \times 10 = -90\ mV$$

and thus

$$v_{in2} = v_{out1} = -90\ mV$$

Then

$$v_{out2} = \left(1 + \frac{90}{10}\right) \times (-90) = -0.9\ V$$

2. We assign R_{LOAD}, v_1, and v_{LOAD} as shown below.

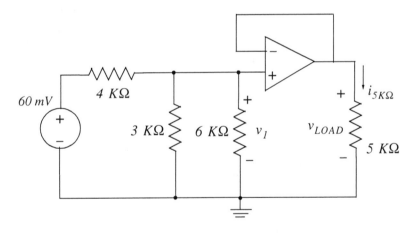

$$3\ K\Omega \parallel 6\ K\Omega = 2\ K\Omega$$

and by the voltage division expression

$$v_1 = \frac{2\ K\Omega}{4\ K\Omega + 2\ K\Omega} \times 60\ mV = 20\ mV$$

and since this is a unity gain amplifier, we get

$$v_{LOAD} = v_1 = 20\ mV$$

Then

$$i_{5K\Omega} = \frac{v_{LOAD}}{R_{LOAD}} = \frac{20\ mV}{5\ K\Omega} = \frac{20 \times 10^{-3}}{5 \times 10^{3}} = 4 \times 10^{-6}\ A = 4\ \mu A$$

3. By superposition

$$v_{out} = v_{out1} + v_{out2} + v_{out3}$$

where

$$v_{out1}\Big|_{\substack{v_{in2} = 0 \\ v_{in3} = 0}} = -\frac{R_f}{R_{in1}} v_{in1}$$

We observe that the minus (−) is a virtual ground and thus there is no current flow in R_{in1} and R_{in2}. Also,

$$v_{out2}\Big|_{\substack{v_{in1} = 0 \\ v_{in3} = 0}} = -\frac{R_f}{R_{in2}} v_{in2}$$

and

$$v_{out3}\Big|_{\substack{v_{in1} = 0 \\ v_{in2} = 0}} = -\frac{R_f}{R_{in3}} (-v_{in3})$$

Then,

$$v_{out} = R_f \left(\frac{v_{in3}}{R_{in3}} - \frac{v_{in2}}{R_{in2}} - \frac{v_{in1}}{R_{in1}} \right)$$

4. We assign voltages v_- and v_+ as shown below.

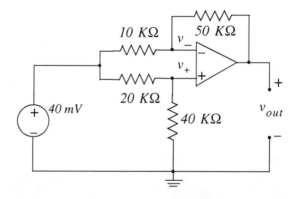

At the minus (−) terminal

$$\frac{v_- - 40 \ mV}{10 \ K\Omega} + \frac{v_- - v_{out}}{50 \ K\Omega} = 0$$

or

$$\frac{6}{50 \times 10^3} v_- - \frac{1}{50 \times 10^3} v_{out} = 4 \times 10^{-6}$$

At the plus (+) terminal

$$\frac{v_+ - 40 \ mV}{20 \ K\Omega} + \frac{v_+}{40 \ K\Omega} = 0$$

or

$$\frac{3}{40 \times 10^3} v_+ = 2 \times 10^{-6}$$

or

$$v_+ = \frac{80 \times 10^{-3}}{3}$$

Since $v_+ = v_-$ we equate the nodal equations and we get

$$\frac{6}{50 \times 10^3}\left(\frac{80 \times 10^{-3}}{3}\right) - \frac{1}{50 \times 10^3} v_{out} = 4 \times 10^{-6}$$

Multiplication by 50×10^3 yields

$$\frac{2 \times 80 \times 10^{-3} \times 50 \times 10^3}{50 \times 10^3} - v_{out} = 4 \times 10^{-6} \times 50 \times 10^3$$

or

$$v_{out} = -40 \ mV$$

Check with (4.21) using MATLAB:

R1=10000; R2=20000; R3=40000; Rf=50000; Vin=40*10^(−3);
Vout=(R1*R3-R2*Rf)*Vin/(R1*(R2+R3))

Vout =

    -0.0400

5. We attach the $5 \ K\Omega$, $15 \ K\Omega$, and R_{LOAD} resistors to the equivalent circuit as shown below. By Thevenin's theorem

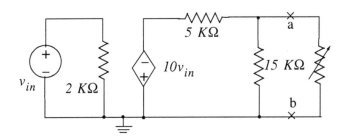

$$v_{TH} = v_{OC} = v_{ab} = \frac{15\ K\Omega}{5\ K\Omega + 15\ K\Omega}(-10v_{in})$$

or

$$v_{TH} = -7.5v_{in}$$

Because the circuit contains a dependent source, we must compute the Thevenin resistance using the relation $R_{TH} = v_{TH}/i_{SC}$ where i_{SC} is found from the circuit below.

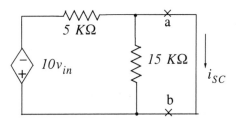

We observe that the short circuit shorts out the *15 KΩ* and thus

$$i_{SC} = \frac{-10v_{in}}{5\ K\Omega} = -2 \times 10^{-3}v_{in}$$

Then

$$R_{TH} = \frac{-7.5v_{in}}{-2 \times 10^{-3}v_{in}} = 3.75\ K\Omega$$

and the Thevenin equivalent circuit is shown below.

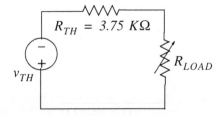

Therefore, for maximum power transfer we must have $R_{LOAD} = R_{TH} = 3.75\ K\Omega$

6. This is a non-inverting op amp whose equivalent circuit is shown below.

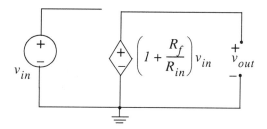

For this circuit $v_{in} = v_{5K\Omega}$ and the value of the VCVS is

$$\left(1 + \frac{R_f}{R_{in}}\right)v_{5K\Omega} = \left(1 + \frac{100}{20}\right)v_{5K\Omega} = 6v_{5K\Omega}$$

Attaching the external resistors to the equivalent circuit above we get the circuit below.

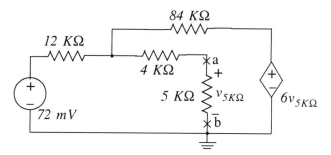

To find the Thevenin equivalent at points a and b we disconnect the $5\ K\Omega$ resistor. When this is done there is no current in the $4\ K\Omega$ and the circuit simplifies to the one shown below.

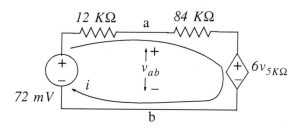

By KVL

$$(12\ K\Omega + 84\ K\Omega)i + 6v_{5K\Omega} = 72\ mV$$

or

$$i = \frac{72\ mV - 6v_{5K\Omega}}{(12\ K\Omega + 84\ K\Omega)}$$

Also

$$v_{TH} = v_{ab} = v_{5K\Omega} = 72 \ mV - (12 \ K\Omega)i = 72 \ mV - 12 \ K\Omega\left(\frac{72 \ mV - 6v_{5K\Omega}}{96 \ K\Omega}\right)$$

$$= 72 \ mV - 9 \ mV + \frac{3}{4}v_{5K\Omega}$$

or

$$v_{5K\Omega} - \frac{3}{4}v_{5K\Omega} = 63 \ mV$$

and thus

$$v_{TH} = v_{ab} = v_{5K\Omega} = 252 \ mV$$

The Thevenin resistance is found from $R_{TH} = v_{OC}/i_{SC}$ where i_{SC} is computed with the terminals a and b shorted making $v_{5K\Omega} = 0$ and the circuit is as shown on the left below. We also perform voltage-source to current-source transformation and we get the circuit on the right below.

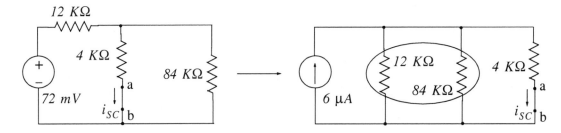

Now

$$12 \ K\Omega \parallel 84 \ K\Omega = 10.5 \ K\Omega$$

and by the current division expression

$$i_{SC} = i_{ab} = \frac{10.5 \ K\Omega}{10.5 \ K\Omega + 4 \ K\Omega} \times 6 \ \mu A = \frac{126}{29} \ \mu A$$

Therefore,

$$R_{TH} = \frac{v_{OC}}{i_{SC}} = \frac{252}{126/29} = 58 \ K\Omega$$

and the Thevenin equivalent circuit with the 5 $K\Omega$ resistor is shown below.

Finally,

$$v_{5K\Omega} = \frac{5}{58 + 5} \times 252 = 20 \ mV$$

7. We assign node voltages v_1 and v_2 as shown below and we write node equations observing that $v_2 = 0$ (virtual ground).

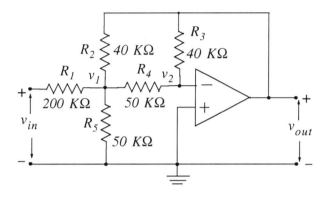

Node 1:

$$\frac{v_1 - v_{in}}{200 \ K\Omega} + \frac{v_1 - v_{out}}{40 \ K\Omega} + \frac{v_1 - 0}{50 \ K\Omega} + \frac{v_1}{50 \ K\Omega} = 0$$

or

$$\left(\frac{1}{200 \ K\Omega} + \frac{1}{40 \ K\Omega} + \frac{1}{50 \ K\Omega} + \frac{1}{50 \ K\Omega} \right) v_1 = \frac{v_{in}}{200 \ K\Omega} + \frac{v_{out}}{40 \ K\Omega}$$

Multiplication of each term by $200 \ K\Omega$ and simplification yields

$$v_1 = \frac{1}{14}(v_{in} + 5v_{out})$$

Node 2:

$$\frac{0 - v_1}{50 \ K\Omega} + \frac{0 - v_{out}}{40 \ K\Omega} = 0$$

or

$$v_1 = -\frac{5}{4}v_{out}$$

Equating the right sides we get

$$\frac{1}{14}(v_{in} + 5v_{out}) = -\frac{5}{4}v_{out}$$

or

$$\frac{37}{28}v_{out} = -\frac{1}{14}v_{in}$$

Simplifying and dividing both sides by v_{in} we get

$$G_v = \frac{v_{out}}{v_{in}} = -\frac{2}{37}$$

8. We assign node voltages v_1 and v_2 as shown below and we write node equations observing that $v_1 = 0$ (virtual ground).

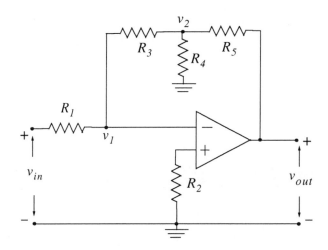

Node 1:

$$\frac{0 - v_{in}}{R_1} + \frac{0 - v_2}{R_3} = 0$$

or

$$v_2 = -\frac{R_3}{R_1}v_{in}$$

Node 2:

$$\frac{v_2 - 0}{R_3} + \frac{v_2}{R_4} + \frac{v_2 - v_{out}}{R_5} = 0$$

or

$$\left(\frac{1}{R_3} + \frac{1}{R_4} + \frac{1}{R_5}\right)v_2 = \frac{v_{out}}{R_5}$$

or

$$v_2 = \frac{1}{R_5/R_3 + R_5/R_4 + 1}v_{out}$$

Equating the right sides we get

$$\frac{1}{R_5/R_3 + R_5/R_4 + 1}v_{out} = -\frac{R_3}{R_1}v_{in}$$

Simplifying and dividing both sides by v_{in} we get

$$G_v = \frac{v_{out}}{v_{in}} = -\frac{1}{R_1}\left[R_5 + R_3\left(\frac{R_5}{R_4} + 1\right)\right]$$

NOTES

Chapter 5

Inductance and Capacitance

This chapter is an introduction to inductance and capacitance, their voltage-current relationships, power absorbed, and energy stored in inductors and capacitors. Procedures for analyzing circuits with inductors and capacitors are presented along with several examples.

5.1 Energy Storage Devices

In the first four chapters we considered resistive circuits only, that is, circuits with resistors and constant voltage and current sources. However, resistance is not the only property that an electric circuit possesses; in every circuit there are two other properties present and these are the inductance and the capacitance. We will see through some examples that will be presented later in this chapter, that inductance and capacitance have an effect on an electric circuit as long as there are changes in the voltages and currents in the circuit.

The effects of the inductance and capacitance properties can best be stated in simple differential equations since they involve the changes in voltage or current with time. We will study inductance first.

5.2 Inductance

Inductance is associated with the magnetic field which is always present when there is an electric current. Thus, when current flows in an electric circuit the conductors (wires) connecting the devices in the circuit are surrounded by a magnetic field. Figure 5.1 shows a simple loop of wire and its magnetic field represented by the small loops.

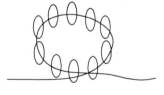

Figure 5.1. Magnetic field around a loop of wire

The direction of the magnetic field (not shown) can be determined by the left-hand rule if conventional current flow is assumed, or by the right-hand rule if electron current flow is assumed. The magnetic field loops are circular in form and are referred to as *lines of magnetic flux*. The unit of magnetic flux is the *weber* (Wb).

In a loosely wound coil of wire such as the one shown in Figure 5.2, the current through the wound coil produces a denser magnetic field and many of the magnetic lines link the coil several times.

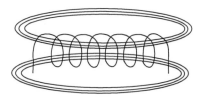

Figure 5.2. Magnetic field around several loops of wire

The magnetic flux is denoted as φ and, if there are N turns and we assume that the flux φ passes through each turn, the total flux, denoted as λ, is called *flux linkage*. Then,

$$\lambda = N\varphi \tag{5.1}$$

Now, we define a linear inductor one in which the flux linkage is proportional to the current through it, that is,

$$\lambda = Li \tag{5.2}$$

where the constant of proportionality L is called *inductance* in webers per ampere.

We also recall Faraday's law of electromagnetic induction which states that

$$v = \frac{d\lambda}{dt} \tag{5.3}$$

and from (5.2) and (5.3),

$$v = L\frac{di}{dt} \tag{5.4}$$

Alternately, the inductance L is defined as the constant which relates the voltage across and the current through a device called *inductor* by the relation of (5.4).

The symbol and the voltage-current[*] designations for the inductor are shown in Figure 5.3.

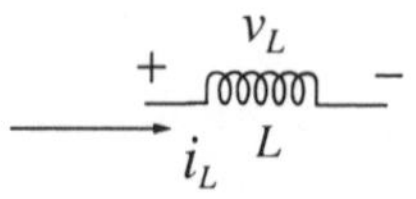

Figure 5.3. Symbol for inductor

---

[*] *In the first four chapters we have used the subscript **LOAD** to denote a voltage across a load, a current through a load, and the resistance of a such load as R_{LOAD} to avoid confusion with the subscript **L** which henceforth will denote inductance. We will continue using the subscript **LOAD** for any load connected to a circuit.*

For an inductor, the voltage-current relationship is

$$v_L = L\frac{di_L}{dt} \tag{5.5}$$

where v_L and i_L have the indicated polarity and direction. Obviously, v_L has a non-zero value only when i_L changes with time.

The unit of inductance is the *Henry* abbreviated as H. Since

$$L = \frac{v_L}{\dfrac{di_L}{dt}} = \frac{volts}{\dfrac{amperes}{seconds}} \tag{5.6}$$

we can say that *one henry is the inductance in a circuit in which a voltage of one volt is induced by a current changing at the rate of one ampere per second.*

By separation of the variables we rewrite (5.5) as

$$di_L = \frac{1}{L}v_L dt \tag{5.7}$$

and integrating both sides we get:

$$\int_{i(t_0)}^{i(t)} di_L = \frac{1}{L}\int_{t_0}^{t} v_L dt$$

or

$$i_L(t) - i_L(t_0) = \frac{1}{L}\int_{t_0}^{t} v_L dt$$

or

$$i_L(t) = \frac{1}{L}\int_{t_0}^{t} v_L dt + i_L(t_0) \tag{5.8}$$

where $i_L(t_0)$, more often denoted as $i_L(0)$, is the current flowing through the inductor at some reference time usually taken as $t = 0$, and it is referred to as the *initial condition.*

We can also express (5.8) as

$$i_L(t) = \frac{1}{L}\int_{-\infty}^{t} v_L dt = \frac{1}{L}\int_{-\infty}^{0} v_L dt + \frac{1}{L}\int_{0}^{t} v_L dt \tag{5.9}$$

where the first integral on the right side represents the initial condition.

Example 5.1

The current $i_L(t)$ passing through a *50 mH* inductor is shown in Figure 5.4.

a. Compute the flux linkage λ at $t = 2, 5, 9$, and *11 ms*

b. Compute and sketch the voltage $v_L(t)$ for the time interval $-\infty < t < 14 \ ms$

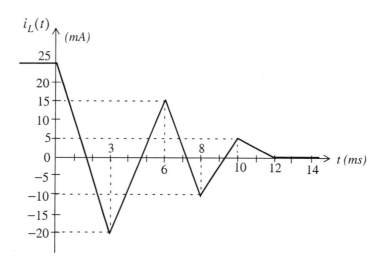

Figure 5.4. Waveform for Example 5.1

Solution:

a. The flux linkage λ is directly proportional to the current; then from (5.1) and (5.2)

$$\lambda = N\varphi = Li$$

Therefore, we need to compute the current i at $t = 2 \ ms$, $t = 5 \ ms$, $t = 9 \ ms$, and $t = 11 \ ms$

For time interval $0 < t < 3 \ ms$, $i = mt + b$ where m is the slope of the straight line segment, and b is the $i - axis$ intercept which, by inspection, is *25 mA*. The slope m is

$$m = \frac{-20 - 25}{3 - 0} = -15$$

and thus

$$i\Big|_{t=0}^{3\ ms} = -15t + 25 \tag{5.10}$$

At $t = 2 \ ms$, (5.10) yields $i = -5 \ mA$. Then, the flux linkage is

$$\lambda = Li = 50 \times 10^{-3} \times (-5) \times 10^{-3}$$

and

$$\lambda\big|_{t = 2\ ms} = -250\ \mu Wb \tag{5.11}$$

For the time interval $3 < t < 6\ ms$, $i = mt + b$ where

$$m = \frac{15 - (-20)}{3 - 0} = \frac{35}{3}$$

and thus

$$i = \frac{35}{3}t + b$$

To find b we use the fact that at $t = 3\ ms$, $i = -20\ mA$ as seen in Figure 5.4. Then,

$$-20 = \frac{35}{3} \times 3 + b$$

from which $b = -55$.

Thus, the straight line equation for the time interval $3 < t < 6\ ms$ is

$$i\big|_{t = 3\ ms}^{6\ ms} = \frac{35}{3}t - 55 \tag{5.12}$$

and therefore at $t = 5\ ms$, $i = 10/3\ mA$, and the flux linkage is

$$\lambda = Li = 50 \times 10^{-3} \times \frac{10}{3} \times 10^{-3}$$

or

$$\lambda\big|_{t = 5\ ms} = \frac{500}{3}\ \mu Wb \tag{5.13}$$

Using the same procedure we find that

$$i\big|_{t = 6\ ms}^{8\ ms} = -12.5t + 90 \tag{5.14}$$

Also,

$$i\big|_{t = 8\ ms}^{10\ ms} = 7.5t - 70 \tag{5.15}$$

and with (5.15),

$$\lambda\big|_{t = 9\ ms} = Li = -125\ \mu Wb \tag{5.16}$$

Likewise,

$$i\Big|_{t = 10\ ms}^{12\ ms} = -2.5t + 30 \tag{5.17}$$

and with (5.17),

$$\lambda\big|_{t = 11\ ms} = Li = 125\ \mu Wb \tag{5.18}$$

b. Since

$$v_L = L\frac{di_L}{dt}$$

to compute and sketch the voltage $v_L(t)$ for the time interval $-\infty < t < 14\ ms$, we only need to differentiate, that is, compute the slope of the straight line segments for this interval. These were found in part (a) as (5.10), (5.12), (5.14), (5.15), and (5.17). Then,

$$slope\big|_{-\infty < t < 0} = 0$$

$$v_L\big|_{-\infty < t < 0} = L \times slope = 0 \tag{5.19}$$

$$slope\big|_{0 < t < 3\ ms} = -15\ mA/ms = -15\ A/s$$

$$v_L\big|_{0 < t < 3\ ms} = L \times slope = 50 \times 10^{-3}\ \frac{v}{A/s} \times (-15\ A/s) = -750\ mV \tag{5.20}$$

$$slope\big|_{3 < t < 6\ ms} = 35/3\ mA/ms = 35/3\ A/s$$

$$v_L\big|_{3 < t < 6\ ms} = L \times slope = 50 \times 10^{-3} \times (35/3) = 583.3\ mV \tag{5.21}$$

$$slope\big|_{6 < t < 8\ ms} = -12.5\ mA/ms = -12.5\ A/s$$

$$v_L\big|_{6 < t < 8\ ms} = L \times slope = 50 \times 10^{-3} \times (-12.5) = -625\ mV \tag{5.22}$$

$$slope\big|_{8 < t < 10\ ms} = 7.5\ mA/ms = 7.5\ A/s$$

$$v_L\big|_{8 < t < 10\ ms} = L \times slope = 50 \times 10^{-3} \times 7.5 = 375\ mV \tag{5.23}$$

$$slope\big|_{10 < t < 12\ ms} = -2.5\ mA/ms = -2.5\ A/s$$

$$v_L\big|_{10 < t < 12\ ms} = L \times slope = 50 \times 10^{-3} \times (-2.5) = -125\ mV \tag{5.24}$$

$$slope\big|_{12 < t < 14\ ms} = 0$$

$$v_L\big|_{12 < t < 14\ ms} = L \times slope = 0 \tag{5.25}$$

We now have all values given by (5.19) through (5.25) to sketch v_L as a function of time. We can do this easily with a spreadsheet such as Excel as shown in Figure 5.5.

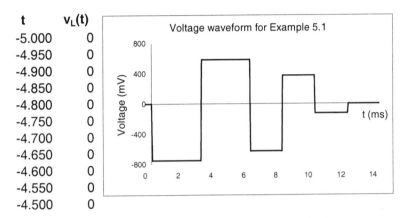

t	$v_L(t)$
-5.000	0
-4.950	0
-4.900	0
-4.850	0
-4.800	0
-4.750	0
-4.700	0
-4.650	0
-4.600	0
-4.550	0
-4.500	0

Figure 5.5. Voltage waveform for Example 5.1

Example 5.2

The voltage across a *50 mH* inductor is as shown on the waveform of Figure 5.6, and it is given that the initial condition is $i_L(t_0) = i_L(0) = 25\ mA$. Compute and sketch the current which flows through this inductor in the interval $-5 < t < 5\ ms$

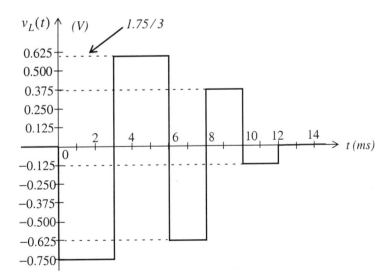

Figure 5.6. Waveform for Example 5.2

Solution:

The current $i_L(t)$ in an inductor is related to the voltage $v_L(t)$ by (5.8) which is repeated here for convenience.

$$i_L(t) = \frac{1}{L} \int_{t_0}^{t} v_L dt + i_L(t_0)$$

where $i_L(t_0) = i_L(0) = 25\ mA$ is the initial condition, that is,

$$i_L\Big|_{-\infty < t < 0} = 25\ mA$$

From the given waveform,

$$v_L\Big|_{0 < t < 3\ ms} = -0.75\ V$$

Then,

$$i_L\Big|_{0 < t < 3\ ms} = \frac{1}{50 \times 10^{-3}} \int_{0}^{3\ ms} (-0.75) dt + 25 \times 10^{-3}$$

$$= 20\left(-0.75 t\Big|_{0}^{3 \times 10^{-3}}\right) + 25 \times 10^{-3} = 20(-2.25 \times 10^{-3}) + 20 \times 0 + 25 \times 10^{-3}$$

$$= -45 \times 10^{-3} + 25 \times 10^{-3} = -20 \times 10^{-3} = -20\ mA$$

that is, the current has dropped linearly from $25\ mA$ at $t = 0$ to $-20\ mA$ at $t = 3\ ms$ as shown in Figure 5.7.

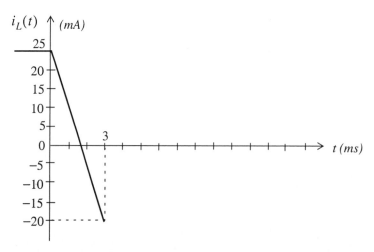

Figure 5.7. Inductor current for $0 < t < 3\ ms$, Example 5.2

The same result can be obtained by graphical integration. Thus,

$$i_L\Big|_{t=3\ ms} = \frac{1}{L}(Area\Big|_{t=0}^{3\ ms}) + initial\ condition$$

$$= 20(-0.750 \times 3 \times 10^{-3}) + 25 \times 10^{-3} = -20\ mA$$

and the value of $i_L\Big|_{t=3\ ms} = -20\ mA$ now becomes our initial condition for the time interval $3 < t < 6\ ms$.

Continuing with graphical integration, we get

$$i_L\Big|_{t=6\ ms} = \frac{1}{L}(Area\Big|_{t=3}^{6\ ms}) + initial\ condition$$

$$= 20\left(\frac{1.75}{3} \times 3 \times 10^{-3}\right) - 20 \times 10^{-3} = 15\ mA$$

and now the current has increased linearly from $-20\ mA$ at $t = 3\ ms$ to $15\ mA$ at $t = 6\ ms$ as shown in Figure 5.8.

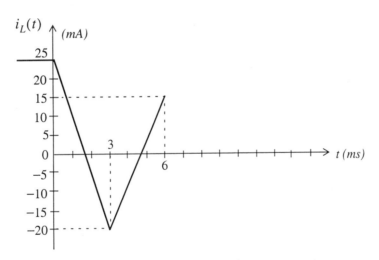

Figure 5.8. Inductor current for $0 < t < 6\ ms$, Example 5.2

For the time interval $6 < t < 8\ ms$, we get

$$i_L\Big|_{t=8\ ms} = \frac{1}{L}(Area\Big|_{t=6}^{8\ ms}) + initial\ condition$$

$$= 20(-0.625 \times 2 \times 10^{-3}) + 15 \times 10^{-3} = -10\ mA$$

Therefore, the current has decreased linearly from $15\ mA$ at $t = 6\ ms$ to $-10\ mA$ at $t = 8\ ms$ as shown in Figure 5.9.

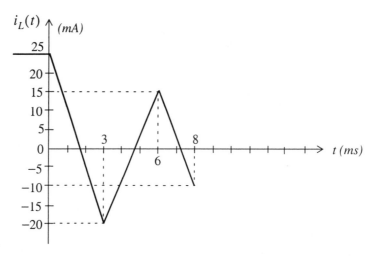

Figure 5.9. Inductor current for $0 < t < 8$ ms, Example 5.2

For the time interval 8 $ms < t < 10$ ms we get

$$i_L\big|_{t = 10\ ms} = \frac{1}{L}\left(Area\big|_{t = 8}^{10\ ms}\right) + initial\ condition$$

$$= 20(0.375 \times 2 \times 10^{-3}) - 10 \times 10^{-3} = 5\ mA$$

that is, the current has increased linearly from -10 mA at $t = 8$ ms to 5 mA at $t = 10$ ms as shown in Figure 5.10.

Finally, for the time interval 10 $ms < t < 12$ ms we get

$$i_L\big|_{t = 12\ ms} = \frac{1}{L}\left(Area\big|_{t = 10}^{12\ ms}\right) + initial\ condition$$

$$= 20(-0.125 \times 2 \times 10^{-3}) + 5 \times 10^{-3} = 0$$

that is, the current has decreased linearly from 5 mA at $t = 10$ ms to 0 mA at $t = 12$ ms and remains at zero for $t > 12$ ms as shown in Figure 5.11.

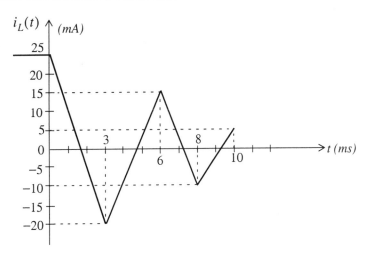

Figure 5.10. Inductor current for 0 < t < 10 ms, Example 5.2

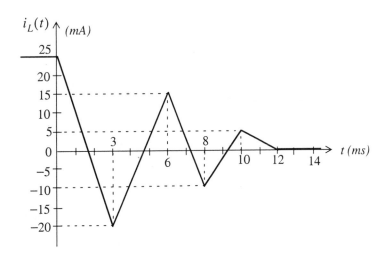

Figure 5.11. Inductor current for 0 < t < 12 ms, Example 5.2

Example 5.2 confirms the well known fact that *the current through an inductor cannot change instantaneously.* This can be observed from the voltage and current waveforms for this and the previous example. We observe that the voltage across the inductor can change instantaneously as shown by the discontinuities at *t = 0, 3, 6, 8, 10,* and *12 ms.* However, the current through the inductor never changes instantaneously, that is, it displays no discontinuities since its value is explicitly defined at all instances of time.

5.3 Power and Energy in an Inductor

Power in an inductor with inductance L is found from

$$p_L = v_L i_L = \left(L\frac{di_L}{dt}\right)i_L = Li_L\frac{di_L}{dt} \tag{5.26}$$

and the energy in an inductor, designated as W_L is the integral of the power, that is,

$$W_L\Big|_{t_0}^t = \int_{t_0}^t p_L dt = L\int_{i(t_0)}^{i(t)} i_L\frac{di_L}{dt}dt = L\int_{i(t_0)}^{i(t)} i_L di_L$$

or

$$W_L\Big|_{t_0}^t = \frac{1}{2}Li_L^2\Big|_{i(t_0)}^{i(t)} = \frac{1}{2}L[i_L^2(t)-i_L^2(t_0)]$$

or

$$W_L(t) - W_L(t_0) = \frac{1}{2}L[i_L^2(t)-i_L^2(t_0)]$$

and letting $i_L = 0$ at $t = 0$, we get the *energy stored in an inductor* as

$$\boxed{W_L(t) = \frac{1}{2}Li_L^2(t)} \tag{5.27}$$

Unlike the resistor which dissipates energy (in the form of heat), the (ideal) inductor is a physical device capable of storing energy in analogy to the potential energy of a stretched spring.

Electric circuits that contain inductors can be simplified if the applied voltage and current sources are constant as shown by the following example.

Example 5.3

For the circuit shown in Figure 5.12, compute v_1, v_2, and v_3, after steady-state[*] conditions have been reached. Then, compute the power absorbed and the energy consumed by the *5 mH* inductor.

Solution:

Since both the voltage and the current sources are constant, the voltages and the currents in all branches of the circuit will be constant after *steady-state conditions* have been reached.

Since

$$v_L = L\frac{di_L}{dt} = L\frac{d}{dt}(constant) = 0$$

---

[*] *By steady state conditions we mean the condition (state) where the voltages and currents, after some transient disturbances, have subsided. Transients will be in Chapter 10.*

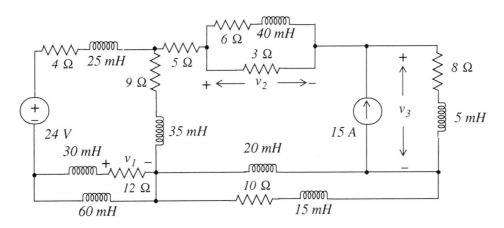

Figure 5.12. Circuit for Example 5.3

then, all voltages across the inductors will be zero and therefore we can replace all inductors by short circuits. The given circuit then reduces to the one shown in Figure 5.13 where the *3 Ω* and *6 Ω* parallel resistors have been combined into a single *2 Ω* resistor.

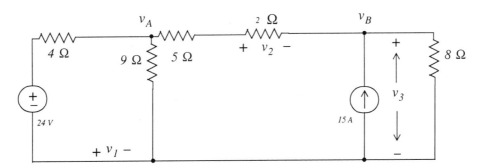

Figure 5.13. Circuit for Example 5.3 after steady-state conditions have been reached

Now, in Figure 5.13, by inspection, $v_1 = 0$ since the *12 Ω* resistor was shorted out by the *60 mH* inductor. To find v_2 and v_2, let us first find v_A and v_B using nodal analysis.

At Node v_A,

$$\frac{v_A - 24}{4} + \frac{v_A}{9} + \frac{v_A - v_B}{5 + 2} = 0$$

or

$$\left(\frac{1}{4} + \frac{1}{9} + \frac{1}{7}\right)v_A - \frac{1}{7}v_B = 6 \tag{5.28}$$

At Node v_B

$$\frac{v_B - v_A}{5 + 2} - 15 + \frac{v_B}{8} = 0$$

or

$$-\frac{1}{7}v_A + \left(\frac{1}{7} + \frac{1}{8}\right)v_B = 15 \tag{5.29}$$

We will use the MATLAB code below to find the solution of (5.28) and (5.29).

```
format rat % Express answers in rational form
G=[1/4+1/9+1/7  -1/7; -1/7  1/7+1/8]; I=[6  15]'; V=G\I;
disp('vA='); disp(V(1)); disp('vB='); disp(V(2))
```

```
vA=
    360/11
vB=
    808/11
```

Therefore,

$$v_A = 360/11 \ V$$

$$v_B = 808/11 \ V$$

$$v_2 = v_A - v_2 = -448/11 \ V$$

$$v_3 = v_2 = 808/11 \ V$$

and

$$p_{5 \ mH} = v_{5 \ mH} \times i_{5 \ mH} = 0 \times i_{5 \ mH} = 0$$

that is,

$$p_{5 \ mH} = 0 \ watts$$

Also,

$$W_{5 \ mH} = \frac{1}{2}Li_{5 \ mH}^2 = \frac{1}{2}L\left(\frac{v_3}{8}\right)^2 = 0.5 \times 5 \times 10^{-3} \times \left(\frac{808/11}{8}\right)^2$$

or

$$W_{5 \ mH} = 0.211 \ J$$

5.4 Combinations of Series and Parallel Inductors

Consider the circuits of figures 5.14 (a) and 5.14 (b) where the source voltage v_S is the same for both circuits. We wish to find an expression for the equivalent inductance which we denote as L_{Seq} in terms of $L_1, L_2, ..., L_N$ in Figure 5.14 (a) so that the current i will be the same for both circuits.

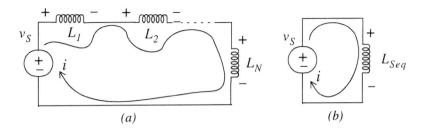

Figure 5.14. *Circuits for derivation of equivalent inductance for inductors in series*

From the circuit of Figure 5.14 (a),

$$L_1 \frac{di}{dt} + L_2 \frac{di}{dt} + \ldots + L_N \frac{di}{dt} = v_S$$

or

$$(L_1 + L_2 + \ldots + L_N) \frac{di}{dt} = v_S \qquad (5.30)$$

From the circuit of Figure 5.14 (b),

$$L_{Seq} \frac{di}{dt} = v_S \qquad (5.31)$$

Equating the left sides of (5.30) and (5.31) we get:

$$\boxed{L_{Seq} = L_1 + L_2 + \ldots + L_N} \qquad (5.32)$$

Thus, inductors in series combine as resistors in series do.

Next, we will consider the circuits of Figures 5.15 (a) and 5.15 (b) where the source current i_S is the same for both circuits. We wish to find an expression for the equivalent inductance which we denote as L_{Peq} in terms of $L_1, L_2, \ldots, L_N$ in Figure 5.15 (a) so that the voltage v will be the same for both circuits.

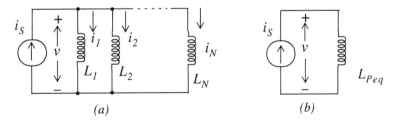

Figure 5.15. *Circuits for derivation of equivalent inductance for inductors in parallel*

From the circuit of Figure 5.15 (a)

$$i_1 + i_2 + \ldots + i_N = i_S$$

or

$$\frac{1}{L_1}\int_{-\infty}^{t} v\,dt + \frac{1}{L_2}\int_{-\infty}^{t} v\,dt + \ldots + \frac{1}{L_N}\int_{-\infty}^{t} v\,dt = i_S$$

or

$$\left(\frac{1}{L_1} + \frac{1}{L_2} + \ldots + \frac{1}{L_N}\right)\int_{-\infty}^{t} v\,dt = i_S \tag{5.33}$$

From the circuit of Figure 5.15 (b)

$$\frac{1}{L_{Peq}}\int_{-\infty}^{t} v\,dt = i_S \tag{5.34}$$

Equating the left sides of (5.33) and (5.34) we get:

$$\boxed{\frac{1}{L_{Peq}} = \frac{1}{L_1} + \frac{1}{L_2} + \ldots + \frac{1}{L_N}} \tag{5.35}$$

and for the special case of two parallel inductors

$$L_{Peq} = \frac{L_1 L_2}{L_1 + L_2} \tag{5.36}$$

Thus, inductors in parallel combine as resistors in parallel do.

Example 5.4

For the network of Figure 5.16, replace all inductors by a single equivalent inductor.

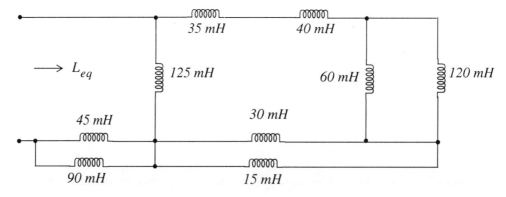

Figure 5.16. Network for Example 5.4

Solution:

Starting at the right end of the network and moving towards the left end, we find that $60 \ mH \parallel 120 \ mH = 40 \ mH$, $30 \ mH \parallel 15 \ mH = 10 \ mH$, $40 \ mH + 35 \ mH = 75 \ mH$, and also $45 \ mH \parallel 90 \ mH = 30 \ mH$. The network then reduces to that shown in Figure 5.17.

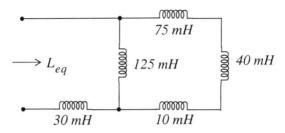

Figure 5.17. First step in combination of inductances

Finally, with reference to Figure 5.17, $(40 \ mH + 35 \ mH + 10 \ mH) \parallel 125 \ mH = 62.5 \ mH$, and $L_{eq} = 30 \ mH + 62.5 \ mH = 92.5 \ mH$ as shown in Figure 5.18.

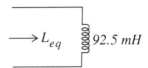

Figure 5.18. Network showing the equivalent inductance of Figure 5.16

5.5 Capacitance

In Section 5.2 we learned that inductance is associated with a *magnetic field* which is created whenever there is current flow. Similarly, *capacitance* is associated with an *electric field*. In a simple circuit we can represent the entire capacitance with a device called *capacitor*, just as we considered the entire inductance to be concentrated in a single inductor. A capacitor consists of two parallel metal plates separated by an air space or by a sheet of some type of insulating material called the *dielectric*.

Now, let us consider the simple series circuit of Figure 5.19 where the device denoted as C, is the standard symbol for a capacitor.

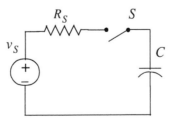

Figure 5.19. Simple circuit to illustrate a charged capacitor

When the switch S closes in the circuit of Figure 5.19, the voltage source will force electrons from its negative terminal through the conductor to the lower plate of the capacitor and it will accumulate *negative charge*. At the same time, electrons which were present in the upper plate of the capacitor will move towards the positive terminal of the voltage source. This action leaves the upper plate of the capacitor deficient in electrons and thus it becomes *positively charged*. Therefore, an *electric field* has been established between the plates of the capacitor.

The distribution of the electric field set up in a capacitor is usually represented by lines of force similar to the lines of force in a magnetic field. However, in an electric field the lines of force start at the positive plate and terminate at the negative plate, whereas magnetic lines of force are always complete loops.

Figure 5.20 shows the distribution of the electric field between the two plates of a capacitor.

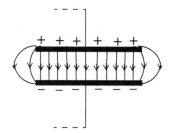

Figure 5.20. Electric field between the plates of a capacitor

We observe that the electric field has an almost uniform density in the area directly between the plates, but it decreases in density beyond the edges of the plates.

The charge q on the plates is directly proportional to the voltage between the plates and the capacitance C is the constant of proportionality. Thus,

$$q = Cv \qquad (5.37)$$

and recalling that the current i is the rate of change of the charge q, we have the relation

$$i = \frac{dq}{dt} = \frac{d}{dt}(Cv)$$

or

$$i_C = C\frac{dv_C}{dt} \qquad (5.38)$$

where i_C and v_C in (5.38) obey the passive sign convention.

The unit of capacitance is the *Farad* abbreviated as F and since

$$C = \frac{i_C}{\frac{dv_C}{dt}} = \frac{amperes}{\frac{volts}{seconds}} \qquad (5.39)$$

we can say that *one farad is the capacitance in a circuit in which a current of one ampere flows when the voltage is changing at the rate of a one volt per second.*

By separation of the variables we rewrite (5.38) as

$$dv_C = \frac{1}{C} i_C dt \qquad (5.40)$$

and integrating both sides we get:

$$\int_{v_C(t_0)}^{v_C(t)} dv_C = \frac{1}{C} \int_{t_0}^{t} i_C dt$$

or

$$v_C(t) - v_C(t_0) = \frac{1}{C} \int_{t_0}^{t} i_C dt$$

or

$$\boxed{v_C(t) = \frac{1}{C} \int_{t_0}^{t} i_C dt + v_C(t_0)} \qquad (5.41)$$

where $v_C(t_0)$ is the initial condition, that is, the voltage across a capacitor at some reference time usually taken as $t = 0$, and denoted as $v_C(0)$.

We can also write (5.41) as

$$v_C(t) = \frac{1}{C} \int_{-\infty}^{t} i_C dt = \frac{1}{C} \int_{-\infty}^{0} i_C dt + \frac{1}{C} \int_{0}^{t} i_C dt$$

where the initial condition is represented by the first integral on the right side.

Example 5.5

The waveform shown in Figure 5.21 represents the current flowing through a *1 µF* capacitor. Compute and sketch the voltage across this capacitor for the time interval $0 < t < 4\ ms$ given that the initial condition is $v_C(0) = 0$.

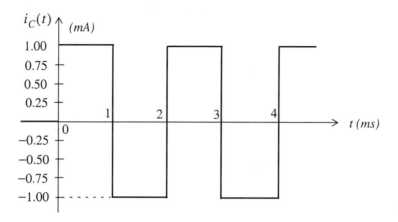

Figure 5.21. Waveform for Example 5.5

Solution:

The initial condition $v_C(0) = 0$, establishes the first point at the coordinates $(0,0)$ on the $v_C(t)$ versus time plot of Figure 5.22.

Next,

$$v_C\big|_{t = 1 \ ms} = \frac{1}{C}\int_0^{1 \times 10^{-3}} i_C \, dt + \underbrace{\frac{v_C(0)}{0}}$$

or

$$v_C\big|_{t = 1 \ ms} = \frac{1}{C}\left(Area\big|_{t = 0}^{1 \times 10^{-3}}\right) = \frac{1}{1 \times 10^{-6}}(1 \times 10^{-3} \times 1 \times 10^{-3}) = 1 \ volt$$

and this value establishes the second point of the straight line segment passing through the origin as shown in Figure 5.22.

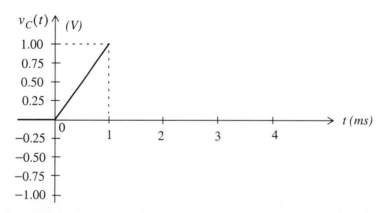

Figure 5.22. Straight line segment for $0 < t < 1 \ ms$ of the voltage waveform for Example 5.5

This value of *1 volt* at $t = 1\ ms$ becomes our initial condition for the time interval $1 < t < 2$. Continuing, we get

$$v_C\big|_{t = 2\ ms} = \frac{1}{C}\left(Area\big|_{t = 0}^{1 \times 10^{-3}}\right) + 1$$

$$= \frac{1}{1 \times 10^{-6}}(-1 \times 10^{-3} \times (2 - 1) \times 10^{-3}) + 1 = 0\ volts$$

Thus, the capacitor voltage then decreases linearly from *1 volt* at $t = 1\ ms$ to *0 volts* at $t = 2\ ms$ as shown in Figure 5.23.

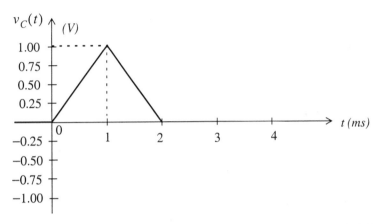

Figure 5.23. Voltage waveform for $0 < t < 2$ ms of Example 5.5

There is no need to calculate the values of the capacitor voltage v_c at $t = 3\ ms$ and at $t = 4\ ms$ because the waveform of the current i_c starts repeating itself at $t = 2\ ms$, and the initial conditions and the areas are the same as before. Accordingly, the capacitor voltage v_c waveform of figure (b) starts repeating itself also as shown in Figure 5.24.

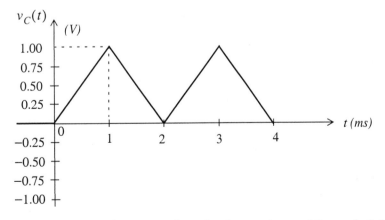

Figure 5.24. Voltage waveform for $0 < t < 4$ ms of Example 5.5

Example 5.5 has illustrated the well known fact that *the voltage across* a capacitor cannot change instantaneously. Referring to the current and voltage waveforms for this example, we observe that the current through the capacitor can change instantaneously as shown by the discontinuities at $t = 1, 2, 3,$ and *4 ms* in Figure 5.21. However, the voltage across the capacitor never changes instantaneously, that is, it displays no discontinuities since its value is explicitly defined at all instances of time as shown in Figure 5.24.

5.6 Power and Energy in a Capacitor

Power in a capacitor with capacitance C is found from

$$p_C = v_C i_C = v_C \left(C \frac{dv_C}{dt} \right)$$

and the energy in a capacitor, denoted as W_C is the integral of the power, that is,

$$W_C \Big|_{t_0}^{t} = \int_{t_0}^{t} p_C dt = C \int_{v(t_0)}^{v(t)} v_C \frac{dv_C}{dt} dt = C \int_{v(t_0)}^{v(t)} v_C dv_c$$

$$= \frac{1}{2} C v_c^2 \Big|_{i(t_0)}^{i(t)} = \frac{1}{2} C [v_c^2(t) - v_c^2(t_0)]$$

or

$$W_C(t) - W_C(t_0) = \frac{1}{2} C [v_c^2(t) - v_c^2(t_0)]$$

and letting $v_C = 0$ at $t = 0$, we get the *energy stored in a capacitor* as

$$\boxed{W_C(t) = \frac{1}{2} C v_C^2(t)} \qquad (5.42)$$

Like an inductor, a capacitor is a physical device capable of storing energy.

It was stated earlier that the current through an inductor and the voltage across a capacitor cannot change instantaneously. These facts can also be seen from the expressions of the energy in an inductor and in a capacitor, equations (5.27) and (5.42) where we observe that if the current in an inductor or the voltage across a capacitor could change instantaneously, then the energies W_L and W_C would also change instantaneously but this is, of course, a physical impossibility.

Example 5.6

In the circuit of figure 5.25, the voltage and current sources are constant.

a. Compute i_{L1} and v_{C2}

b. Compute the power and energy in the 2 μF capacitor.

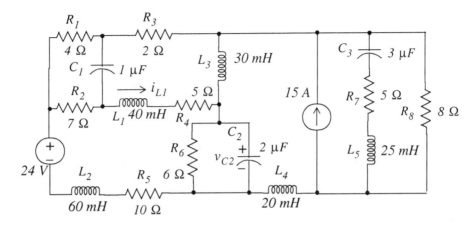

Figure 5.25. Circuit for Example 5.6

Solution:

a. The voltage and current sources are constant; thus, after steady-state conditions have been reached, the voltages across the inductors will be zero and the currents through the capacitors will be zero. Therefore, we can replace the inductors by short circuits and the capacitors by open circuits and the given circuit reduces to that shown in Figure 5.26.

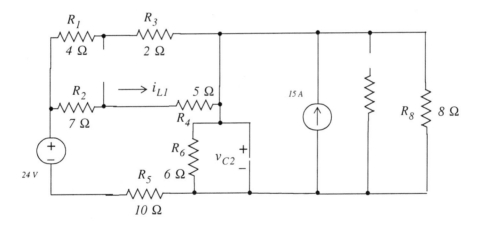

Figure 5.26. First simplification of the circuit of Example 5.6

We can simplify the circuit of figure 5.26 by first exchanging the *15 A* current source and resistor R_8 for a voltage source of $15 \times 8 = 120\ V$ in series with R_8 as shown in Figure 5.27. We also combine the series-parallel resistors R_1 through R_4. Thus, $R_{eq} = (4 + 2) \parallel (7 + 5) = 4\ \Omega$. But now we observe that the branch in which the current i_{L1} flows has disappeared; however, this presents no problem since we can apply the current division expression once i, shown in Figure 5.27, is found. The simplified circuit then is

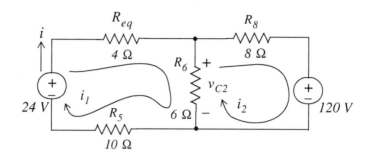

Figure 5.27. Final simplification of the circuit of Example 5.6

We can apply superposition here. Instead, we will write two mesh equations and we will solve using MATLAB. These in matrix form are

$$\begin{bmatrix} 20 & -6 \\ -6 & 14 \end{bmatrix} \begin{bmatrix} i_1 \\ i_2 \end{bmatrix} = \begin{bmatrix} 24 \\ -120 \end{bmatrix}$$

Solution using MATLAB:

format rat; R=[20 −6; −6 14]; V=[24 −120]'; I=R\V; disp('i1='); disp(I(1)); disp('i2='); disp(I(2))

```
i1=
 -96/61
i2=
-564/61
```

Therefore, with reference to the circuit of Figure 5.28

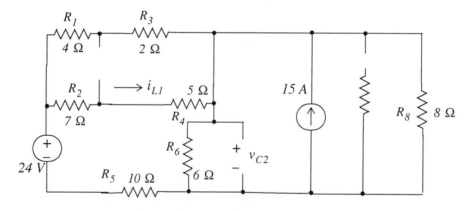

Figure 5.28. Circuit for computation of i_{L1} and v_{C2} for Example 5.6

$$i_{L1} = \frac{(4+2)}{(4+2)+(7+5)} \times \left(-\frac{96}{61}\right) = -\frac{32}{61} = -0.525 \ A$$

and

$$v_{C2} = 6\left(-\frac{96}{61} + \frac{564}{61}\right) = \frac{2808}{61} = 46.03 \ V$$

b.

$$p_{2 \ \mu F} = v_{2 \ \mu F} \times i_{2 \ \mu F} = v_{C2} \times 0 = 0$$

and

$$W_{2 \ \mu F} = \frac{1}{2}Cv_{2 \ \mu F}^2 = 0.5 \times 2 \times 10^{-6} \times \left(\frac{2808}{61}\right)^2 = 2 \ mJ$$

5.7 Capacitance Combinations

Consider the circuits of figures 5.29 (a) and 5.29 (b) in which the source voltage v_S is the same for both circuits. We want to find an expression for the equivalent capacitance which we denote as C_{Seq} in terms of $C_1, C_2, ..., C_N$ in Figure 5.29 (a) so that the current i will be the same in both circuits.

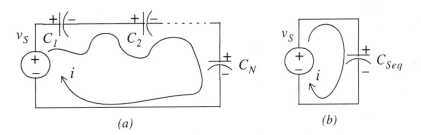

Figure 5.29. Circuits for derivation of equivalent capacitance for capacitors in series

From the circuit of Figure 5.29 (a),

$$v_{C1} + v_{C2} + ... + v_{CN} = v_S$$

or

$$\frac{1}{C_1}\int_{-\infty}^{t} i dt + \frac{1}{C_2}\int_{-\infty}^{t} i dt + ... + \frac{1}{C_N}\int_{-\infty}^{t} i dt = v_S$$

or

$$\left(\frac{1}{C_1} + \frac{1}{C_2} + ... + \frac{1}{C_N}\right)\int_{-\infty}^{t} i dt = v_S \qquad (5.43)$$

From the circuit of Figure 5.29 (b)

$$\frac{1}{C_{Seq}} \int_{-\infty}^{t} i \, dt = v_S \qquad (5.44)$$

Equating the left sides of (5.43) and (5.44) we get:

$$\boxed{\frac{1}{C_{Seq}} = \frac{1}{C_1} + \frac{1}{C_2} + \dots + \frac{1}{C_N}} \qquad (5.45)$$

and for the special case of two capacitors in series

$$C_{Seq} = \frac{C_1 C_2}{C_1 + C_2} \qquad (5.46)$$

Thus capacitors in series combine as resistors in parallel do.

Next, we will consider the circuits of figures 5.30 (a) and 5.30 (b) where the source current i_S is the same for both circuits. We wish to find an expression for the equivalent capacitance which we denote as C_{Peq} in terms of $C_1, C_2, \dots, C_N$ in Figure 5.30 (a) so that the voltage v will be the same in both circuits.

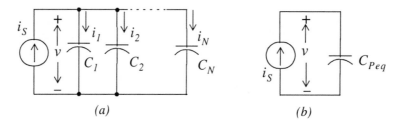

(a)   (b)

Figure 5.30. Circuits for derivation of equivalent capacitance for capacitors in parallel

From the circuit of Figure 5.30 (a),

$$i_1 + i_2 + \dots + i_N = i_S$$

or

$$C_1 \frac{dv}{dt} + C_2 \frac{dv}{dt} + \dots + C_N \frac{dv}{dt} = i_S$$

or

$$(C_1 + C_2 + \dots + C_N) \frac{dv}{dt} = i_S \qquad (5.47)$$

From the circuit of Figure 5.30 (b),

$$C_{Peq} \frac{dv}{dt} = i_S \qquad (5.48)$$

Equating the left sides of (5.47) and (5.48) we get:

$$C_{Peq} = C_1 + C_2 + \ldots + C_N \qquad (5.49)$$

Thus, capacitors in parallel combine as resistors in series do.

Example 5.7

For the network of Figure 5.31, replace all capacitors by a single equivalent capacitor.

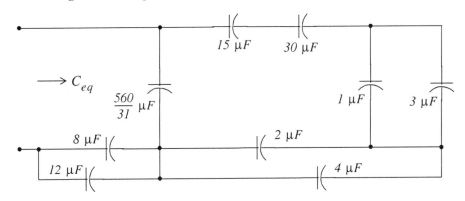

Figure 5.31. Network for Example 5.7

Solution:

Starting at the right of the network and moving towards the left, we find that $3~\mu F \parallel 1~\mu F = 4~\mu F$, $2~\mu F \parallel 4~\mu F = 6~\mu F$, $15~\mu F$ in series with $30\mu F = 10\mu F$, and $8~\mu F \parallel 12~\mu F = 20~\mu F$. The network then reduces to that shown in Figure 5.32.

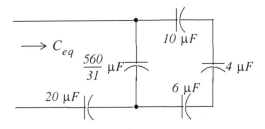

Figure 5.32. First step in combination of capacitances

Next, the series combination of $10, 4,$ and $6\mu F$ capacitors yields $60/31~\mu F$ and $60/31~\mu F \parallel 560/31~\mu F = 20~\mu F$. Finally, the series combination of $20~\mu F$ and $20~\mu F$ yields $C_{eq} = 10~\mu F$ as shown in Figure 5.33.

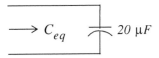

Figure 5.33. Network showing the equivalent inductance of Figure 5.16

5.8 Nodal and Mesh Equations in General Terms

In Examples 5.3 and 5.6 the voltage and current sources were constant and therefore, the steady-state circuit analysis could be performed by nodal, mesh or any other method of analysis as we learned in Chapter 3. However, if the voltage and current sources are time-varying quantities we must apply KCL or KVL in general terms as illustrated by the following example.

Example 5.8

Write nodal and mesh equations for the circuit shown in Figure 5.34.

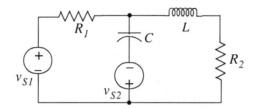

Figure 5.34. Circuit for Example 5.8

Solution:

Nodal Analysis:

We assign nodes as shown in Figure 5.35. Thus, we need $N - 1 = 5 - 1 = 4$ nodal equations.

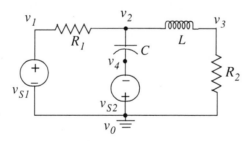

Figure 5.35. Nodal analysis for the circuit of Example 5.8

At Node 1:

$$v_1 = v_{S1}$$

At Node 2:

$$\frac{v_2 - v_1}{R_1} + C\frac{d}{dt}(v_2 - v_4) + \frac{1}{L}\int_{-\infty}^{t} (v_2 - v_3)dt = 0$$

At Node 3:

$$\frac{1}{L}\int_{-\infty}^{t} (v_3 - v_2)dt + \frac{v_3}{R_2} = 0$$

At Node 4:

$$v_4 = -v_{S2}$$

Mesh Analysis:

We need $M = B - 1 = 6 - 5 + 1 = 2$ mesh equations. Thus, we assign currents i_1 and i_2 as shown in Figure 5.36.

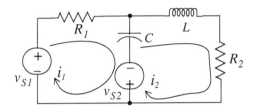

Figure 5.36. Mesh analysis for the circuit of Example 5.8

For Mesh 1:

$$R_1 i_1 + \frac{1}{C}\int_{-\infty}^{t} (i_1 - i_2)dt - v_{S1} - v_{S2} = 0$$

For Mesh 2:

$$L\frac{d}{dt}i_2 + R_2 i_2 + v_{S2} + \frac{1}{C}\int_{-\infty}^{t} (i_2 - i_1)dt = 0$$

In both the nodal and mesh equations, the initial conditions are included in the limits of integration. Alternately, we can add the initial condition terms and replace the lower limit of integration $-\infty$ with zero in the integrodifferential equations above.

5.9 Summary

- Inductance is associated with a magnetic field which is created whenever there is current flow.

- The magnetic field loops are circular in form and are called lines of magnetic flux. The unit of magnetic flux is the weber (Wb).

- The magnetic flux is denoted as φ and, if there are N turns and we assume that the flux φ passes through each turn, the total flux, denoted as λ, is called flux linkage. Then, $\lambda = N\varphi$

- For an inductor, the voltage-current relationship is $v_L = L(di_L/dt)$

- The unit of inductance is the Henry abbreviated as H.

- Unlike the resistor which dissipates energy (in the form of heat), the (ideal) inductor is a physical device capable of storing energy in analogy to the potential energy of a stretched spring.

- The energy stored in an inductor is $W_L(t) = (1/2)Li_L^2(t)$

- The current through an inductor cannot change instantaneously.

- In circuits where the applied voltage source or current source are constants, after steady-state conditions have been reached, an inductor behaves like a short circuit.

- Inductors in series combine as resistors in series do.

- Inductors in parallel combine as resistors in parallel do.

- Capacitance is associated with an electric field.

- A capacitor consists of two parallel metal plates separated by an air space or by a sheet of some type of insulating material called the dielectric.

- The charge q on the plates of a capacitor is directly proportional to the voltage between the plates and the capacitance C is the constant of proportionality. Thus, $q = Cv$

- In a capacitor, the voltage-current relationship is $i_C = C(dv_C/dt)$

- The unit of capacitance is the Farad abbreviated as F.

- Like an inductor, a capacitor is a physical device capable of storing energy.

- The energy stored in a capacitor is $W_C(t) = (1/2)Cv_C^2(t)$

- The voltage across a capacitor cannot change instantaneously.

- In circuits where the applied voltage source or current source are constants, after steady-state conditions have been reached, a capacitor behaves like an open circuit.

- Capacitors in series combine as resistors in parallel do.

- Capacitors in parallel combine as resistors in series do.

- In a circuit that contains inductors and/or capacitors, if the applied voltage and current sources are time-varying quantities, the nodal and mesh equations are, in general, integrodifferential equations.

5.10 Exercises

Multiple Choice

1. The unit of inductance is the

    A. Farad

    B. Ohm

    C. mH

    D. Weber

    E. None of the above

2. The unit of capacitance is the

    A. μF

    B. Ohm

    C. Farad

    D. Coulomb

    E. None of the above

3. Faraday's law of electromagnetic induction states that

    A. $\lambda = N\varphi$

    B. $\lambda = Li$

    C. $v = L(di/dt)$

    D. $v = d\lambda/dt$

    E. None of the above

4. In an electric field of a capacitor, the lines of force

    A. are complete loops

    B. start at the positive plate and end at the negative plate

    C. start at the negative plate and end at the positive plate

    D. are unpredictable

    E. None of the above

5. The energy in an inductor is

   A. $(1/2)(Li^2)$

   B. $(1/2)(Lv^2)$

   C. $v_L i_L$

   D. dissipated in the form of heat

   E. None of the above

6. The energy in a capacitor is

   A. $(1/2)(Ci^2)$

   B. $(1/2)(Cv^2)$

   C. $v_C i_C$

   D. dissipated in the form of heat

   E. None of the above

7. In an inductor

   A. the voltage cannot change instantaneously

   B. the current cannot change instantaneously

   C. neither the voltage nor the current can change instantaneously

   D. both the voltage and the current can change instantaneously

   E. None of the above

8. In a capacitor

   A. the voltage cannot change instantaneously

   B. the current cannot change instantaneously

   C. neither the voltage nor the current can change instantaneously

   D. both the voltage and the current can change instantaneously

   E. None of the above

9. In the circuit of Figure 5.37 after steady-state conditions have been established, the current i_L through the inductor will be

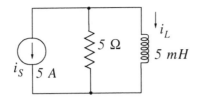

Figure 5.37. Circuit for Question 9

A. *0 A*

B. *∞ A*

C. *2.5 A*

D. *5 A*

E. None of the above

10. In the circuit of Figure 5.38 after steady-state conditions have been established, the voltage v_C across the capacitor will be

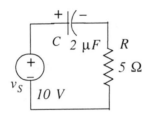

Figure 5.38. Circuit for Question 10

A. *0 V*

B. *∞ V*

C. *−10 V*

D. *10 V*

E. None of the above

Problems

1. The current i_L flowing through a 10 mH inductor is shown by the waveform of Figure 5.39.

   a. Compute and sketch the voltage v_L across this inductor for $t > 0$

   b. Compute the first time after $t = 0$ when the power p_L absorbed by this inductor is $p_L = 50 \ \mu w$ Answer: $t = 5 \ ms$

c. Compute the first time after $t = 0$ when the power p_L absorbed by this inductor is $p_L = -50\ \mu w$ Answer: $t = 25\ ms$

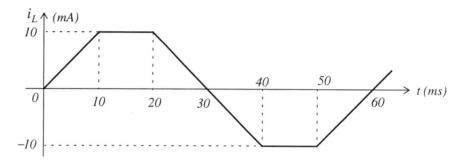

Figure 5.39. Waveform for Problem 1

2. The current i_C flowing through a *1 μF* capacitor is given as $i_C(t) = cos\,100t\ mA$, and it is known that $v_C(0) = 0$

   a. Compute and sketch the voltage v_C across this capacitor for $t > 0$

   b. Compute the first time after $t = 0$ when the power p_C absorbed by this capacitor is $p_C = 5\ mw$. Answer: *7.85 ms*

   c. Compute the first time after $t = 0$ when the power p_C absorbed by this capacitor is $p_C = -5\ mw$. Answer: *23.56 ms*

3. For the network of Figure 5.40, compute the total energy stored in the series combination of the resistor, capacitor, and inductor at $t = 10\ ms$ if:

   a. $i(t) = 0.1e^{-100t}\ mA$ and it is known that $v_C(0) = -10\ V$. Answer:*3.4 mJ*

   b. $i(t) = 0.5\,cos\,5t\ mA$ and it is known that $v_C(0) = 0$. Answer:*50 μJ*

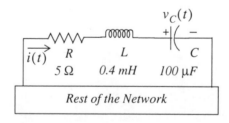

Figure 5.40. Network for Problem 3

4. For the circuit of Figure 5.41, compute the energy stored in the *5 mH* inductor at $t = 1\ s$ given that $i(0) = 0$. Answer: *1 mJ*

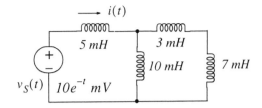

Figure 5.41. Circuit for Problem 4

5. For the circuit of Figure 5.42, replace all capacitors with an equivalent capacitance C_{eq} and then compute the energy stored in C_{eq} at $t = 1\ ms$ given that $v_C(0) = 0$ in all capacitors.

   Answer: *10 pJ*

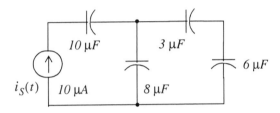

Figure 5.42. Circuit for Problem 5

6. Write nodal equations for the circuit of Figure 5.43.

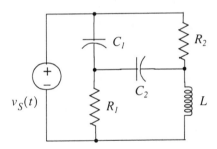

Figure 5.43. Circuit for Problem 6

7. Write mesh equations for the circuit of Figure 5.44.

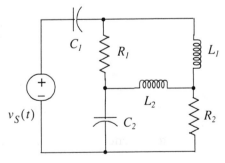

Figure 5.44. Circuit for Problem 7

5.11 Answers to Exercises

Multiple Choice

1. E Henry

2. C

3. D

4. B

5. A

6. B

7. B

8. A

9. E −5 A

10. D

Problems

1. a.   In an inductor the voltage and current are related by $v_L = L(di_L/dt) = L \times slope$. Thus, we

   need to compute the slope of each segment of the given waveform and multiply it by L.

$$v_L\Big|_0^{10\ ms} = L \times slope = L\frac{\Delta i_L}{\Delta t} = 10 \times 10^{-3} \times \frac{10 \times 10^{-3}\ A}{10 \times 10^{-3}\ s} = 10\ mV$$

Likewise,

$$v_L\Big|_{10}^{20\ ms} = L \times slope = L \times 0 = 0\ mV$$

$$v_L\Big|_{20}^{40\ ms} = L \times slope = 10 \times 10^{-3} \times \frac{[-10-(10)] \times 10^{-3}\ A}{(40-20) \times 10^{-3}\ s} = -10\ mV$$

$$v_L\Big|_{40}^{50\ ms} = L \times slope = L \times 0 = 0\ mV$$

$$v_L\Big|_0^{10\ ms} = L \times slope = 10 \times 10^{-3} \times \frac{[0-(-10)] \times 10^{-3}\ A}{(60-50) \times 10^{-3}\ s} = 10\ mV$$

The current, voltage, and power waveforms are shown below.

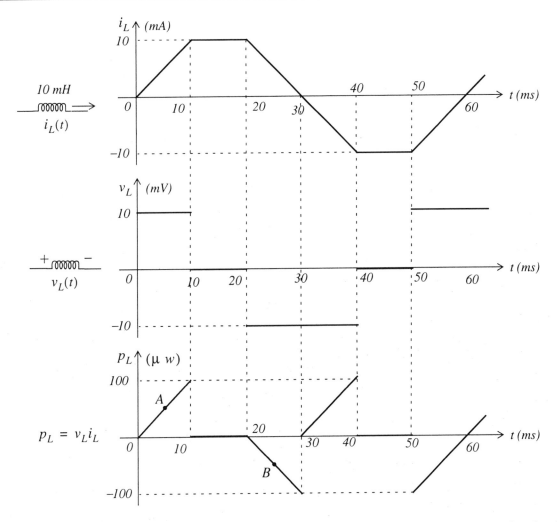

b. From the power waveform above, we observe that $p_L = v_L i_L = 50 \ \mu w$ occurs for the first time at point A where $t = 5 \ ms$

c. From the power waveform above, we observe that $p_L = v_L i_L = -50 \ \mu w$ occurs for the first time at point A where $t = 25 \ ms$

2. a. For this problem $C = 1 \ \mu F = 10^{-6} \ F$ and the current i_C is a sinusoid given as $i_C(t) = \cos 100t \ mA$ as shown below. The voltage $v_C(t)$ across this capacitor is found from

$$v_C(t) = \frac{1}{C} \int_0^t i_C \ d\tau + v_C(0) = 10^6 \int_0^t (10^{-3}) \cos 100\tau d\tau + 0$$

$$= 10^3 \int_0^t \cos 100\tau d\tau = \frac{10^3}{100} \sin 100\tau \Big|_0^t = 10 \sin 100t$$

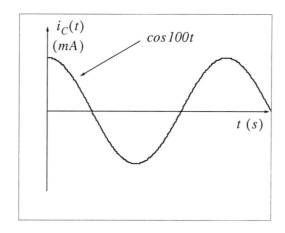

and the waveform of $v_C(t)$ is shown below.

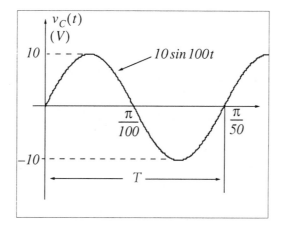

Now, $\omega T = 2\pi$ or $\omega = 2\pi/T$. Then, $10\sin\frac{2\pi}{T}t = 10\sin 100t$ or $2\pi/T = 100$ and $T = 2\pi/100$ or $T = \pi/50$

b. Since $v_C(t)$ is a sine function and $i_C(t)$ a cosine function, the first time after zero that their product will be positive is in the interval $0 < t < \pi/200$ where we want $p_C = v_C i_C = 5\ mw$ or

$$p_C = (10\sin 100t)(10^{-3}\cos 100t) = 5\times 10^{-3}\ w$$

or

$$p_C = (10\sin 100t)(\cos 100t) = 5\ w$$

Recalling that

$$sin2x = 2sinx cos x$$

it follows that

$$p_C = 5sin200t = 5 \ w$$

or

$$sin200t = 1$$

or

$$t = \frac{sin^{-1}1}{200} = \frac{\pi/2}{200} = \frac{\pi}{400} = 0.00785 \ s = 7.85 \ ms$$

c. The time where $p_C = -5 \ mw$ will occur for the first time is $7.85 \ ms$ after $t = \pi/200 \ s$ or after $t = 1000\pi/200 \ ms = 5\pi \ ms$. Therefore, $p_C = -5 \ mw$ will occur for the first time at $t = 7.85 + 5\pi = 7.85 + 15.71 = 23.56 \ ms$

3. a. There is no energy stored in the resistor; it is dissipated in the form of heat. Thus, the total energy is stored in the capacitor and the inductor, that is,

$$W_T = W_L + W_C = \frac{1}{2}Li_L^2 + \frac{1}{2}Cv_C^2$$

where

$$i_L = i(t) = 0.1e^{-100t}$$

and

$$v_C(t) = \frac{1}{C}\int_0^t i_C \ d\tau + v_C(0) = 10^4 \int_0^t 0.1e^{-100\tau}d\tau - 10$$

$$= \frac{10^4 \times 0.1}{-100}e^{-100\tau}\Big|_0^t - 10 = 10e^{-100\tau}\Big|_t^0 = 10 - 10e^{-100t} - 10$$

or

$$v_C(t) = -10e^{-100t}$$

Then,

$$W_T\Big|_{t = 10 \ ms} = \frac{1}{2} \times 0.4 \times 10^{-3} \times (0.1e^{-100t})^2 + \frac{1}{2} \times 10^{-4} \times (-10e^{-100t})^2$$

$$= 2.5 \times 10^{-4}[(0.1e^{-1})^2 + (-10e^{-1})^2] = 3.4 \ mJ$$

We've used MATLAB as a calculator to obtain the answer, that is,

```
WT=2.5*10^(−4)*((0.1*exp(−1))^2+((−10)*exp(−1))^2);
fprintf(' \n'); fprintf('WT=%7.4f J',WT); fprintf(' \n')
```

```
WT= 0.0034 J
```

b. For this part

$$i_L = i(t) = 0.5\cos 5t \ mA$$

and

$$v_C(t) = \frac{1}{C}\int_0^t i_C \, d\tau + v_C(0) = 10^4 \times 10^{-4} \int_0^t 5\cos 5\tau d\tau + 0$$

$$= \sin 5\tau \Big|_0^t = \sin 5t$$

Then,

$$W_T = W_L + W_C = \frac{1}{2} \times 0.4 \times 10^{-4} \times (0.5\cos 5t)^2 + \frac{1}{2} \times 10^{-4} \times (\sin 5t)^2$$

$$= 0.5 \times 10^{-4} \left[\underbrace{\cos^2 5t + \sin^2 5t}_{1} \right] = 0.05 \ mJ = 50 \ \mu J$$

We observe that the total power is independent of time.

4. Starting with the right side and proceeding to the left, the series-parallel combination of $7 + 3 = 10$, $10 \parallel 10 = 5$, and $5 + 5 = 10 \ mH$ reduces the given circuit to the one shown below.

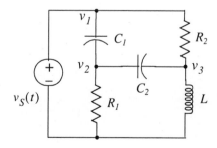

The current $i_L(t)$ is

$$i_L(t) = \frac{1}{L}\int_0^t v_S \, d\tau + i_L(0) = \frac{1}{10 \times 10^{-3}} \times 10 \times 10^{-3} \int_0^t e^{-\tau} d\tau = -e^{-\tau}\Big|_0^t = e^{-\tau}\Big|_t^0 = 1 - e^{-t}$$

Then,

$$W_{5 \ mH}\Big|_{t=1 \ s} = \frac{1}{2} \times 5 \times 10^{-3}(1 - e^{-t})^2\Big|_{t=1 \ s} = 2.5 \times 10^{-3} \times (1 - e^{-1})^2 \approx 1 \ mJ$$

6. We assign node voltages v_1, v_2, and v_3 as shown below.

Then,

$$v_1 = v_S$$

$$C_1\frac{d}{dt}(v_2 - v_1) + C_2\frac{d}{dt}(v_2 - v_3) + \frac{v_2}{R_1} = 0$$

$$\frac{v_3 - v_1}{R_2} + C_2\frac{d}{dt}(v_3 - v_2) + \frac{1}{L}\int_{-\infty}^{t} v_3 \, d\tau = 0$$

7. We assign mesh currents i_1, i_2, and i_3 as shown below.

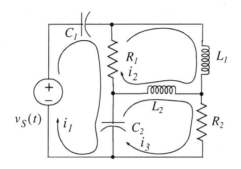

Then,

$$\frac{1}{C_1}\int_{-\infty}^{t} i_1 \, d\tau + R_1(i_1 - i_2) + \frac{1}{C_2}\int_{-\infty}^{t} (i_1 - i_3) \, d\tau = v_S$$

$$R_1(i_2 - i_1) + L_1\frac{di_2}{dt} + L_2\frac{d}{dt}(i_2 - i_3) = 0$$

$$\frac{1}{C_2}\int_{-\infty}^{t} (i_3 - i_1) \, d\tau + L_2\frac{d}{dt}(i_3 - i_2) + R_2 i_3 = 0$$

NOTES

Chapter 6

Sinusoidal Circuit Analysis

This chapter is an introduction to circuits in which the applied voltage or current are sinusoidal. The time and frequency domains are defined and phasor relationships are developed for resistive, inductive and capacitive circuits. Reactance, susceptance, impedance and admittance are also defined. It is assumed that the reader is familiar with sinusoids and complex numbers. If not, it is strongly recommended that Appendix B is reviewed thoroughly before reading this chapter.

6.1 Excitation Functions

The applied voltages and currents in electric circuits are generally referred to as *excitations* or *driving functions*, that is, we say that a circuit is "excited" or "driven" by a constant, or a sinusoidal, or an exponential function of time. Another term used in circuit analysis is the word *response*; this may be the voltage or current in the "load" part of the circuit or any other part of it. Thus the response may be anything we define it as a response. Generally, the response is the voltage or current at the output of a circuit, but we need to specify what the output of a circuit is.

In Chapters 1 through 4 we considered circuits that consisted of excitations (active sources) and resistors only as the passive devices. We used various methods such as nodal and mesh analyses, superposition, Thevenin's and Norton's theorems to find the desired response such as the voltage and/or current in any particular branch. The circuit analysis procedure for these circuits is the same for DC and AC circuits. Thus, if the excitation is a constant voltage or current, the response will also be some constant value; if the excitation is a sinusoidal voltage or current, the response will also be sinusoidal with the same frequency but different amplitude and phase.

In Chapter 5 we learned that when the excitation is a constant and steady-state conditions are reached, an inductor behaves like a short circuit and a capacitor behaves like an open circuit. However, when the excitation is a time-varying function such as a sinusoid, inductors and capacitors behave entirely different as we will see in our subsequent discussion.

6.2 Circuit Response to Sinusoidal Inputs

We can apply the circuit analysis methods which we have learned in previous chapters to circuits where the voltage or current sources are sinusoidal. To find out how easy (or how difficult) the procedure becomes, we will consider the simple series circuit of Example 6.1.

Example 6.1

For the circuit shown in Figure 6.1, derive an expression for $v_C(t)$ in terms of V_p, R, C, and ω where the subscript p is used to denote the *peak* or *maximum* value of a time varying function, and the sine symbol inside the circle denotes that the excitation is a sinusoidal function.

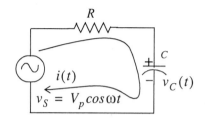

Figure 6.1. Circuit for Example 6.1

Solution:

By KVL,

$$v_R + v_C = v_S \tag{6.1}$$

where

$$v_R = Ri = Ri_C$$

and

$$i_C = C\frac{dv_C}{dt}$$

Then,

$$v_R = RC\frac{dv_C}{dt}$$

and by substitution into (6.1) we get

$$RC\frac{dv_C}{dt} + v_C = v_S = V_p \cos\omega t \tag{6.2}$$

As we know, differentiation (and integration) of a sinusoid of radian frequency ω results in another sinusoid of the same frequency ω. Accordingly, the solution of (6.2) must have the form

$$v_C(t) = A\cos(\omega t + \theta) \tag{6.3}$$

where the amplitude A and phase angle θ are constants to be determined from the circuit parameters of V_p, R, C, and ω. Substitution of (6.3) into (6.2) yields

$$-A\omega RC\sin(\omega t + \theta) + A\cos(\omega t + \theta) = V_p\cos\omega t \tag{6.4}$$

and recalling that

$$sin(x + y) = sinx cosy + cosx siny$$

and

$$cos(x + y) = cosx cosy - sinx siny$$

we rewrite (6.4) as

$$-A\omega RC \sin\omega t \cos\theta - A\omega RC \cos\omega t \sin\theta + A\cos\omega t \cos\theta - A\sin\omega t \sin\theta = V_p \cos\omega t$$

Collecting sine and cosine terms, equating like terms and, after some more tedious work, solving for amplitude A and phase angle θ we get:

$$v_C(t) = \frac{V_p}{\sqrt{1+(\omega RC)^2}} \cos(\omega t - \tan^{-1}(\omega RC)) \tag{6.5}$$

Obviously, analyzing circuits with sinusoidal excitations when they contain capacitors and/or inductors, using the above procedure is impractical. We will see on the next section that the *complex excitation function* greatly simplifies the procedure of analyzing such circuits. Complex numbers are discussed in Appendix B.

The complex excitation function does not imply complexity of a circuit; it just entails the use of complex numbers. We should remember also that when we say that the imaginary part of a complex number is some value, there is nothing "imaginary" about this value. In other words, the imaginary part is just as "real" as the real part of the complex number but it is defined on a different axis. Thus we display the real part of a complex function on the axis of the reals (usually the x-axis), and the imaginary part on the imaginary axis or the y-axis.

6.3 The Complex Excitation Function

We recall that the derivatives and integrals of sinusoids always produce sinusoids of the *same frequency* but different amplitude and phase since the cosine and sine functions are 90 degrees out-of-phase. Thus, if

$$v(t) = A\cos(\omega t + \theta)$$

then

$$\frac{dv}{dt} = -\omega A \sin(\omega t + \theta)$$

and if

$$i(t) = B\sin(\omega t + \phi)$$

then

$$\frac{di}{dt} = \omega B \cos(\omega t + \phi)$$

Let us consider the network of Figure 6.2 which consists of resistors, inductors and capacitors, and it is driven (excited) by a sinusoidal voltage source $v_S(t)$.

Figure 6.2. General presentation of a network showing excitation and load

Let us also define the voltage across the load as $v_{LOAD}(t)$ [*] as the response. As we know from Chapter 5, the nodal and mesh equations for such circuits are integrodifferential equations, and it is shown in differential equations textbooks[†] that the *forced response* or *particular solution* of these circuits have the form

$$v_{LD}(t) = A\cos(\omega t) + B\sin(\omega t)$$

We also know from Euler's identity that

$$A\cos\omega t + jA\sin\omega t = Ae^{j\omega t} \qquad (6.6)$$

and therefore, the real component is the response due to $\cos\omega t$ and the imaginary component is the response to $\sin\omega t$ We will use Example 6.2 to illustrate the ease by which we can obtain the response of a circuit, which is excited by a sinusoidal source, using the complex function $Ae^{j\omega t}$ approach. In this text, we will represent all sinusoidal variations in terms of the cosine function.

Example 6.2

Repeat Example 6.1, that is, find the capacitor voltage $v_C(t)$ for the circuit of Figure 6.3 using the complex excitation method.

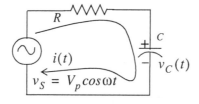

Figure 6.3. Circuit for Example 6.2

---

[*] *Some textbooks denote the voltage across and the current through the load as v_L and i_L respectively. As we stated previously, in this text, we use the v_{LOAD} and i_{LOAD} notations to avoid confusion with the voltage v_L across and the current i_L through an inductor.*

[†] *This topic is also discussed in Circuit Analysis II with MATLAB® Applications by this author*

Solution:

Since

$$cos\omega t = Re\{e^{j\omega t}\}$$

we let the excitation be

$$v_S(t) = V_p e^{j\omega t}$$

and thus the response will have the form

$$v_C(t) = V_C e^{j(\omega t + \varphi)}$$

As in Example 6.1,

$$RC\frac{dv_C}{dt} + v_C = V_p e^{j\omega t} \qquad (6.7)$$

or

$$RC\frac{d}{dt}(V_C e^{j(\omega t + \varphi)}) + V_C e^{j(\omega t + \varphi)} = V_p e^{j\omega t}$$

or

$$(j\omega RC + 1)V_C e^{j(\omega t + \varphi)} = V_p e^{j\omega t}$$

The last expression above shows that radian frequency ω is the same for the response as it is for the excitation; therefore we only need to be concerned with the magnitude and the phase angle of the response. Accordingly, we can eliminate the radian frequency ω by dividing both sides of that expression by $e^{j\omega t}$ and thus the input-output (excitation-response) relation reduces to

$$(j\omega RC + 1)V_C e^{j\varphi} = V_p$$

from which

$$V_C e^{j\varphi} = \frac{V_p}{j\omega RC + 1} = \frac{V_p}{\sqrt{1 + \omega^2 R^2 C^2}\, e^{j[tan^{-1}(\omega RC)]}} = \frac{V_p}{\sqrt{1 + \omega^2 R^2 C^2}}\, e^{-j[tan^{-1}(\omega RC)]}$$

This expression above shows the response as a function of the maximum value of the excitation, its radian frequency and the circuit constants R and C.

If we wish to express the response in complete form, we simply multiply both sides by $e^{j\omega t}$ and we get

$$V_C e^{j(\omega t + \varphi)} = \frac{V_p}{\sqrt{1 + \omega^2 R^2 C^2}}\, e^{j[\omega t - tan^{-1}(\omega RC)]}$$

Finally, since the excitation is the real part of the complex excitation, we use Euler's identity on both sides and equating reals parts, we get

$$v_C(t) = V_C \cos(\omega t + \phi) = \frac{V_p}{\sqrt{1 + \omega^2 R^2 C^2}} \cos[\omega t - \tan^{-1}(\omega RC)]$$

The first part of the above procedure where the excitation-response relation is simplified to amplitude and phase relationship is known as *time-domain* to *frequency-domain transformation;* the second part where the excitation-response is put back to its sinusoidal form is known as *frequency-domain to time-domain transformation.* For brevity, we will denote the time domain as the $t-domain$, and the frequency domain as the $j\omega-domain$.

If a sinusoid is given in terms of the sine function, we must first convert it to a cosine function. Thus,

$$m(t) = A\sin(\omega t + \theta) = A\cos(\omega t + \theta - 90°) \qquad (6.8)$$

and in the $j\omega-domain$ it is expressed as

$$M = Ae^{j(\theta - 90°)} = A\angle(\theta - 90°) \qquad (6.9)$$

where M represents a *phasor* (rotating vector) voltage V or current I^*.

In summary, the $t-domain$, to $j\omega-domain$ transformation procedure is as follows:

1. Express the given sinusoid as a cosine function

2. Express the cosine function as the real part of the complex excitation using Euler's identity

3. Extract the magnitude and phase angle from it.

Example 6.3

Transform the sinusoid $v(t) = 10\sin(100t - 60°)$ to its equivalent $j\omega-domain$ expression.

Solution:

For this example, we have

$$v(t) = 10\sin(100t - 60°) = 10\cos(100t - 60° - 90°)$$

or

$$v(t) = 10\cos(100t - 150°) = Re\left\{10e^{j(100t - 150°)}\right\}$$

Since the $j\omega - domain$ contains only the amplitude and phase, we extract these from the bracketed term on the right side of the above expression, and we get the phasor V as

$$V = 10e^{-j150^{\circ}} = 10\angle -150^{\circ}$$

The $j\omega - domain$ to $t - domain$ transformation procedure is as follows:

1. Convert the given phasor from polar to exponential form

2. Add the radian frequency ω multiplied by t to the exponential form

3. Extract the real part from it.

Example 6.4

Transform the phasor $I = 120\angle -90^{\circ}$ to its equivalent time-domain expression.

Solution:

First, we express the given phasor in exponential form, that is,

$$I = 120\angle -90^{\circ} = 120e^{-j90^{\circ}}$$

Next, adding the radian frequency ω multiplied by t to the exponent of the above expression we get

$$i(t) = 120e^{j(\omega t - 90^{\circ})}$$

and finally we extract the real part from it. Then,

$$i(t) = Re\left\{120e^{j(\omega t - 90^{\circ})}\right\} = 120\cos(\omega t - 90^{\circ}) = 120\sin\omega t$$

We can add, subtract, multiply and divide sinusoids of the same frequency using phasors as illustrated by the following example.

Example 6.5

It is given that $i_1(t) = 10\cos(120\pi t + 45^{\circ})$ and $i_2(t) = 5\sin(120\pi t - 45^{\circ})$. Compute the sum $i(t) = i_1(t) + i_2(t)$.

Solution:

As a first step, we express $i_2(t)$ as a cosine function, that is,

$$i_2(t) = 5\sin(120\pi t - 45°) = 5\cos(120\pi t - 45° - 90°) = 5\cos(120\pi t - 135°)$$

Next, we perform the $t - domain$ to $j\omega - domain$ transformation and we obtain the phasors

$$I_1 = 10\angle 45° \text{ and } I_2 = 5\angle -135°$$

and by addition,

$$I = I_1 + I_2 = 10\angle 45° + 5\angle -135° = 10\left(\frac{\sqrt{2}}{2} + j\frac{\sqrt{2}}{2}\right) + 5\left(-\frac{\sqrt{2}}{2} - j\frac{\sqrt{2}}{2}\right)$$

or

$$I = 5\left(\frac{\sqrt{2}}{2} + j\frac{\sqrt{2}}{2}\right) = 5\angle 45°$$

and finally transforming the phasor I into the $t - domain$, we get

$$i(t) = 5\cos(120\pi t + 45°)$$

Also, for brevity, in our subsequent discussion we will designate resistive, inductive and capacitive circuits as R, L, and C respectively.

6.4 Phasors in R, L, and C Circuits

The circuit analysis of circuits containing R, L, and C devices, and which are excited by sinusoidal sources, is considerably simplified with the use of phasor voltages and phasor currents which we will represent by the boldface capital letters V and I respectively. We will now derive V and I phasor relationships in the $j\omega - domain$. *We must always remember that phasor quantities exist only in the $j\omega - domain$.*

1. V and I phasor relationship in R branches

Consider circuit 6.4 (a) below where the load is purely resistive. We know from Ohm's law that $v_R(t) = Ri_R(t)$ where the resistance R is a constant. We will show that this relationship also holds for the phasors V_R and I_R shown in circuit 6.4 (b), that is, we will prove that

$$V_R = RI_R$$

Proof:

In circuit 6.4 (a) we let $v_R(t)$ be a complex voltage, that is,

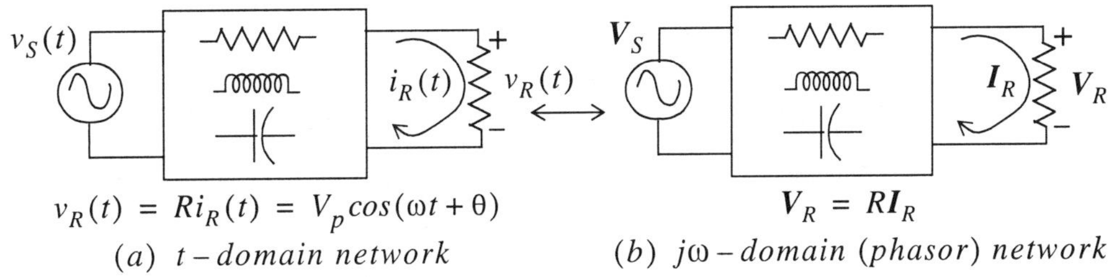

$$v_R(t) = Ri_R(t) = V_p\cos(\omega t + \theta)$$

$$V_R = RI_R$$

(a) t – domain network   *(b) jω – domain (phasor) network*

Figure 6.4. Voltage across a resistive load in t – domain and jω – domain

$$V_p e^{j(\omega t + \phi)} = V_p\cos(\omega t + \phi) + jV_p\sin(\omega t + \phi) \tag{6.10}$$

and since R is a constant, it will produce a current of the same frequency ω and the same phase ϕ [*]
whose form will be

$$I_p e^{j(\omega t + \phi)} = I_p\cos(\omega t + \phi) + jI_p\sin(\omega t + \phi)$$

and by Ohm's law,

$$V_p e^{j(\omega t + \phi)} = RI_p e^{j(\omega t + \phi)} \tag{6.11}$$

Transforming (6.11) to the $j\omega – domain$, we obtain the phasor relationship

$$V_p e^{j\phi} = RI_p e^{j\phi} \quad \text{or} \quad V_p\angle\phi = RI_p\angle\phi$$

Since the phasor current I is *in-phase* with the voltage V (both I and V have the same phase ϕ), we
let

$$V_p\angle\phi = V_R \quad \text{and} \quad I_p\angle\phi = I_R$$

and it follows that

$$\boxed{V_R = RI_R} \tag{6.12}$$

Therefore, the phasor V and I relationship in resistors, obeys Ohm's law also, and the current
through a resistor is always in–phase with the voltage across that resistor.

Example 6.6

For the network of Figure 6.5, find $i_R(t)$ when $v_R(t) = 40\sin(377t - 75°)$.

---

[*] *The phase will be the same since neither differentiation nor integration is performed here.*

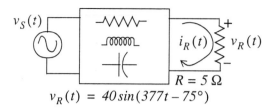

Figure 6.5. Voltage across the resistive load of Example 6.6

Solution:

We first perform the $t-domain$ to $j\omega-domain$ i.e., $v_R(t) \Leftrightarrow V_R$ transformation as follows:

$$v_R(t) = 40\sin(377t - 75°) = 40\cos(377t - 165°) \Leftrightarrow V_R = 40\angle-165°$$

Then,

$$I_R = \frac{V_R}{R} = \frac{40\angle-165°}{5} = 8\angle-165° \text{ A}$$

Therefore,

$$\underbrace{I_R = 8\angle-165° \text{ A}}_{j\omega-domain} \Leftrightarrow \underbrace{i_R(t) = 8\cos(377t - 165°) = 8\sin(377t - 75°) \text{ A}}_{t-domain}$$

Alternately, since the resistance R is a constant, we can compute $i_R(t)$ directly from the $t-domain$ expression for $v_R(t)$, that is,

$$i_R(t) = \frac{v_R(t)}{R} = \frac{40\sin(377t - 75°)}{5} = 8\sin(377t - 75°) \text{ A}$$

2. *V* and *I* phasor relationship in *L* branches

Consider circuit 6.6 (a) below where the load is purely inductive.

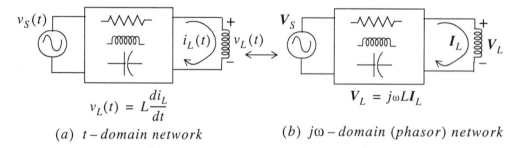

$$v_L(t) = L\frac{di_L}{dt}$$

$$V_L = j\omega LI_L$$

(a) t – domain network   *(b) jω – domain (phasor) network*

Figure 6.6. Voltage across an inductive load in t – domain and jω – domain

We will prove that the relationship between the phasors V_L and I_L shown in circuit 6.6 (b) is

$$V_L = j\omega L I_L \tag{6.13}$$

Proof:

In circuit 6.6 (a) we let $v_L(t)$ be a complex voltage, that is,

$$V_p e^{j(\omega t + \phi)} = V_p \cos(\omega t + \phi) + jV_p \sin(\omega t + \phi) \tag{6.14}$$

and recalling that if $x(t) = \sin(\omega t + \phi)$ then $dx/dt = \omega\cos(\omega t + \phi)$, that is, differentiation (or integration) does not change the radian frequency ω or the phase angle ϕ, the current through the inductor will have the form

$$I_p e^{j(\omega t + \phi)} = I_p \cos(\omega t + \phi) + jI_p \sin(\omega t + \phi) \tag{6.15}$$

and since

$$v_L(t) = L\frac{di_L}{dt}$$

then,

$$V_p e^{j(\omega t + \phi)} = L\frac{d}{dt}(I_p e^{j(\omega t + \phi)}) = j\omega L I_p e^{j(\omega t + \phi)} \tag{6.16}$$

Next, transforming (6.16) to the $j\omega - domain$, we obtain the phasor relationship

$$V_p e^{j\phi} = j\omega L I_p e^{j\phi} \quad \text{or} \quad V_p \angle\phi = j\omega L I_p \angle\phi$$

and letting

$$V_p \angle\phi = V_L \quad \text{and} \quad I_p \angle\phi = I_L$$

we get

$$\boxed{V_L = j\omega L I_L} \tag{6.17}$$

The presence of the j operator in (6.17) indicates that the voltage across an inductor leads the current through it by $90°$.

Example 6.7

For the network of Figure 6.7, find $i_L(t)$ when $v_L(t) = 40\sin(2t - 75°)$.

Solution:

We first perform the $t - domain$ to $j\omega - domain$ i.e., $v_L(t) \Leftrightarrow V_L$ transformation as follows:

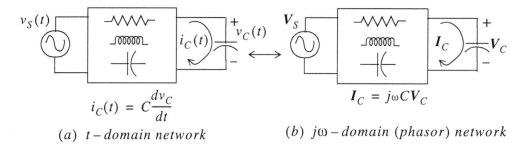

$$v_L(t) = 40\sin(2t - 75°)$$

Figure 6.7. Voltage across the inductive load of Example 6.7

$$v_L(t) = 40\sin(2t - 75°) = 40\cos(2t - 165°) \Leftrightarrow V_L = 40\angle{-165°} \ mV$$

and

$$I_L = \frac{V_L}{j\omega L} = \frac{(40\angle{-165°}) \times 10^{-3}}{j10 \times 10^{-3}} = \frac{40\angle{-165°}}{10\angle{90°}} = 4\angle{-255°} = 4\angle{105°} \ A$$

Therefore,

$$\underbrace{I_L = 4\angle{105°} \ A}_{j\omega - domain} \Leftrightarrow \underbrace{i_L(t) = 4\cos(2t + 105°) = 4\sin(2t - 165°) \ A}_{t - domain}$$

3. *V* and *I* phasor relationship in *C* branches

Consider circuit 6.8 (a) below where the load is purely capacitive.

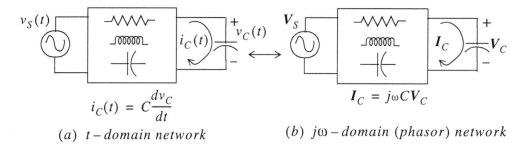

$$i_C(t) = C\frac{dv_C}{dt}$$

$$I_C = j\omega CV_C$$

$$(a) \ \ t - domain \ network \qquad\qquad (b) \ \ j\omega - domain \ (phasor) \ network$$

Figure 6.8. Voltage across a capacitive load in t – domain and jω – domain

We will prove that the relationship between the phasors V_C and I_C shown in the network of Figure 6.8 (b) is

$$I_C = j\omega CV_C \tag{6.18}$$

Proof:

In circuit 6.8 (a) we let $v_C(t)$ be a complex voltage, that is,

$$V_p e^{j(\omega t + \phi)} = V_p \cos(\omega t + \phi) + jV_p \sin(\omega t + \phi)$$

then the current through the capacitor will have the form

$$I_p e^{j(\omega t + \phi)} = I_p \cos(\omega t + \phi) + jI_p \sin(\omega t + \phi)$$

and since

$$i_C(t) = \frac{dv_C}{dt}$$

then

$$I_p e^{j(\omega t + \phi)} = C\frac{d}{dt}(V_p e^{j(\omega t + \phi)}) = j\omega C V_p e^{j(\omega t + \phi)} \tag{6.19}$$

Next, transforming (6.19) to the $j\omega - domain$, we obtain the phasor relationship

$$I_p e^{j\phi} = j\omega C V_p e^{j\phi} \quad or \quad I_p \angle\phi = j\omega C V_p \angle\phi$$

and letting

$$I_p \angle\phi = I_C \quad and \quad V_p \angle\phi = V_C$$

we get

$$\boxed{I_C = j\omega C V_C} \tag{6.20}$$

The presence of the j operator in (6.26) indicates that the current through a capacitor leads the voltage across it by $90°$.

Example 6.8

For the circuit shown below, find $i_C(t)$ when $v_C(t) = 170\cos(60\pi t - 45°)$.

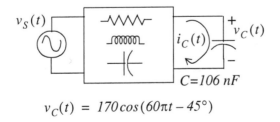

$$v_C(t) = 170\cos(60\pi t - 45°)$$

Figure 6.9. Voltage across the capacitive load of Example 6.8

Solution:

We first perform the $t - domain$ to $j\omega - domain$ i.e., $v_C(t) \Leftrightarrow V_C$ transformation as follows:

$$v_C(t) = 170\cos(60\pi t - 45°) \Leftrightarrow V_C = 170\angle-45°$$

Then,

$$I_C = j\omega C V_C = j \times 60\pi \times 106 \times 10^{-9} \times 170\angle-45° = 1\angle 90° \times 3.4 \times 10^{-3} \times 1\angle-45°$$

$$= 3.4 \times 10^{-3}\angle 45° = 3.4\angle 45° \ mA$$

Therefore,

$$\underbrace{I_C = 3.4\angle 45° \ mA}_{j\omega - domain} \Leftrightarrow \underbrace{i_C(t) = 3.4\cos(60\pi + 45°) \ mA}_{t - domain}$$

6.5 Impedance

Consider the $t-domain$ circuit in Figure 6.10 (a) and its equivalent phasor circuit shown in Figure 6.10 (b).

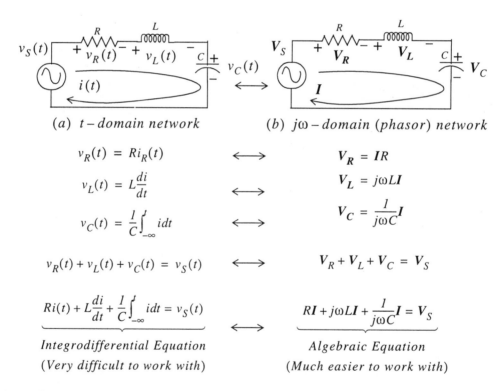

$$(a) \ \ t-domain \ network \qquad\qquad (b) \ \ j\omega-domain \ (phasor) \ network$$

$$v_R(t) = Ri_R(t) \qquad\longleftrightarrow\qquad V_R = IR$$

$$v_L(t) = L\frac{di}{dt} \qquad\longleftrightarrow\qquad V_L = j\omega LI$$

$$v_C(t) = \frac{1}{C}\int_{-\infty}^{t} idt \qquad\longleftrightarrow\qquad V_C = \frac{1}{j\omega C}I$$

$$v_R(t) + v_L(t) + v_C(t) = v_S(t) \qquad\longleftrightarrow\qquad V_R + V_L + V_C = V_S$$

$$\underbrace{Ri(t) + L\frac{di}{dt} + \frac{1}{C}\int_{-\infty}^{t} idt = v_S(t)}_{\substack{Integrodifferential \ Equation \\ (Very \ difficult \ to \ work \ with)}} \qquad\longleftrightarrow\qquad \underbrace{RI + j\omega LI + \frac{1}{j\omega C}I = V_S}_{\substack{Algebraic \ Equation \\ (Much \ easier \ to \ work \ with)}}$$

Figure 6.10. The $t-domain$ and $j\omega-domain$ relationships in a series RLC circuit

The last equation of the phasor circuit may be written as

$$\left(R + j\omega L + \frac{1}{j\omega C}\right)I = V_S$$

and dividing both sides of (6.26) by I we obtain the *impedance* which, by definition, is

$$\text{Impedance} = Z = \frac{\text{Phasor Voltage}}{\text{Phasor Current}} = \frac{V_S}{I} = R + j\omega L + \frac{1}{j\omega C} \qquad (6.21)$$

Expression (6.21) is referred to as *Ohm's law for AC Circuits.*

Like resistance, the unit of impedance is the Ohm (Ω).

We can express the impedance Z as the sum of a real and an imaginary component as follows:

Since

$$\frac{1}{j} = \frac{1}{j} \cdot \frac{j}{j} = \frac{j}{j^2} = -j$$

then

$$\frac{1}{j\omega C} = -j\frac{1}{\omega C}$$

and thus

$$Z = R + j\left(\omega L - \frac{1}{\omega C}\right) \qquad (6.22)$$

We can also express (6.22) in polar form as

$$Z = \sqrt{R^2 + \left(\omega L - \frac{1}{\omega C}\right)^2} \angle \tan^{-1}\left(\omega L - \frac{1}{\omega C}\right)/R \qquad (6.23)$$

We must remember that *the impedance is not a phasor*, it is a complex quantity whose real part is the resistance R and the imaginary part is $\omega L - 1/\omega C$, that is,

$$Re\{Z\} = R \quad \text{and} \quad Im\{Z\} = \omega L - \frac{1}{\omega C} \qquad (6.24)$$

The imaginary part of the impedance Z is called *reactance* and it is denoted with the letter X. The two components of reactance are the *inductive reactance* X_L and the *capacitive reactance* X_C, i.e.,

$$X = X_L + X_C = \omega L - 1/\omega C \qquad (6.25)$$

$$X_L = \omega L \qquad (6.26)$$

$$X_C = 1/\omega C \qquad (6.27)$$

The unit of the inductive and capacitive reactances is also the Ohm (Ω).

In terms of reactances, the impedance can be expressed as

$$Z = R + jX = R + j(X_L - X_C) = \sqrt{R^2 + (X_L - X_C)^2} \angle tan^{-1}[(X_L - X_C)/R] \qquad (6.28)$$

By a procedure similar to that of Chapter 2, we can show that impedances combine as resistances do.

Example 6.9

For the circuit below, find the current $i(t)$ given that $v_S(t) = 100\cos(100t - 30°)$.

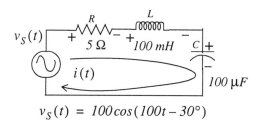

$$v_S(t) = 100\cos(100t - 30°)$$

Figure 6.11. Circuit for Example 6.9

Solution:

If we attempt to solve this problem in the time-domain directly, we will need to solve an integrodifferential equation. But as we now know, a much easier solution is with the transformation of the given circuit to a phasor circuit. Here, $\omega = 100 \ rad/s$ and thus

$$j\omega L = jX_L = j100 \times 0.1 = j10 \ \Omega$$

and

$$\frac{1}{j\omega C} = -j\frac{1}{\omega C} = -jX_C = -j\frac{1}{10^2 \times 10^2 \times 10^{-6}} = -j100$$

Also,

$$V_S = 100\angle -30°$$

and the phasor circuit is as shown in Figure 6.12.

$$V_S = ZI_S$$

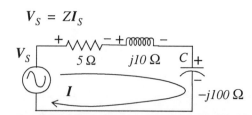

Figure 6.12. Phasor circuit for Example 6.9

From the phasor circuit of Figure 6.12

$$Z = 5 + j10 - j100 = 5 - j90 = \sqrt{5^2 + 90^2} \angle tan^{-1}(-90/5) = 90.14 \angle -86.82°$$

and

$$I = \frac{V_S}{Z} = \frac{100 \angle -30°}{90.14 \angle -86.82°} = 1.11 \angle [-30° - (-86.82°)]$$

Therefore,

$$I = 1.11 \angle 56.82° \Leftrightarrow i(t) = 1.11 \cos(100t + 56.82)$$

6.6 Admittance

Consider the $t-domain$ circuit in Figure 6.13 (a) and its equivalent phasor circuit shown in Figure 6.13 (b).

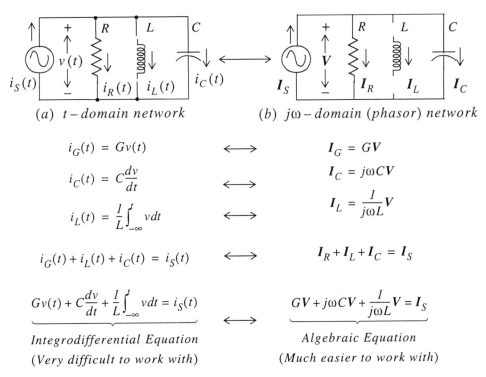

$$i_G(t) = Gv(t) \qquad \longleftrightarrow \qquad I_G = GV$$

$$i_C(t) = C\frac{dv}{dt} \qquad \longleftrightarrow \qquad I_C = j\omega C V$$

$$i_L(t) = \frac{1}{L}\int_{-\infty}^{t} vdt \qquad \longleftrightarrow \qquad I_L = \frac{1}{j\omega L}V$$

$$i_G(t) + i_L(t) + i_C(t) = i_S(t) \qquad \longleftrightarrow \qquad I_R + I_L + I_C = I_S$$

$$\underbrace{Gv(t) + C\frac{dv}{dt} + \frac{1}{L}\int_{-\infty}^{t} vdt = i_S(t)}_{} \qquad \longleftrightarrow \qquad \underbrace{GV + j\omega C V + \frac{1}{j\omega L}V = I_S}_{}$$

Integrodifferential Equation   *Algebraic Equation*

(Very difficult to work with)   *(Much easier to work with)*

Figure 6.13. The t – domain and jω – domain relationships in a parallel RLC circuit

The last equation of the phasor circuit may be written as

$$\left(G + \frac{1}{j\omega L} + j\omega C \right)V = I_S \tag{6.29}$$

Dividing both sides of (6.29) by V, we obtain the *admittance*, that is, by definition

$$Admittance = Y = \frac{Phasor\ Current}{Phasor\ Voltage} = \frac{I_S}{V} = G + \frac{1}{j\omega L} + j\omega C = \frac{1}{Z} \qquad (6.30)$$

Here we observe that the admittance Y is the reciprocal of the impedance Z as conductance G is the reciprocal of the resistance R.

Like conductance, the unit of admittance is the siemens or mho (Ω^{-1}).

As with the impedance Z, we can express the admittance Y as the sum of a real component and an imaginary component as follows:

$$Y = G + j\left(\omega C - \frac{1}{\omega L}\right) \qquad (6.31)$$

and in polar form

$$Y = \sqrt{G^2 + \left(\omega C - \frac{1}{\omega L}\right)^2} \angle tan^{-1}\left[\left(\omega C - \frac{1}{\omega L}\right)/G\right] \qquad (6.32)$$

Like the impedance Z, *the admittance Y it is not a phasor;* it is a complex quantity whose real part is the conductance G and the imaginary part is $\omega C - \frac{1}{\omega L}$, that is,

$$Re\{Y\} = G \quad and \quad Im\{Y\} = \omega C - \frac{1}{\omega L} \qquad (6.33)$$

The imaginary part of the admittance Y is called *susceptance* and it is denoted with the letter B. The two components of susceptance are the *capacitive susceptance* B_C and the *inductive susceptance* B_L, that is,

$$B = B_C + B_L = \omega C - 1/\omega L \qquad (6.34)$$

$$B_C = \omega C \qquad (6.35)$$

$$B_L = 1/\omega L \qquad (6.36)$$

The unit of the susceptances B_C and B_L is also the siemens (Ω^{-1}).

In terms of susceptances, the admittance Y can be expressed as

$$Y = G + jB = G + j(B_C - B_L) = \sqrt{G^2 + (B_C - B_L)^2} \angle tan^{-1}[(B_C - B_L)/G] \qquad (6.37)$$

By a procedure similar to that of Chapter 2, we can show that admittances combine as conductances do.

Duality is a term meaning that there is a similarity in which some quantities are related to others. The dual quantities we have encountered thus far are listed in Table 6.1.

TABLE 6.1 Dual quantities

Series	Parallel
Voltage	Current
Resistance	Conductance
Thevenin	Norton
Inductance	Capacitance
Reactance	Susceptance
Impedance	Admittance

Example 6.10

Consider the series and parallel networks shown in Figure 6.14. How should their real and imaginary terms be related so that they will be equivalent?

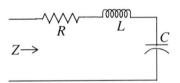 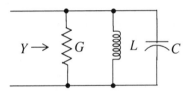

Figure 6.14. Networks for Example 6.10

Solution:

For these circuits to be equivalent, their impedances Z or admittances Y must be equal. Therefore,

$$Y = \frac{1}{Z} = \frac{1}{R+jX} = G+jB = \frac{1}{R+jX} \cdot \frac{R-jX}{R-jX} = \frac{R-jX}{R^2+X^2} = \frac{R}{R^2+X^2} - j\frac{X}{R^2+X^2}$$

and equating reals and imaginaries we get

$$G = \frac{R}{R^2+X^2} \qquad \text{and} \qquad B = \frac{-X}{R^2+X^2} \qquad (6.38)$$

Relation (6.38) is worth memorizing.

Example 6.11

Compute Z and Y for the network of Figure 6.15.

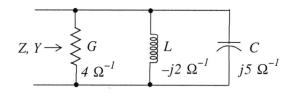

Figure 6.15. Network for Example 6.11

Solution:

Since this is a parallel network, it is easier to compute the admittance Y first. Thus,

$$Y = G + \frac{1}{j\omega L} + j\omega C = 4 - j2 + j5 = 4 + j3 = 5\angle 36.9°$$

Since the impedance is the reciprocal of admittance, it follows that

$$Z = \frac{1}{Y} = \frac{1}{5\angle 36.9°} = 0.2\angle -36.9° = 0.16 - j0.12$$

Example 6.12

Compute Z and Y for the circuit shown below. Verify your answers with MATLAB.

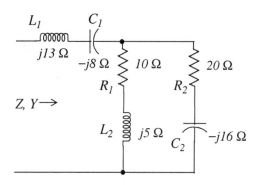

Figure 6.16. Network for Example 6.12

Solution:

Let the given network be represented as shown in Figure 6.17 where $Z_1 = j13 - j8 = j5$, $Z_2 = 10 + j5$, and $Z_3 = 20 - j16$

Then,

$$Z = Z_1 + \frac{Z_2 Z_3}{Z_2 + Z_3} = j5 + \frac{(10 + j5)(20 - j16)}{10 + j5 + 20 - j16} = j5 + \frac{(11.18\angle 26.6°)(25.61\angle -38.7°)}{31.95\angle -20.1°}$$

$$= j5 + 8.96\angle 8° = j5 + 8.87 + j1.25 = 8.87 + j6.25 = 10.85\angle 35.2°$$

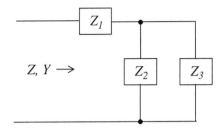

Figure 6.17. Simplified network for Example 6.12

and

$$Y = \frac{1}{Z} = \frac{1}{10.85 \angle 35.2°} = 0.092 \angle -35.2° = 0.0754 - j0.531$$

Check with MATLAB:

z1=j*5; z2=10+j*5; z3=20−j*16; z=z1+(z2*z3/(z2+z3))   % Find impedance z

z =
   8.8737+ 6.2537i

y=1/z   % Find admittance y

y =
   0.0753- 0.0531i

6.7 Summary

- Excitations or driving functions refer to the applied voltages and currents in electric circuits.

- A response is anything we define it as a response. Typically response is the voltage or current in the "load" part of the circuit or any other part of it.

- If the excitation is a constant voltage or current, the response will also be some constant value.

- If the excitation is a sinusoidal voltage or current, in general, the response will also be sinusoidal with the same frequency but with different amplitude and phase.

- If the excitation is a time-varying function such as a sinusoid, inductors and capacitors do not behave like short circuits and open circuits respectively as they do when the excitation is a constant and steady-state conditions are reached. They behave entirely different.

- Circuit analysis in circuits where the excitation is a time-varying quantity such as a sinusoid is very difficult and thus impractical in the $t - domain$.

- The complex excitation function greatly simplifies the procedure of analyzing such circuits when excitation is a time-varying quantity such as a sinusoid.

- The procedure where the excitation-response relation is simplified to amplitude and phase relationship is known as time-domain to frequency-domain transformation.

- The procedure where the excitation-response is put back to its sinusoidal form is known as frequency-domain to time-domain transformation.

- For brevity, we denote the time domain as the $t-domain$, and the frequency domain as the $j\omega-domain$.

- If a sinusoid is given in terms of the sine function, it is convenient to convert it to a cosine function using the identity $m(t) = A\sin(\omega t + \theta) = A\cos(\omega t + \theta - 90°)$ before converting it to the $j\omega-domain$.

- The $t-domain$ to $j\omega-domain$ transformation procedure is as follows:

  1. Express the given sinusoid as a cosine function

  2. Express the cosine function as the real part of the complex excitation using Euler's identity

  3. Extract the magnitude and phase angle from it.

- The $j\omega-domain$ to $t-domain$ transformation procedure is as follows:

  1. Convert the given phasor from polar to exponential form

  2. Add the radian frequency ω multiplied by t to the exponential form

  3. Extract the real part from it.

- The circuit analysis of circuits containing R, L, and C devices, and which are excited by sinusoidal sources, is considerably simplified with the use of phasor voltages and phasor currents which we represent by the boldface capital letters V and I respectively.

- Phasor quantities exist only in the $j\omega-domain$

- In the $j\omega-domain$ the current through a resistor is always in–phase with the voltage across that resistor

- In the $j\omega-domain$ the current through an inductor lags the voltage across that inductor by 90°

- In the $j\omega-domain$ the current through a capacitor leads the voltage across that capacitor by 90°

- In the $j\omega-domain$ the impedance Z is defined as

$$Impedance = Z = \frac{Phasor\ Voltage}{Phasor\ Current} = \frac{V_S}{I} = R + j\omega L + \frac{1}{j\omega C}$$

- Like resistance, the unit of impedance is the Ohm (Ω).

- Impedance is a complex quantity whose real part is the resistance R, and the imaginary part is $\omega L - 1/\omega C$, that is,

$$Re\{Z\} = R \quad \text{and} \quad Im\{Z\} = \omega L - \frac{1}{\omega C}$$

- In polar form the impedance is expressed as

$$Z = \sqrt{R^2 + \left(\omega L - \frac{1}{\omega C}\right)^2} \angle tan^{-1}\left(\omega L - \frac{1}{\omega C}\right)/R$$

- The imaginary part of the impedance Z is called reactance and it is denoted with the letter X. The two components of reactance are the inductive reactance X_L and the capacitive reactance X_C, i.e.,

$$X = X_L - X_C = \omega L - \frac{1}{\omega C}$$

- The unit of the inductive and capacitive reactances is also the Ohm (Ω).

- In the $j\omega - domain$ the admittance Y is defined as

$$Admittance = Y = \frac{Phasor\ Current}{Phasor\ Voltage} = \frac{I_S}{V} = G + \frac{1}{j\omega L} + j\omega C = \frac{1}{Z}$$

- The admittance Y is the reciprocal of the impedance Z as conductance G is the reciprocal of the resistance R.

- The unit of admittance is the siemens or mho (Ω^{-1}).

- The admittance Y is a complex quantity whose real part is the conductance G and the imaginary part is $\omega C - \frac{1}{\omega L}$, that is,

$$Re\{Y\} = G \quad \text{and} \quad Im\{Y\} = \omega C - \frac{1}{\omega L}$$

- The imaginary part of the admittance Y is called susceptance and it is denoted with the letter B. The two components of susceptance are the capacitive susceptance B_C and the inductive suscep-tance B_L, that is,

$$B = B_C - B_L = \omega C - \frac{1}{\omega L}$$

- In polar form the admittance is expressed as

$$Y = \sqrt{G^2 + \left(\omega C - \frac{1}{\omega L}\right)^2} \angle tan^{-1}\left[\left(\omega C - \frac{1}{\omega L}\right)/G\right]$$

- The unit of the susceptances B_C and B_L is also the siemens (Ω^{-1}).

- Admittances combine as conductances do.

- In phasor circuit analysis, conductance is not necessarily the reciprocal of resistance, and susceptance is not the negative reciprocal of reactance. Whenever we deal with resistance and reactance we must think of devices in series, and when we deal with conductance and susceptance we must think of devices in parallel. However, the admittance is always the reciprocal of the impedance

- The ratio V/I of the phasor voltage to the phasor current exists only in the $j\omega - domain$ and it is not the ratio $v(t)/i(t)$ in the $t - domain$. Although the ratio $v(t)/i(t)$ could yield some value, this value is not impedance. Similarly, the ratio $i(t)/v(t)$ is not admittance.

- Duality is a term meaning that there is a similarity in which some quantities are related to others.

6.8 Exercises

Multiple Choice

1. Phasor voltages and phasor currents can be used in the $t-domain$ if a circuit contains

    A. independent and dependent sources with resistors only

    B. independent and dependent sources with resistors and inductors only

    C. independent and dependent sources with resistors and capacitors only

    D. independent and dependent sources with resistors, inductors, and capacitors

    E. none of the above

2. If the excitation in a circuit is a single sinusoidal source with amplitude A, radian frequency ω, and phase angle θ, and the circuit contains resistors, inductors, and capacitors, all voltages and all currents in that circuit will be of the same

    A. amplitude A but different radian frequency ω and different phase angle θ

    B. radian frequency ω but different amplitude A and different phase angle θ

    C. phase angle θ but different amplitude A and different radian frequency ω

    D. amplitude A same radian frequency ω and same phase angle θ

    E. none of the above

3. The sinusoid $v(t) = 120\sin(\omega t + 90°)$ in the $j\omega-domain$ is expressed as

    A. $V = 120e^{j(\omega t + 90°)}$

    B. $V = 120e^{j\omega t}$

    C. $V = 120e^{j90°}$

    D. $V = 120e^{j0°}$

    E. none of the above

4. A series RLC circuit contains two voltage sources with values $v_1(t) = 100\cos(10t + 45°)$ and $v_2(t) = 200\sin(5t-60°)$. We can transform this circuit to a phasor equivalent to find the current by first replacing these with a single voltage source $v(t) = v_1(t) + v_2(t)$ whose value is

    A. $v(t) = 300\cos(15t-15°)$

    B. $v(t) = 100\cos(5t + 105°)$

C. $v(t) = 150\cos(7.5t-15°)$

D. $v(t) = 150\cos(7.5t + 15°)$

E. none of the above

5. The equivalent impedance Z_{eq} of the network of Figure 6.18 is

A. $1 + j1$

B. $1 - j1$

C. $-j1$

D. $2 + j0$

E. none of the above

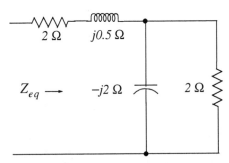

Figure 6.18. Network for Questions 5 and 6

6. The equivalent admittance Y_{eq} of the network of Figure 6.18 is

A. $4-j1.5$

B. $\dfrac{16}{73} + j\dfrac{6}{73}$

C. $\dfrac{12}{37} + j\dfrac{2}{37}$

D. $2 - j2$

E. none of the above

7. The resistance of a coil is $R = 1.5\ \Omega$ and the inductance of that coil is $L = 5.3\ mH$. If a current of $i(t) = 4\cos\omega t\ A$ flows through that coil and operates at the frequency of $f = 60Hz$, the phasor voltage V across that coil is

A. $10\angle 53.1°\ V$

B. $6\angle 0° \ V$

C. $5.3 \times 10^{-3}\angle 90° \ V$

D. $6.8\angle 45° \ V$

E. none of the above

8. A resistor with value $R = 5 \ \Omega$ is in series with a capacitor whose capacitive reactance at some particular frequency ω is $-jX_C = -5 \ \Omega$. A phasor current with value $I = 8\angle 0° \ A$ is flowing through this series combination. The $t-domain$ voltage across this series combination is

A. $80\cos\omega t$

B. $80\sin\omega t$

C. $56.6\cos(\omega t - 45°)$

D. $56.6\cos(\omega t + 45°)$

E. none of the above

9. A conductance with value $G = 0.3 \ \Omega^{-1}$ is in parallel with a capacitor whose capacitive suscep-tance at some particular frequency ω is $jB_C = j0.3 \ \Omega^{-1}$. A phasor voltage with value $V = 10\angle 0°$ is applied across this parallel combination. The $t-domain$ total current through this parallel combination is

A. $3\cos\omega t + j3\sin\omega t$

B. $3\cos\omega t - j3\sin\omega t$

C. $5\sin(\omega t + 53.2°)$

D. $5\cos(\omega t + 53.2°)$

E. none of the above

10. If the phasor $I = je^{j(\pi/2)}$, then in the $t-domain$ $i(t)$ is

A. $\cos(\omega t + \pi/2)$

B. $\sin(\omega t + \pi/2)$

C. $-\cos\omega t$

D. $-\sin\omega t$

E. none of the above

Problems

1. Express the sinusoidal voltage waveform of Figure 6.18 as $v(t) = A\cos(\omega t + \theta)$, that is, find A, ω, and θ. Answer: $v(t) = 2\cos(1000t + 36.1°)$

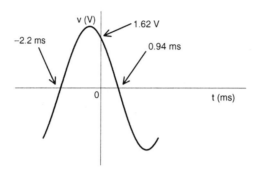

Figure 6.19. Circuit for Problem 1

2. The current $i(t)$ through a device decays exponentially as shown by the waveform of Figure 6.20, and two values are known as indicated. Compute $i(1)$, that is, the current at $t = 1\ ms$

Answers: $i(t) = 50e^{-750t}\ mA$, $23.62\ mA$

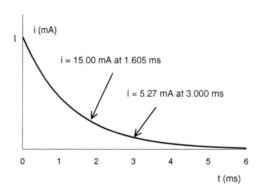

Figure 6.20. Circuit for Problem 2

3. At what frequency f is the network of Figure 6.21 operating if it is known that $v_S = 120\cos\omega t\ V$ and $i = 12\cos(\omega t - 36.9°)\ A$? Answer: $f = 5.533\ KHz$

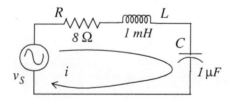

Figure 6.21. Circuit for Problem 3

4. In the circuit of Figure 6.22, $v_S = V\cos(2000t + \theta)$ V and the symbols V and A inside the circles denote an AC *voltmeter*[*] and *ammeter* respectively. Assume that the ammeter has negligible internal resistance. The variable capacitor C is adjusted until the voltmeter reads 25 V and the ammeter 5 A. Find the value of the capacitor. Answer: $C = 89.6\ \mu F$

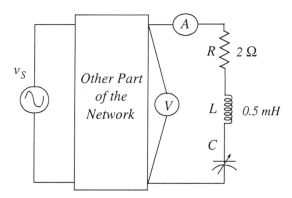

Figure 6.22. Network for Problem 4

5. For the circuit shown on Figure 6.23, is it possible to adjust the variable resistor R_1 and the variable capacitor C so that Z_{IN} and Y_{IN} have the same numerical value regardless of the operating frequency? If so, what are these values? Answer: Yes, if $C = 1$ F and $R_1 = 1\ \Omega$

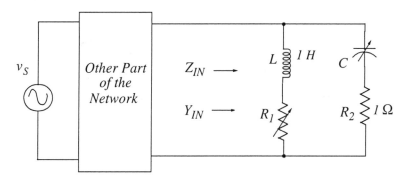

Figure 6.23. Network for Problem 5

---

[*] *Voltmeters and Ammeters are discussed in Chapter 8. For this exercise, it will suffice to say that these instruments indicate the magnitude (absolute) values of voltage and current.*

6.9 Answers to Exercises

Multiple Choice

1.E Phasors exist in the $j\omega - domain$ only

2. B

3. D

4. E The voltage sources $v_1(t)$ and $v_1(t)$ operate at different frequencies. Therefore, to find the current we must apply superposition.

5. E $3 - j0.5$ This value is obtained with the MATLAB code z1=2+0.5j; z2=2*(−2j)/(2−2j); z=z1+z2

   z = 3.0000-0.5000i

6. C

7. A $\omega = 2\pi f = 2\pi \times 60 = 377\ r/s$, $jX_L = j\omega L = j \times 377 \times 5.3 \times 10^{-3} = j2\ \Omega$

   $Z = 1.5 + j2 = 2.5\angle53.13°$, $V = ZI = 2.5\angle53.13° \times 4\angle0° = 10\angle53.13°$

8. C

9. D

10. C

Problems

1. The $t - axis$ crossings define half of the period T. Thus, $T/2 = 2.2 + 0.94 = 3.14\ ms$, and one period is $T = 6.28\ ms$. The frequency is $f = 1/T = 10^3/6.28 = 10^3/2\pi$. Then, $\omega = 2\pi f$ or $\omega = 2\pi \times 10^3/2\pi = 1000\ r/s$

   Next, we find the phase angle θ from the figure above observing that $\pi/2 + \theta = 2.2\ ms$

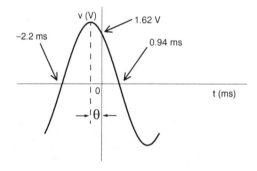

   or

$$\theta = 2.2ms - \frac{\pi}{2} = 2.2 \times 10^{-3} \; s \times \frac{2\pi \; rad}{6.28 \times 10^{-3} \; s} \times \frac{180°}{\pi \; rad} - \frac{\pi}{2}$$

$$= \frac{2.2 \times 2 \times 180°}{6.28} - \frac{\pi}{2} = 126.1° - 90° = 36.1°$$

Finally, we find the amplitude A by observing that at $t = 0$, $v = 1.62 \; V$, that is,

$$v(0) = 1.62 = A\cos(0 + 36.1°)$$

or

$$A = \frac{1.62}{\cos 36.1°} = 2 \; V$$

Therefore,

$$v(t) = 2\cos(1000t + 36.1°)$$

2. The decaying exponential has the form $i(t) = Ae^{-\alpha t} \; mA$ where the time is in ms and thus for this problem we need to compute the values of A and α using the given values. Then,

$$i\big|_{t = 1.605 \; ms} = 15 \; mA = Ae^{-(1.605 \times 10^{-3})\alpha}$$

and

$$i\big|_{t = 3.000 \; ms} = 5.27 \; mA = Ae^{-(3.000 \times 10^{-3})\alpha}$$

Division of the first equation by the second yields

$$\frac{Ae^{-(1.605 \times 10^{-3})\alpha}}{Ae^{-(3.000 \times 10^{-3})\alpha}} = \frac{15 \; mA}{5.27 \; mA}$$

or

$$e^{-(1.605 \times 10^{-3})\alpha + (3.000 \times 10^{-3})\alpha} = \frac{15}{5.27}$$

or

$$e^{1.395 \times 10^{-3}\alpha} = \frac{15}{5.27}$$

or

$$ln\left(\frac{15}{5.27}\right) = 1.395 \times 10^{-3}\alpha$$

or

$$\alpha = \frac{ln(15/5.27) \times 10^3}{1.395} = 750$$

and thus

$$i(t) = Ae^{-750t} \; mA$$

To find the value of A we make use of the fact that $i|_{t\,=\,3\,ms} = 5.27\ mA$. Then,

$$5.27 = Ae^{-750 \times 3 \times 10^{-3}}$$

or

$$A = \frac{5.27 \times 10^{-3}}{e^{-2.25}}$$

or

$$A = 0.050\ A = 50mA$$

Therefore,

$$i(t) = 50e^{-750t}\ mA$$

and

$$i|_{t\,=\,1\,ms} = 50e^{-750 \times 10^{-3}} = 23.62\ mA$$

3. The equivalent phasor circuit is shown below.

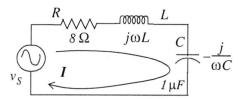

In the $j\omega - domain$ $V_S = 120\angle 0°\ V$, $I = 12\angle{-36.9°}\ A$, $j\omega L = j10^{-3}\omega$, and $-j/\omega C = -j10^{6}/\omega$
Then,

$$Z = \frac{V_S}{I} = \frac{120\angle 0°\ V}{12\angle{-36.9°}\ A} = 10\angle 36.9°$$

and

$$|Z| = 10 = \sqrt{R^2 + (\omega L - 1/\omega C)^2}$$

or

$$R^2 + (\omega L - 1/\omega C)^2 = 100$$

or

$$8^2 + (\omega L - 1/\omega C)^2 = 100$$

or

$$(\omega L - 1/\omega C)^2 = 36$$

or

$$\omega L - 1/\omega C = 6$$

or

$$\omega^2 - \frac{6}{L}\omega - \frac{1}{LC} = 0$$

or

$$\omega^2 - 6 \times 10^3 \omega - 10^9 = 0$$

Solving for ω and ignoring the negative value, we get

$$\omega = \frac{6 \times 10^3 + \sqrt{36 \times 10^6 + 4 \times 10^9}}{2} = 34,765 \ r/s$$

and

$$f = \frac{\omega}{2\pi} = \frac{34,765 \ r/s}{2\pi} = 5,533 \ Hz = 5.533 \ KHz$$

Check: $j\omega L = j34.765$, $-j/\omega C = -j28.765$

$$Z = R + j(\omega L - 1/(\omega C)) = 8 + j(34.765 - 28.765) = 8 + j6 = 10\angle 36.9°$$

and

$$I = \frac{120\angle 0°}{10\angle 36.9°} = 12\angle -36.9°$$

4. Since the instruments read absolute values, we are only need to be concerned the magnitudes of the phasor voltage, phasor current, and impedance. Thus,

$$|V| = |Z||I| = 25 = \sqrt{R^2 + (\omega L - 1/\omega C)^2} \times 5$$

or

$$|V|^2 = 25^2 = [R^2 + (\omega L - 1/\omega C)^2] \times 25 = \left[4 + \left(1 - \frac{5 \times 10^{-4}}{C} \right)^2 \right] \times 25$$

$$= 100 + 25 - \frac{250 \times 10^{-4}}{C} + \frac{625 \times 10^{-8}}{C^2} = 625$$

and after simplification we get

$$500C^2 + 250 \times 10^{-4} C - 625 \times 10^{-8} = 0$$

Using MATLAB, we get

p=[500 250*10^(−4) −625*10^(−8)]; r=roots(p)

and this yields $C = 89.6 \ \mu F$

The second root of this polynomial is negative and thus it is discarded.

5. We group the series devices as shown below.

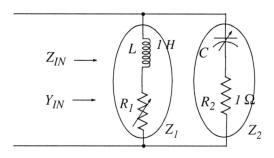

Thus $Z_1 = R_1 + j\omega$, $Z_2 = 1 - j/(\omega C)$, and

$$Z_{IN} = \frac{Z_1 \cdot Z_2}{Z_1 + Z_2} = \frac{(R_1 + j\omega)(1 - j/\omega C)}{R_1 + j\omega + 1 - j/\omega C}$$

and at any frequency ω

$$Y_{IN} = \frac{1}{Z_{IN}} = \frac{R_1 + j\omega + 1 - j/\omega C}{(R_1 + j\omega)(1 - j/\omega C)}$$

Therefore, if the condition $Y_{IN} = Z_{IN}$ is to hold for all frequencies, the right sides of Z_{IN} and Y_{IN} must be equal, that is,

$$\frac{(R_1 + j\omega)(1 - j/\omega C)}{R_1 + j\omega + 1 - j/\omega C} = \frac{R_1 + j\omega + 1 - j/\omega C}{(R_1 + j\omega)(1 - j/\omega C)}$$

$$[(R_1 + j\omega)(1 - j/\omega C)]^2 = [R_1 + j\omega + 1 - j/\omega C]^2$$

$$(R_1 + j\omega)(1 - j/\omega C) = R_1 + j\omega + 1 - j/\omega C$$

$$R_1 - j\frac{R_1}{\omega C} + j\omega + \frac{1}{C} = R_1 + 1 + j\left(\omega - \frac{1}{\omega C}\right)$$

$$\left(R_1 + \frac{1}{C}\right) + j\left(\omega - \frac{R_1}{\omega C}\right) + j\omega = (R_1 + 1) + j\left(\omega - \frac{1}{\omega C}\right)$$

Equating reals and imaginaries we get

$$R_1 + \frac{1}{C} = R_1 + 1 \qquad \omega - \frac{R_1}{\omega C} = \omega - \frac{1}{\omega C}$$

From the first equation above we get $C = 1 \ F$ and by substitution of this value into the second equation we get $R_1 = 1 \ \Omega$

Chapter 7

This chapter begins with the application of nodal analysis, mesh analysis, superposition, and Thevenin's and Norton's theorems in phasor circuits. Then, phasor diagrams are introduced, and the input-output relationships for an RC low-pass filter and an RC high-pass filter are developed.

7.1 Nodal Analysis

The procedure of analyzing a phasor[*] circuit is the same as in Chapter 3, except that in this chapter we will be using phasor quantities. The following example illustrates the procedure.

Example 7.1

Use nodal analysis to compute the phasor voltage $V_{AB} = V_A - V_B$ for the circuit of Figure 7.1.

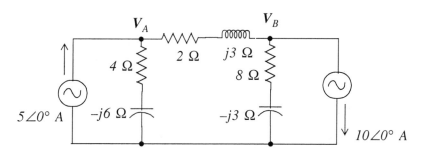

Figure 7.1. Circuit for Example 7.1

Solution:

As before, we choose a reference node as shown in Figure 7.2, and we write nodal equations at the other two nodes A and B. Also, for convenience, we designate the devices in series as Z_1, Z_2, and Z_3 as shown, and then we write the nodal equations in terms of these impedances.

$$Z_1 = 4 - j6 = 7.211\angle{-56.3°}$$

$$Z_2 = 2 + j3 = 3.606\angle{56.3°}$$

$$Z_3 = 8 - j3 = 8.544\angle{-20.6°}$$

---

[*] *A phasor is a rotating vector*

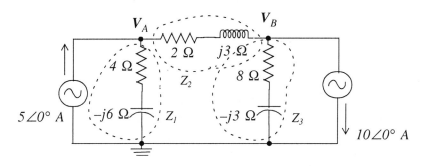

Figure 7.2. Nodal analysis for the circuit for Example 7.1

By application of KCL at V_A,

$$\frac{V_A}{Z_1} + \frac{V_A - V_B}{Z_2} = 5\angle 0°$$

or

$$\left(\frac{1}{Z_1} + \frac{1}{Z_2}\right)V_A - \frac{1}{Z_2}V_B = 5\angle 0° \qquad (7.1)$$

or

$$\left(\frac{Z_1 + Z_2}{Z_1 Z_2}\right)V_A - \frac{1}{Z_2}V_B = 5\angle 0°$$

and by substitution for Z_1 and Z_2 we get

$$\frac{4 - j6 + 2 + j3}{(7.211\angle -56.3)(3.606\angle 56.3°)}V_A - \frac{1}{3.606\angle 56.3°}V_B = 5\angle 0°$$

$$\frac{6 - j3}{26.0\angle 0°}V_A - (0.277\angle -56.3°)V_B = 5\angle 0°$$

$$\frac{6.708\angle -26.6°}{26\angle 0°}V_A - (0.277\angle -56.3°)V_B = 5\angle 0°$$

$$(0.258\angle -26.6°)V_A - (0.277\angle -56.3°)V_B = 5\angle 0° \qquad (7.2)$$

Next, at V_B:

$$\frac{V_B - V_A}{Z_2} + \frac{V_B}{Z_3} = -10\angle 0°$$

or

$$-\frac{1}{Z_2}V_A + \left(\frac{1}{Z_2} + \frac{1}{Z_3}\right)V_B = -10\angle 0° \qquad (7.3)$$

In matrix form (7.1) and (7.3) are

$$\begin{bmatrix} \left(\dfrac{1}{Z_1} + \dfrac{1}{Z_2}\right) & -\dfrac{1}{Z_2} \\ -\dfrac{1}{Z_2} & \left(\dfrac{1}{Z_2} + \dfrac{1}{Z_3}\right) \end{bmatrix} \begin{bmatrix} V_A \\ V_B \end{bmatrix} = \begin{bmatrix} 5 \\ -10 \end{bmatrix} \qquad (7.4)$$

We will follow a step-by-step procedure to solve these equations using Cramer's rule, and we will use MATLAB®[*] to verify the results.

We rewrite (7.3) as

$$-\frac{1}{Z_2}V_A + \left(\frac{Z_2 + Z_3}{Z_2 Z_3}\right)V_B = 10\angle 180°$$

$$-\frac{1}{3.606\angle 56.3°}V_A + \frac{2 + j3 + 8 - j3}{(3.606\angle 56.3°)(8.544\angle -20.6°)}V_B = 10\angle 180°$$

$$-(0.277\angle -56.3°)V_A + \frac{10}{30.810\angle 35.7°}V_B = 10\angle 180°$$

$$-(0.277\angle -56.3°)V_A + (0.325\angle -35.7°)V_B = 10\angle 180° \qquad (7.5)$$

and thus with (7.2) and (7.5) the system of equations is

$$\begin{aligned} (0.258\angle -26.6°)V_A - (0.277\angle -56.3°)V_B &= 5\angle 0° \\ -(0.277\angle -56.3°)V_A + (0.325\angle -35.7°)V_B &= 10\angle 180° \end{aligned} \qquad (7.6)$$

We find V_A and V_B from

$$V_A = \frac{D_1}{\Delta} \qquad (7.7)$$

and

$$V_B = \frac{D_2}{\Delta} \qquad (7.8)$$

The determinant Δ is

---

* *If unfamiliar with MATLAB, please refer to Appendix A*

$$\Delta = \begin{vmatrix} (0.258\angle{-26.6°}) & -(0.277\angle{-56.3°}) \\ -(0.277\angle{-56.3°}) & (0.325\angle{-35.7°}) \end{vmatrix}$$

$$= (0.258\angle{-26.6°}) \cdot (0.325\angle{-35.7°}) - (0.277\angle{-56.3°}) \cdot (-0.277\angle{-56.3})$$

$$= (0.084\angle{-62.3°}) - (0.077\angle{-112.6}) - (0.039 - j0.074) - (-0.023 - j0.071)$$

$$= (0.062 - j0.003 = 0.062\angle{-2.8°})$$

Also,

$$D_1 = \begin{vmatrix} 5\angle{0°} & -(0.277\angle{-56.3°}) \\ 10\angle{180°} & (0.325\angle{-35.7°}) \end{vmatrix}$$

$$= (5\angle{0°})(0.325\angle{-35.7°}) - (10\angle{180°})[-(0.277\angle{-56.3°})]$$

$$= (1.625\angle{-35.7°} + 2.770\angle{123.7°}) = 1.320 - j0.948 + (-1.537 + j2.305)$$

$$= -0.217 + j1.357 = 1.374\angle{99.1°}$$

and

$$D_2 = \begin{vmatrix} (0.258\angle{-26.6°}) & 5\angle{0°} \\ -(0.277\angle{-56.3°}) & 10\angle{180°} \end{vmatrix}$$

$$= (0.258\angle{-26.6°})(10\angle{180°}) - (-0.277\angle{-56.3°})(5\angle{0°})$$

$$= 2.580\angle{153.4°} + 1.385\angle{-56.3°} = (-2.307 + j1.155 + 0.769 - j1.152)$$

$$= -1.358 + j0.003 = 1.358\angle{179.9°}$$

Therefore, by substitution into (7.7) and (7.8), we get

$$V_A = \frac{D_1}{\Delta} = \frac{1.374\angle{99.1°}}{0.062\angle{-2.8°}} = 22.161\angle{101.9°} = -4.570 + j21.685$$

and

$$V_B = \frac{D_2}{\Delta} = \frac{1.358\angle{179.9°}}{0.062\angle{-2.8°}} = 24.807\angle{-177.3°} = -24.780 - j1.169$$

Finally,

$$V_{AB} = V_A - V_B = -4.570 + j21.685 - (-24.780 - j1.169)$$

$$= 20.21 + j22.85 = 30.5\angle{48.5°}$$

Check with MATLAB:

```
z1=4–j*6; z2=2+j*3; z3=8–j*3;          % Define z1, z2 and z3
Z=[1/z1+1/z2 –1/z2; –1/z2 1/z2+1/z3];  % Elements of matrix Z
I=[5 –10]';                             % Column vector I
V=Z\I; Va=V(1,1); Vb=V(2,1); Vab=Va–Vb; % Va = V(1), Vb = V(2) are also acceptable
% With fprintf only the real part of each parameter is processed so we will use disp
fprintf(' \n'); disp('Va = '); disp(Va); disp('Vb = '); disp(Vb); disp('Vab = '); disp(Vab);
fprintf(' \n');
```

```
Va =  -4.1379 + 19.6552i

Vb = -22.4138 - 1.0345i

Vab = 18.2759 + 20.6897i
```

These values differ by about 10% from the values we obtained with Cramer's rule where we rounded the values to three decimal places. MATLAB performs calculations with accuracy of 15 decimal places, although it only displays four decimal places in the short (default) number display format. Accordingly, we should accept the MATLAB values as more accurate.

7.2 Mesh Analysis

Again, the procedure of analyzing a phasor circuit is the same as in Chapter 3 except that in this chapter we will be using phasor quantities. The following example illustrates the procedure.

Example 7.2

For the circuit of Figure 7.3, use mesh analysis to find the voltage V_{10A}, that is, the voltage across the $10\angle 0°$ current source.

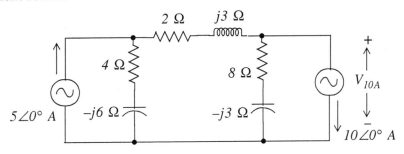

Figure 7.3. Circuit for Example 7.2

Solution:

As in the previous example, for convenience, we denote the passive devices in series as $Z_1, Z_2,$ and Z_3, and we write mesh equations in terms of these impedances. The circuit then is as shown in Figure 7.4 with the mesh currents assigned in a clockwise direction.

We observe that the voltage across the $10\angle 0°$ current source is the same as the voltage across the $8\ \Omega$ and $-j3\ \Omega$ series combination.

By inspection, for Mesh 1,

$$I_1 = 5\angle 0° \tag{7.9}$$

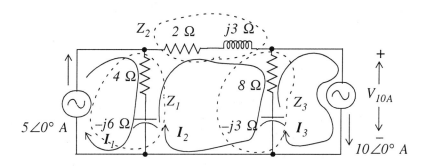

Figure 7.4. Mesh analysis for the circuit of Example 7.2

By application of KVL around Mesh 2,

$$-Z_1I_1 + (Z_1 + Z_2 + Z_3)I_2 - Z_3I_3 = 0$$

or

$$-(4-j6)I_1 + (14-j6)I_2 - (8-j3)I_3 = 0 \tag{7.10}$$

Also, by inspection for Mesh 3,

$$I_3 = 10\angle 0° \tag{7.11}$$

and in matrix form, (7.9), (7.10), and (7.11) are written as

$$\begin{bmatrix} 1 & 0 & 0 \\ -(4-j6) & (14-j6) & -(8-j3) \\ 0 & 0 & 1 \end{bmatrix} \begin{bmatrix} I_1 \\ I_2 \\ I_3 \end{bmatrix} = \begin{bmatrix} 5 \\ 0 \\ 10 \end{bmatrix} \tag{7.12}$$

We use MATLAB for the solution of 7.12.[*]

```
Z=[1 0 0; –(4–j*6) 14–j*6 –(8–j*3); 0 0 1];
V=[5 0 10]';
I=Z\V; i1=I(1); i2=I(2); i3=I(3); fprintf(' \n');
disp('i1 = '); disp(i1); disp('i2 = '); disp(i2); disp('i3 = '); disp(i3); fprintf(' \n');

i1 = 5     i2 = 7.5862 - 1.0345i     i3 = 10
```

Therefore, the voltage across the $10\angle 0°$ A current source is

$$V_{10A} = Z_3(I_2 - I_3) = (8-j3)(7.586 - j1.035 - 10) = -22.417 - j1.038$$

We observe that this is the same value as that of the voltage V_B in the previous example.

---

[*] *As we experienced with Example 7.1, the computation of phasor voltages and currents becomes quite tedious. Accordingly, in our subsequent discussion we will use MATLAB for the solution of simultaneous equations with complex coefficients.*

7.3 Application of Superposition Principle

As we know from Chapter 3, the superposition principle is most useful when a circuit contains two or more independent voltage or current sources. The following example illustrates the application of the superposition principle in phasor circuits.

Example 7.3

Use the superposition principle to find the phasor voltage across capacitor C_2 in the circuit of Figure 7.5.

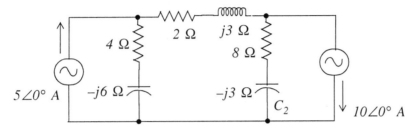

Figure 7.5. Circuit for Example 7.3

Solution:

Let the phasor voltage across C_2 due to the $5\angle 0°\ A$ current source acting alone be denoted as V'_{C2}, and that due to the $10\angle 0°\ A$ current source as V''_{C2}. Then,

$$V_{C2} = V'_{C2} + V''_{C2}$$

With the $5\angle 0°\ A$ current source acting alone, the circuit reduces to that shown in Figure 7.6.

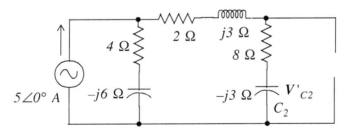

Figure 7.6. Circuit for Example 7.3 with the $5\angle 0°\ A$ current source acting alone

By application of the current division expression, the current I'_{C2} through C_2 is

$$I'_{C2} = \frac{4-j6}{4-j6+2+j3+8-j3}5\angle 0° = \frac{7.211\angle -56.3°}{15.232\angle -23.2°}5\angle 0° = 2.367\angle -33.1°$$

The voltage across C_2 with the $5\angle 0°$ current source acting alone is

$$V'_{C2} = (-j3)(2.367\angle-33.1°) = (3\angle-90°)(2.367\angle-33.1°)$$
$$= 7.102\angle-123.1° = -3.878 - j5.949$$

(7.13)

Next, with the $10\angle0°$ A current source acting alone, the circuit reduces to that shown in Figure 7.7.

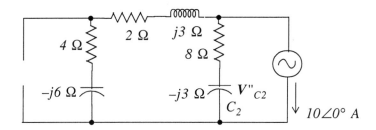

Figure 7.7. Circuit for Example 7.3 with the $10\angle0°$ A current source acting alone

and by application of the current division expression, the current I''_{C2} through C_2 is

$$I''_{C2} = \frac{4 - j6 + 2 + j3}{4 - j6 + 2 + j3 + 8 - j3}(-10\angle0°)$$
$$= \frac{6.708\angle-26.6°}{15.232\angle-23.2°}10\angle180° = 4.404\angle176.6°$$

The voltage across C_2 with the $10\angle0°$ current source acting alone is

$$V''_{C2} = (-j3)(4.404\angle176.6°) = (3\angle-90°)(4.404\angle176.6°)$$
$$= (13.213\angle86.6 = 0.784 + j13.189)$$

(7.14)

Addition of (7.13) with (7.14) yields

$$V_{C2} = V'_{C2} + V''_{C2} = -3.878 - j5.949 + 0.784 + j13.189$$

or

$$V_{C2} = -3.094 + j7.240 = 7.873\angle113.1°$$

(7.15)

7.4 Thevenin's and Norton's Theorems

These two theorems also offer a very convenient method in analyzing phasor circuits as illustrated by the following example.

Example 7.4

For the circuit of Figure 7.8, apply Thevenin's theorem to compute I_X and then draw Norton's equivalent circuit.

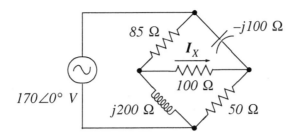

Figure 7.8. Circuit for Example 7.4

Solution:

With the *100* Ω resistor disconnected, the circuit reduces to that shown in Figure 7.9.

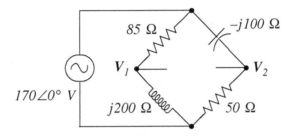

Figure 7.9. Circuit for Example 7.4 with the 100 Ω resistor disconnected

By application of the voltage division expression,

$$V_1 = \frac{j200}{85 + j200} 170\angle 0° = \frac{200\angle 90°}{217.31\angle 67°} 170\angle 0° = 156.46\angle 23° = 144 + j61.13 \qquad (7.16)$$

and

$$V_2 = \frac{50}{50 - j100} 170\angle 0° = \frac{50}{111.8\angle(-63.4)°} 170\angle 0° = 76\angle 63.4° = 34 + j68 \qquad (7.17)$$

Then, from (7.16) and (7.17),

$$V_{TH} = V_{OC} = V_{12} = V_1 - V_2 = 144 + j61.13 - (34 + j68)$$

or

$$V_{TH} = 110 - j6.87 = 110.21\angle -3.6° \qquad (7.18)$$

Next, we find the Thevenin equivalent impedance Z_{TH} by shorting the *170*∠0° *V* voltage source. The circuit then reduces to that shown in Figure 7.10.

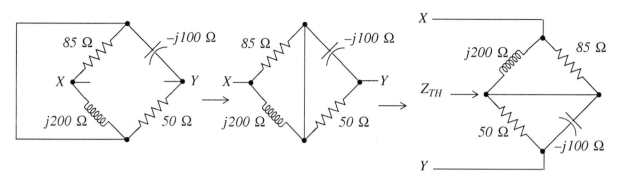

Figure 7.10. Circuit for Example 7.4 with the voltage source shorted

We observe that the parallel combinations $j200 \parallel 85$ and $50 \parallel j100$ are in series as shown in Figure 7.11.

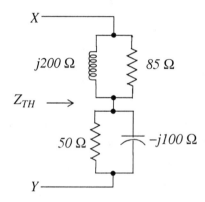

Figure 7.11. Network for the computation of Z_{TH} for Example 7.4

From Figure 7.11,

$$Z_{TH} = \frac{85 \times j200}{85 + j200} + \frac{50 \times (-j100)}{50 - j100}$$

and with MATLAB,

Zth=85*200j/(85+200j) + 50*(−100j)/(50−100j)

```
Zth =
 1.1200e+002 + 1.0598e+001i
```

or

$$Z_{TH} = 112.0 + j10.6 = 112.5 \angle 5.4° \; \Omega$$

The Thevenin equivalent circuit is shown in Figure 7.12.

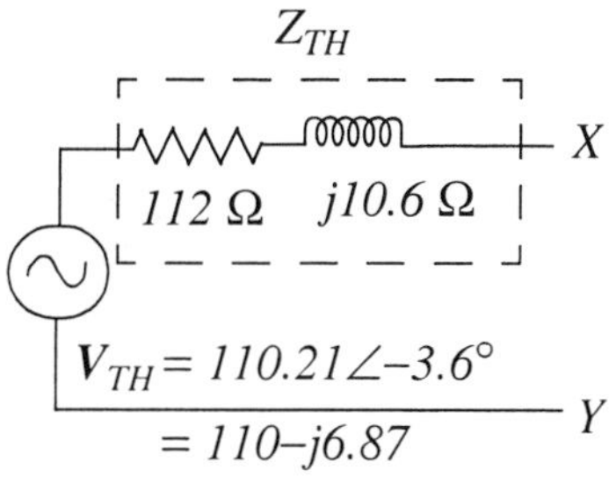

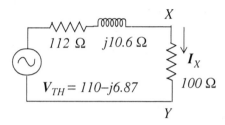

Figure 7.12. Thevenin equivalent circuit for Example 7.4

With the *100 Ω* resistor connected at X-Y, the circuit becomes as shown in Figure 7.13.

Figure 7.13. Simplified circuit for computation of I_X in Example 7.4

We find I_X using MATLAB:

```
Vth=110–6.87j;  Zth=112+10.6j; Ix=Vth/(Zth+100);
fprintf(' \n'); disp('Ix = '); disp(Ix); fprintf(' \n');

Ix = 0.5160 - 0.0582i
```

that is,

$$I_X = \frac{V_{TH}}{Z_{TH} + 100\ \Omega} = 0.516 - j0.058 = 0.519\angle{-6.4°}\ A \qquad (7.19)$$

The same answer is found in Example C.18 of Appendix *C* where we applied nodal analysis to find I_X.

Norton's equivalent is obtained from Thevenin's circuit by exchanging V_{TH} and its series Z_{TH} with I_N in parallel with Z_N as shown in Figure 7.14. Thus,

$$I_N = \frac{V_{TH}}{Z_{TH}} = \frac{110.21\angle{-3.6°}}{112.5\angle{5.4°}} = 0.98\angle{-9°}\ A$$

and

$$Z_N = Z_{TH} = 112.5\angle{5.4°}\ \Omega$$

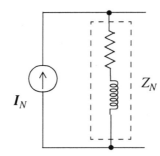

Figure 7.14. Norton equivalent circuit for Example 7.4

7.5 Phasor Analysis in Amplifier Circuits

Other circuits such as those who contain op amps and op amp equivalent circuits can be analyzed using any of the above methods.

Example 7.5

Compute $i_X(t)$ for the circuit below where $v_{in}(t) = 2\cos(30000\omega t)\ V$.

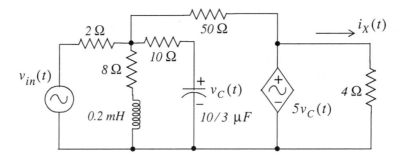

Figure 7.15. Circuit for Example 7.5

Solution:

As a first step, we perform the $t-domain$, to $j\omega-domain$ transformation. Thus,

$$jX_L = j\omega L = j0.2 \times 10^{-3} \times 30 \times 10^3 = j6$$

and

$$-jX_C = -j\frac{1}{\omega C} = -j\frac{1}{30 \times 10^3 \times \dfrac{10}{3} \times 10^{-6}} = -j10$$

Also,

$$V_{IN} = 2\angle 0°$$

and the phasor circuit is shown in Figure 5.16.

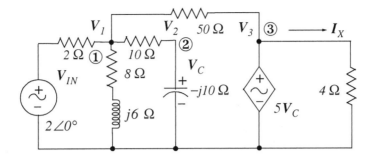

Figure 7.16. Phasor circuit for Example 7.5

At Node ①:

$$\frac{V_1 - 2\angle 0°}{2} + \frac{V_1}{8 + j6} + \frac{V_1 - V_C}{10} + \frac{V_1 - 5V_C}{50} = 0 \qquad (7.20)$$

and since

$$\frac{1}{8 + j6} = \frac{1}{8 + j6} \cdot \frac{8 - j6}{8 - j6} = \frac{8 - j6}{100} = \frac{4}{50} - j\frac{3}{50}$$

the nodal equation of (7.20) simplifies to

$$\left(\frac{35}{50} - j\frac{3}{50}\right) V_1 - \frac{1}{5} V_C = 1\angle 0° \qquad (7.21)$$

At Node ②:

$$\frac{V_C - V_1}{10} + \frac{V_C}{-j10} = 0$$

or

$$-\frac{1}{10} V_1 + \left(\frac{1}{10} + j\frac{1}{10}\right) V_C = 0 \qquad (7.22)$$

At Node ③:

$$V_3 = 5V_C$$

We use MATLAB to solve (7.21) and (7.22).

```
G=[35/50 −j*3/50; −1/5 1/10+j*1/10]; I=[1 0]'; V=G\I;
Ix=5*V(2,1)/4;                         % Multiply Vc by 5 and divide by 4 to get current Ix
magIx=abs(Ix); theta=angle(Ix)*180/pi;    % Convert current Ix to polar form
fprintf(' \n'); disp(' Ix = ' ); disp(Ix);...
fprintf('magIx = %4.2f A \t', magIx); fprintf('theta = %4.2f deg \t', theta);...
fprintf(' \n'); fprintf(' \n');

 Ix = 2.1176 - 1.7546i    magIx = 2.75 A    theta = -39.64 deg
```

Therefore,

$$I = 2.75\angle{-39.6°} \iff i(t) = 2.75\cos(30000t - 39.6°)$$

Example 7.6

Compute the phasor V_{out} for the op amp circuit of Figure 7.17.

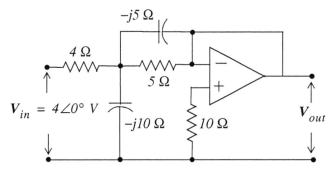

Figure 7.17. Circuit for Example 7.6

Solution:

We assign phasor voltages V_1 and V^+ as shown in Figure 7.18, and we apply KCL at these nodes, while observing that $V_{out} = V^+$

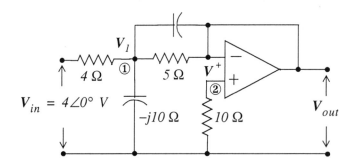

Figure 7.18. Application of KCL for the circuit of Example 7.6

At Node ①:

$$\frac{V_1 - 4\angle{0°}}{4} + \frac{V_1 - V_{out}}{-j5} + \frac{V_1 - V_{out}}{5} + \frac{V_1}{-j10} = 0$$

or

$$\left(\frac{9}{20} + j\frac{3}{10}\right)V_1 - \left(\frac{1}{5} + j\frac{1}{5}\right)V_{out} = 1\angle{0°} \tag{7.23}$$

At Node ②,

$$V_2 = V^+ = V_{out}$$

and thus,

$$\frac{V_{out}}{10} + \frac{V_{out} - V_1}{5} + \frac{V_{out} - V_1}{-j5} = 0$$

or

$$-\left(\frac{1}{5} + j\frac{1}{5}\right)V_1 + \left(\frac{3}{10} + j\frac{1}{5}\right)V_{out} = 0 \qquad (7.24)$$

Solving (7.23) and (7.24) with MATLAB we get:

```
format rat
G=[9/20+j*3/10 −1/5−j*1/5; −1/5−j*1/5 3/10+j*1/5]; I=[1 0]'; V=G\I;
fprintf(' \n');disp('V1 = '); disp(V(1,1)); disp('Vout = '); disp(V(2,1));
format short
magV=abs(V(2,1)); thetaV=angle(V(2,1))*180/pi;
fprintf('magIx = %5.3f A \t', magIx); fprintf('theta = %4.2f deg \t', theta);...
fprintf(' \n'); fprintf(' \n')
```

```
V1 = 68/25 - 24/25i    Vout = 56/25 - 8/25i

magIx = 2.750 A        theta = -39.64 deg
```

Therefore,

$$V_{out} = 2.263 \angle{-8.13°} \qquad (7.25)$$

7.6 Phasor Diagrams

A *phasor diagram* is a sketch showing the magnitude and phase relationships among the phasor voltages and currents in phasor circuits. The procedure is best illustrated with the examples below.

Example 7.7

Compute and sketch all phasor quantities for the circuit of Figure 7.19.

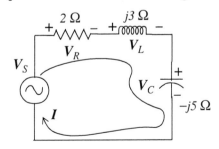

Figure 7.19. Circuit for Example 7.7

Solution:

Since this is a series circuit, the phasor current I is common to all circuit devices. Therefore, we assign to this phasor current the value $I = 1\angle 0°$ and use it as our reference as shown in the phasor diagram of Figure 7.20. Then,

$$V_R = (2 \ \Omega)(1\angle 0°) = 2\angle 0° \ V$$

$$V_L = (j3 \ \Omega)(1\angle 0°) = j3 = 3\angle 90° \ V$$

$$V_C = (-j5 \ \Omega)(1\angle 0°) = -j5 = 5\angle -90° \ V$$

$$V_S = V_R + (V_L + V_C) = 2 - j2 = 2\sqrt{2}\angle -45°$$

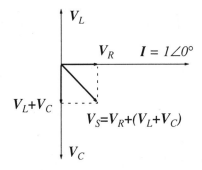

Figure 7.20. Phasor diagram for the circuit of Example 7.7

Example 7.8

Compute and sketch all phasor quantities for the circuit of Figure 7.21.

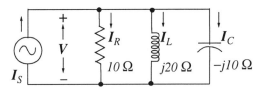

Figure 7.21. Circuit for Example 7.8

Solution:

Since this is a parallel circuit, the phasor voltage V is common to all circuit devices. Therefore let us assign this phasor voltage the value $V = 1\angle 0°$ and use it as our reference phasor as shown in the phasor diagram of Figure 7.22. Then,

$$I_R = 1\angle 0° / 10 = 100\angle 0° \ mA$$

$$I_L = 1\angle 0° / j20 = 1\angle 0° / 20\angle 90° = 50\angle -90° \ m$$

$$I_C = 1\angle 0° / (-j10) = 1\angle 0° / 10\angle -90° = 100\angle 90° \ mA$$

$$I_C + I_L = 50\angle 90° \ mA$$

$$I_S = I_R + (I_C + I_L) = 100 + j50 = 111.8\angle 26.6°$$

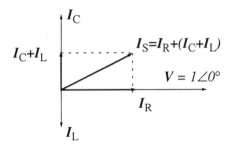

Figure 7.22. Phasor diagram for Example 7.8

We can draw a phasor diagram for other circuits that are neither series nor parallel by assigning any phasor quantity as a reference.

Example 7.9

Compute and sketch all phasor voltages for the circuit of Figure 7.23. Then, use MATLAB to plot these quantities in the $t-domain$.

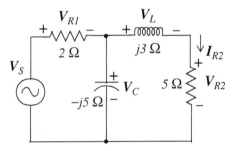

Figure 7.23. Circuit for Example 7.9

Solution:

We will begin by selecting $I_{R2} = 1\angle 0°$ A as our reference as shown on the phasor diagram of Figure 7.24. Then,

$$V_{R2} = 5 \ \Omega \times I_{R2} = 5 \times 1\angle 0° = 5\angle 0°$$

$$V_L = j3 \ \Omega \times I_{R2} = 3\angle 90° \times 1\angle 0° = 3\angle 90°$$

$$V_C = V_L + V_{R2} = 5\angle 0° + 3\angle 90° = 5 + j3 = 5.83\angle 31°$$

$$V_{R1} = 2\,\Omega \times I_{R1} = 2(I_C + I_{R2}) = 2\left(\frac{V_C}{-j5} + I_{R2}\right) = 2\left(\frac{5.83\angle31°}{5\angle-90°} + 5\angle0°\right)$$

$$= 2.33\angle121° + 10\angle0° = -1.2 + j2 + 10 = 8.8 + j2 = 9\angle12.8°$$

and

$$V_S = V_{R1} + V_C = 8.8 + j2 + 5 + j3 = 13.8 + j5 = 14.7\angle20°$$

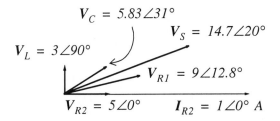

Figure 7.24. Phasor diagram for Example 7.9

Now, we can transform these phasors into time-domain quantities and use MATLAB to plot them. We will use the voltage source as a reference with the value $V_S = 1\angle0°$, and we will apply nodal analysis with node voltages V_1, V_2, and V_3 assigned as shown in Figure 7.25.

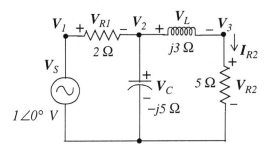

Figure 7.25. Circuit for Example 7.9 with the voltage source taken as reference

The node equations are shown below in matrix form.

$$\underbrace{\begin{bmatrix} 1 & 0 & 0 \\ -\dfrac{1}{2} & \left(\dfrac{1}{2} + \dfrac{1}{-j5} + \dfrac{1}{j3}\right) & -\dfrac{1}{j3} \\ 0 & -\dfrac{1}{j3} & \left(\dfrac{1}{j3} + \dfrac{1}{5}\right) \end{bmatrix}}_{G} \underbrace{\begin{bmatrix} V_1 \\ V_2 \\ V_3 \end{bmatrix}}_{V} = \underbrace{\begin{bmatrix} 1 \\ 0 \\ 0 \end{bmatrix}}_{I}$$

The MATLAB code is as follows:

```
% Enter the non-zero values of the G matrix
G(1,1)=1;
G(2,1)=-1/2;
G(2,2)=1/2-1/5j+1/3j;
G(2,3)=-1/3j;
G(3,2)=-1/3j;
G(3,3)=1/3j+1/5;
%
% Enter all values of the I matrix
I=[1  0  0]';
%
% Compute node voltages
V=G\I;
%
VR1=V(1)-V(2);
VL=V(2)-V(3);
% Compute magnitudes and phase angles of voltages
magV1=abs(V(1)); magV2=abs(V(2)); magV3=abs(V(3));
phaseV1=angle(V(1))*180/pi; phaseV2=angle(V(2))*180/pi; phaseV3=angle(V(3))*180/pi;
magVR1=abs(VR1); phaseVR1=angle(VR1)*180/pi;
magVL=abs(VL); phaseVL=angle(VL)*180/pi;
%
% Denote radian frequency as w and plot wt for 0 to 2*pi range
wt=linspace(0,2*pi);
V1=magV(1)*cos(wt-phaseV(1));
V2=magV(2)*cos(wt-phaseV(2));
V3=magV(3)*cos(wt-phaseV(3));
VR1t=magVR1*cos(wt-phaseVR1);
VLt=magVL*cos(wt-phaseVL);
%
% Convert wt to degrees
deg=wt*180/pi;
%
% Print phasor voltages, magnitudes, and phase angles
fprintf(' \n');
% With fprintf only the real part of each parameter is processed so we will use disp
disp('V1 = '); disp(V(1)); disp('V2 = '); disp(V(2)); disp('V3 = '); disp(V(3));
disp('VR1 = '); disp(VR1); disp('VL = '); disp(VL);
fprintf('magV1 = %4.2f V \t', magV1); fprintf('magV2 = %4.2f V \t', magV2);
fprintf('magV3 = %4.2f V', magV3); fprintf(' \n'); fprintf(' \n');
fprintf('phaseV1 = %4.2f deg \t', phaseV1);
fprintf('phaseV2 = %4.2f deg \t', phaseV2); fprintf('phaseV3 = %4.2f deg', phaseV3);
fprintf(' \n'); fprintf(' \n');
fprintf('magVR1 = %4.2f V \t', magVR1); fprintf('phaseVR1 = %4.2f deg ', phaseVR1);
fprintf(' \n'); fprintf(' \n');
fprintf('magVL = %4.2f V \t', abs(VL)); fprintf('phaseVL = %4.2f deg ', phaseVL);
fprintf(' \n');
%
plot(deg,V1,deg,V2,deg,V3,deg,VR1t,deg,VLt)
fprintf(' \n');

V1 = 1

V2 = 0.7503 - 0.1296i
```

```
V3 = 0.4945 - 0.4263i
VR1 = 0.2497 + 0.1296i
VL = 0.2558 + 0.2967i
magV1 = 1.00 V          magV2 = 0.76 V          magV3 = 0.65 V
phaseV1 = 0.00 deg      phaseV2 = -9.80 deg     phaseV3 = -40.76 deg
magVR1 = 0.28 V         phaseVR1 = 27.43 deg
magVL = 0.39 V          phaseVL = 49.24 deg
```

and with these values we have

$$v_S(t) = v_1(t) = \cos\omega t \quad v_2(t) = 0.76\cos(\omega t - 9.8°) \quad v_3(t) = 0.65\cos(\omega t - 40.8°)$$

$$v_{R1}(t) = 0.28\cos(\omega t + 27.4°) \quad v_L(t) = 0.39\cos(\omega t + 49.2°)$$

These are plotted with MATLAB as shown in Figure 7.26.

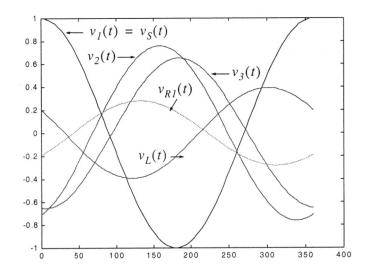

Figure 7.26. The t − domain plots for Example 7.9

7.7 Electric Filters

The characteristics of electric filters were introduced in Chapter 4 but are repeated below for convenience.

Analog filters are defined over a continuous range of frequencies. They are classified as *low-pass, high-pass, band-pass* and *band-elimination (stop-band)*. Another, less frequently mentioned filter, is the *all-pass* or *phase shift* filter. It has a constant amplitude response but is phase varies with frequency. This is discussed in *Signals and Systems with MATLAB Applications*, ISBN 0-9709511-3-2, by this author.

The ideal amplitude characteristics of each are shown in Figure 7.27. The ideal characteristics are not physically realizable; we will see that practical filters can be designed to approximate these characteristics. In this section we will derive the passive RC low and high-pass filter characteristics and those of an active low-pass filter using phasor analysis.

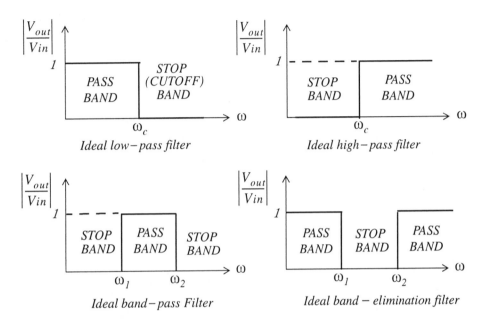

Figure 7.27. Amplitude characteristics of the types of filters

A *digital filter*, in general, is a computational process, or algorithm that converts one sequence of numbers representing the input signal into another sequence representing the output signal. Accordingly, a digital filter can perform functions as differentiation, integration, estimation, and, of course, like an analog filter, it can filter out unwanted bands of frequency. Digital filters are discussed in *Signals and Systems with MATLAB Applications* by this author, Orchard Publications.

7.8 Basic Analog Filters

An analog filter can also be classified as *passive* or *active*. Passive filters consist of passive devices such as resistors, capacitors and inductors. Active filters are, generally, operational amplifiers with resistors and capacitors connected to them externally. We can find out whether a filter, passive or active, is a low-pass, high-pass, etc., from its the frequency response that can be obtained from its transfer function. The procedure is illustrated with the examples that follow.

Example 7.10

Derive expressions for the magnitude and phase responses of the series RC network of Figure 7.28, and sketch their characteristics.

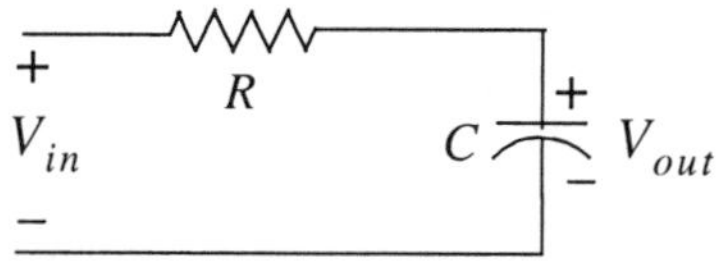

Figure 7.28. Series RC network for Example 7.10

Solution:

By the voltage division expression,

$$V_{out} = \frac{1/j\omega C}{R + 1/j\omega C} V_{in}$$

and denoting the ratio V_{out}/V_{in} as $G(j\omega)$, we get

$$G(j\omega) = \frac{V_{out}}{V_{in}} = \frac{1}{1 + j\omega RC} = \frac{1}{(\sqrt{1 + \omega^2 R^2 C^2})\angle atan(\omega RC)}$$

$$= \frac{1}{\sqrt{1 + \omega^2 R^2 C^2}}\angle -atan(\omega RC) \qquad (7.26)$$

The magnitude of (7.26) is

$$\boxed{|G(j\omega)| = \left|\frac{V_{out}}{V_{in}}\right| = \frac{1}{\sqrt{1 + \omega^2 R^2 C^2}}} \qquad (7.27)$$

and the phase angle θ, also known as the *argument*, is

$$\boxed{\theta = arg\{G(j\omega)\} = arg\left(\frac{V_{out}}{V_{in}}\right) = -atan(\omega RC)} \qquad (7.28)$$

We can obtain a quick sketch for the magnitude $|G(j\omega)|$ versus ω by evaluating (7.27) at $\omega = 0$, $\omega = 1/RC$, and $\omega \rightarrow \infty$. Thus,

as $\omega \rightarrow 0$, $|G(j\omega)| \cong 1$

for $\omega = 1/RC$, $|G(j\omega)| = 1/\sqrt{2} = 0.707$

and as $\omega \rightarrow \infty$, $|G(j\omega)| \cong 0$

To obtain a smooth curve, we will use a spreadsheet such as Microsoft Excel to plot $|G(j\omega)|$ versus radian frequency for several values of ω. This is shown in Figure 7.29 where, for convenience, we let

$RC = 1$. The plot shows that this circuit is an approximation, although not a good one, to the amplitude characteristics of a low-pass filter.

ω	$\lvert G(j\omega)\rvert$
0.000	1
0.020	0.9998
0.040	0.9992
0.060	0.9982
0.080	0.99682
0.100	0.99504
0.120	0.99288
0.140	0.99034
0.160	0.98744
0.180	0.98418
0.200	0.98058
0.220	0.97664

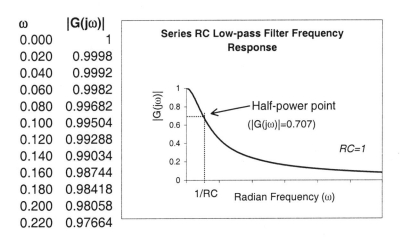

Figure 7.29. Amplitude characteristics of a series RC low-pass filter

We can also obtain a quick sketch for the phase angle, i.e., $\theta = arg\{G(j\omega)\}$ versus ω by evaluating of (11.3) at $\omega = 0$, $\omega = 1/RC$, $\omega = -1/RC$, $\omega \to -\infty$ and $\omega \to \infty$. Thus,

as $\omega \to 0$, $\theta \cong -atan\,0 \cong 0°$

for $\omega = 1/RC$, $\theta = -atan\,1 = -45°$

for $\omega = -1/RC$, $\theta = -atan(-1) = 45°$

as $\omega \to -\infty$, $\theta = -atan(-\infty) = 90°$

and as $\omega \to \infty$, $\theta = -atan(\infty) = -90°$

We use Excel to plot θ versus radian frequency for several values of ω. This is shown in Figure 7.31 where, again for convenience, we let $RC = 1$

Example 7.11

The network of Figure 7.30 is also a series RC circuit, where the positions of the resistor and capacitor have been interchanged. Derive expressions for the magnitude and phase responses, and sketch their characteristics.

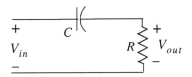

Figure 7.30. RC network for Example 7.11

ω	θ
-12.00	85.24
-11.98	85.23
-11.96	85.22
-11.94	85.21
-11.92	85.20
-11.90	85.20
-11.88	85.19
-11.86	85.18
-11.84	85.17
-11.82	85.16
-11.80	85.16
-11.78	85.15
-11.76	85.14
-11.74	85.13

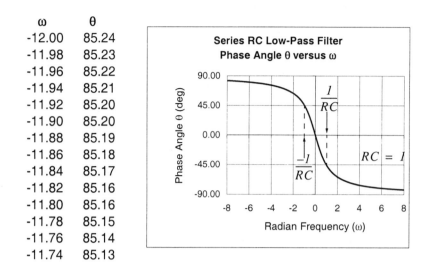

Figure 7.31. Phase characteristics of a series RC low-pass filter

Solution:

$$V_{out} = \frac{R}{R + 1/j\omega C} V_{in}$$

or

$$G(j\omega) = \frac{V_{out}}{V_{in}} = \frac{j\omega RC}{1 + j\omega RC} = \frac{j\omega RC + \omega^2 R^2 C^2}{1 + \omega^2 R^2 C^2} = \frac{\omega RC(j + \omega RC)}{1 + \omega^2 R^2 C^2}$$

$$= \frac{\omega RC\sqrt{1 + \omega^2 R^2 C^2}\angle atan\left(\frac{1}{\omega RC}\right)}{1 + \omega^2 R^2 C^2} = \frac{1}{\sqrt{1 + \frac{1}{\omega^2 R^2 C^2}}}\angle atan\left(\frac{1}{\omega RC}\right) \tag{7.29}$$

The magnitude of (7.29) is

$$\boxed{|G(j\omega)| = \frac{1}{\sqrt{1 + \frac{1}{\omega^2 R^2 C^2}}}} \tag{7.30}$$

and the phase angle or argument, is

$$\boxed{\theta = arg\{G(j\omega)\} = atan\left(\frac{1}{\omega RC}\right)} \tag{7.31}$$

We can obtain a quick sketch for the magnitude $|G(j\omega)|$ versus ω by evaluating (7.30) at $\omega = 0$, $\omega = 1/RC$, and $\omega \to \infty$. Thus,

as $\omega \to 0$, $|G(j\omega)| \cong 0$

for $\omega = 1/RC$, $|G(j\omega)| = 1/\sqrt{2} = 0.707$

and as $\omega \to \infty$, $|G(j\omega)| \cong 1$

Figure 7.32 shows $|G(j\omega)|$ versus radian frequency for several values of ω where $RC = 1$. The plot shows that this circuit is an approximation, although not a good one, to the amplitude characteristics of a high-pass filter.

ω	$\|G(j\omega)\|$
0.000	1E-07
0.020	0.02
0.040	0.03997
0.060	0.05989
0.080	0.07975
0.100	0.0995
0.120	0.11915
0.140	0.13865
0.160	0.15799
0.180	0.17715
0.200	0.19612
0.220	0.21486

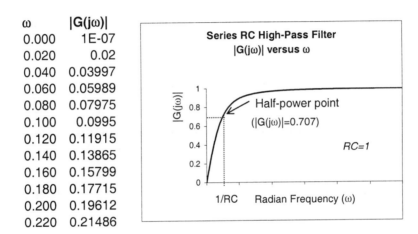

Figure 7.32. Amplitude characteristics of a series RC high-pass filter

We can also obtain a quick sketch for the phase angle, i.e., $\theta = arg\{G(j\omega)\}$ versus ω, by evaluating (7.31) at $\omega = 0$, $\omega = 1/RC$, $\omega = -1/RC$, $\omega \to -\infty$, and $\omega \to \infty$. Thus,

as $\omega \to 0$, $\theta \cong -atan0 \cong 0°$

for $\omega = 1/RC$, $\theta = -atan1 = -45°$
for $\omega = -1/RC$, $\theta = -atan(-1) = 45°$

as $\omega \to -\infty$, $\theta = -atan(-\infty) = 90°$

and as $\omega \to \infty$, $\theta = -atan(\infty) = -90°$

Figure 7.33 shows the phase angle θ versus radian frequency for several values of ω, where $RC = 1$

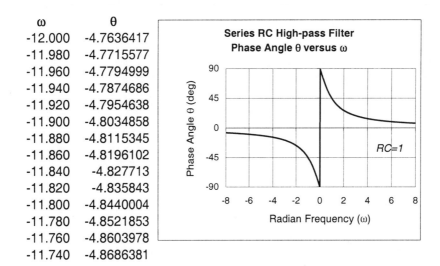

ω	θ
-12.000	-4.7636417
-11.980	-4.7715577
-11.960	-4.7794999
-11.940	-4.7874686
-11.920	-4.7954638
-11.900	-4.8034858
-11.880	-4.8115345
-11.860	-4.8196102
-11.840	-4.827713
-11.820	-4.835843
-11.800	-4.8440004
-11.780	-4.8521853
-11.760	-4.8603978
-11.740	-4.8686381

Figure 7.33. Phase characteristics of an RC high-pass filter

7.9 Active Filter Analysis

We can analyze active filters, such as those we discussed in Chapter 4, using phasor circuit analysis.

Example 7.12

Compute the approximate cut-off frequency of the circuit of Figure 7.34 which is known as a *Multiple Feed Back (MFB) active low-pass filter.*

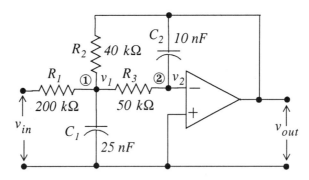

Figure 7.34. Low-pass filter for Example 7.12

Solution:

We write the phasor circuit nodal equations as follows:

At Node ①:

$$\frac{v_1 - v_{in}}{R_1} + \frac{v_1}{1/(j\omega C_1)} + \frac{v_1 - v_{out}}{R_2} + \frac{v_1 - v_2}{R_3} = 0 \qquad (7.32)$$

At node ②:

$$\frac{v_2 - v_1}{R_3} = \frac{C_2}{1/(j\omega C_2)} \tag{7.33}$$

and since $v_2 = 0$ (virtual ground), relation (7.33) reduces to

$$v_1 = (-j\omega R_3 C_2)v_{out} \tag{7.34}$$

and by substitution of (7.34) into (7.32), rearranging, and collecting like terms, we get:

$$\left[\left(\frac{1}{R_1} + \frac{1}{R_2} + \frac{1}{R_3} + j\omega C_1\right)(-j\omega R_3 C_2) - \frac{1}{R_2}\right]v_{out} = \frac{1}{R_1}v_{in} \tag{7.35}$$

or

$$\frac{v_{out}}{v_{in}} = \frac{1}{R_1\left[\left(\frac{1}{R_1} + \frac{1}{R_2} + \frac{1}{R_3} + j\omega C_1\right)(-j\omega R_3 C_2) - \frac{1}{R_2}\right]} \tag{7.36}$$

By substitution of given values of resistors and capacitors, we get

$$\frac{v_{out}}{v_{in}} = \frac{1}{2 \times 10^5\left[\left(\frac{1}{20 \times 10^3} + j2.5 \times 10^{-8}\omega\right)(-j5 \times 10^4 \times 10^{-8}\omega) - \frac{1}{4 \times 10^4}\right]}$$

or

$$|G(j\omega)| = \frac{v_{out}}{v_{in}} = \frac{-1}{2.5 \times 10^{-6}\omega^2 - j5 \times 10^{-3}\omega + 5} \tag{7.37}$$

and now we can use MATLAB to find and plot the magnitude of (7.37) with the following code.

```
w=1:10:10000; Gjw=-1./(2.5.*10.^(-6).*w.^2-5.*j.*10.^(-3).*w+5);
semilogx(w,abs(Gjw)); grid; hold on
xlabel('Radian Frequency w'); ylabel('|Vout/Vin|');
title('Magnitude Vout/Vin vs. Radian Frequency')
```

The plot is shown in Figure 7.35 where we see that the cutoff frequency occurs at about $700 \ rad/s$. We observe that the half-power point for this plot is $0.2 \times 0.707 = 0.141$.

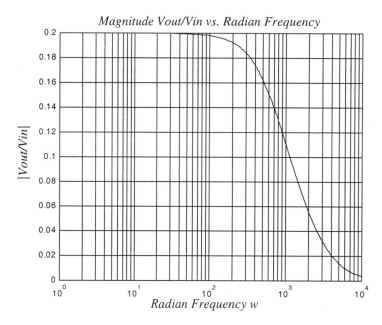

Figure 7.35. Plot for the magnitude of the low-pass filter circuit of Example 7.12

7.10 Summary

- In Chapter 3 we were concerned with constant voltage and constant current sources, resistances and conductances. In this chapter we were concerned with alternating voltage and alternating current sources, impedances, and admittances.

- Nodal analysis, mesh analysis, the principle of superposition, Thevenin's theorem, and Norton's theorem can also be applied to phasor circuits.

- The use of complex numbers make the phasor circuit analysis much easier.

- MATLAB can be used very effectively to perform the computations since it does not require any special procedures for manipulation of complex numbers.

- Whenever a branch in a circuit contains two or more devices in series or two or more devices in parallel, it is highly recommended that they are grouped and denoted as z_1, z_2, and so on before writing nodal or mesh equations.

- Phasor diagrams are sketches that show the magnitude and phase relationships among several phasor voltages and currents. When constructing a phasor diagram, the first step is to select one phasor as a reference, usually with zero phase angle, and all other phasors must be drawn with the correct relative angles.

- The RC low-pass and RC high-pass filters are rudimentary types of filters and are not used in practice. They serve as a good introduction to electric filters.

7.11 Exercises

Multiple Choice

1. In the circuit of Figure 7.36 the phasor voltage V is

   A. $2 + j0 \ V$

   B. $1 + j0 \ V$

   C. $1 - j0 \ V$

   D. $1 + j \ V$

   E. none of the above

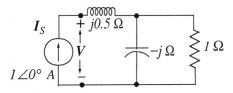

Figure 7.36. Circuit for Question 1

2. In the circuit of Figure 7.37 the phasor current I is

   A. $0 + j2 \ A$

   B. $0 - j2 \ A$

   C. $1 + j0 \ A$

   D. $2 + j2 \ A$

   E. none of the above

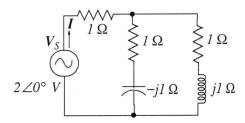

Figure 7.37. Circuit for Question 2

3. In the circuit of Figure 7.38 the voltage across the capacitor C_2 is

   A. $8 \times 10^{-4} \sin(2000t + 90°) \ V$

   B. $\sqrt{50} \cos(2000t - 45°) \ V$

C. $\sqrt{50}\cos(2000t + 45°)$ V

D. $\sqrt{50}\cos(2000t + 90°)$ V

E. none of the above

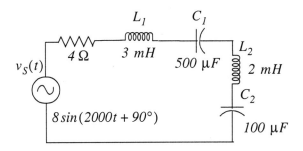

Figure 7.38. Circuit for Question 3

4. In the circuit of Figure 7.39 the current $i_C(t)$ through the capacitor is

A. $4\sin 2000t$

B. $4\sin(2000t + 180°)$

C. $\sqrt{32}\cos(2000t - 45°)$

D. $\sqrt{32}\cos(2000t + 90°)$

E. none of the above

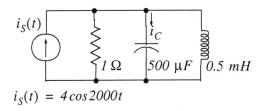

$i_S(t) = 4\cos 2000t$

Figure 7.39. Circuit for Question 4

5. The Thevenin equivalent voltage V_{TH} at terminals A and B in the circuit of Figure 7.40 is

A. $10\angle{-90°}$ V

B. $10\angle{-53.13°}$ V

C. $10\angle{53.13°}$ V

D. $10\angle{-45°}$ V

E. none of the above

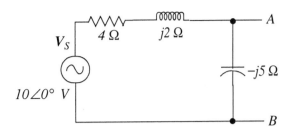

Figure 7.40. Circuit for Questions 5 and 6

6. The Thevenin equivalent impedance Z_{TH} at terminals A and B in the circuit of Figure 7.40 is

A. $2 + j4 \ \Omega$

B. $4 + j2 \ \Omega$

C. $4 - j2 \ \Omega$

D. $-j5 \ \Omega$

E. none of the above

7. In the circuit of Figure 7.41 the phasor voltage V_C is

A. $5\angle -90° \ V$

B. $5\angle -45° \ V$

C. $4\angle -53.1° \ V$

D. $4\angle 53.1° \ V$

E. none of the above

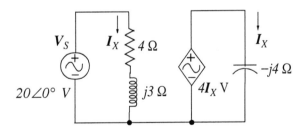

Figure 7.41. Circuit for Question 7

8. In the circuit of Figure 7.42 the phasor voltage $V_{R_{5\ \Omega}}$ is

A. $20 + j0 \ V$

B. $0 + j20 \ V$

C. $20 + j20$ V

D. $80 - j80$ V

E. none of the above

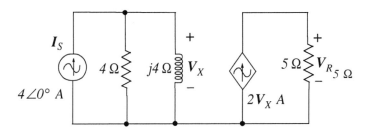

Figure 7.42. Circuit for Question 8

9. In the circuit of Figure 7.43 the phasor voltage $V_{OUT\,2}$ is

A. $2 + j0$ V

B. $4 + j0$ V

C. $4 - j0$ V

D. $1 + j1$ V

E. none of the above

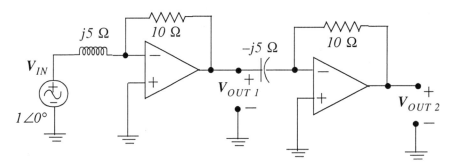

Figure 7.43. Circuit for Question 9

10. In the circuit of Figure 7.44 the $t - domain$ voltage $v_{AB}(t)$ is

A. $1.89 \cos(\omega t + 45°)$ V

B. $0.53 \cos(\omega t - 45°)$ V

C. $2 \cos \omega t$ V

D. $0.5 \cos(\omega t + 53.1°)$ V

E. none of the above

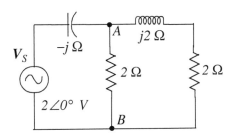

Figure 7.44. Circuit for Question 10

Problems

1. For the circuit of Figure 7.45, $i_S(t) = 2\cos 1000t\ A$. Compute $v_{AB}(t)$ and $i_C(t)$.

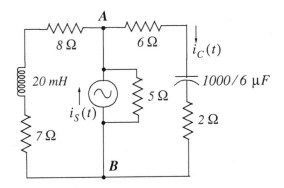

Figure 7.45. Circuit for Problem 1

2. Write nodal equations and use MATLAB to compute $i_C(t)$ for the circuit of Figure 7.46 given that $v_S(t) = 12\cos(1000t + 45°)\ V$.

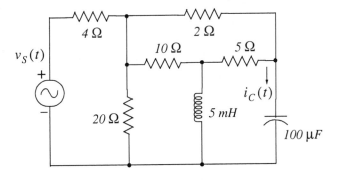

Figure 7.46. Circuit for Problem 2

3. Write mesh equations and use MATLAB to compute $i_L(t)$ for the circuit of Figure 7.47 given that $v_S(t) = 100\cos(10000t + 60°)\ V$.

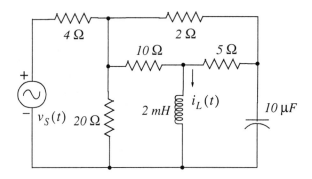

Figure 7.47. Circuit for Problem 3

4. For the circuit of Figure 7.48, it is given that $v_{S1}(t) = 40\cos(5000t + 60°) \; V$ and $v_{S2}(t) = 60\sin(5000t + 60°) \; V$. Use superposition to find $v_C(t)$.

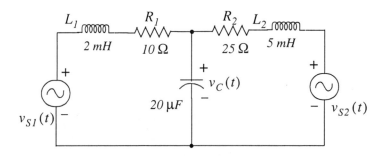

Figure 7.48. Circuit for Problem 4

5. For the circuit of Figure 7.49, find $v_C(t)$ if $v_{S1} = 15 \; V$, $v_{S2}(t) = 20\cos 1000t \; V$, and $i_S(t) = 4\cos 2000t \; A$. Plot $v_C(t)$ using MATLAB or Excel.

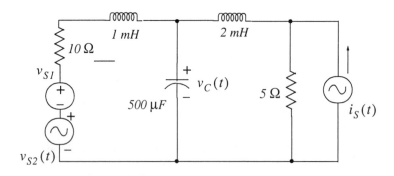

Figure 7.49. Circuit for Problem 5

6. For the circuit of Figure 7.50, find the value of Z_{LD} which will receive maximum power.

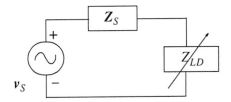

Figure 7.50. Circuit for Problem 6

7. For the circuit of Figure 7.51, to what value should the load impedance Z_{LD} be adjusted so that it will receive maximum power from the voltage source?

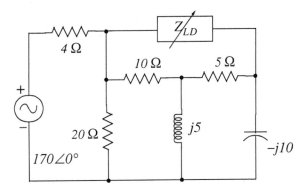

Figure 7.51. Circuit for Problem 6

8. For the circuit of Figure 7.52, draw a phasor diagram which shows the voltage and current in each branch.

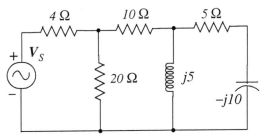

Figure 7.52. Circuit for Problem 8

9. For the op amp circuit of Figure 7.53, $v_{in}(t) = 3\cos 1000t$ V. Find $v_{out}(t)$.

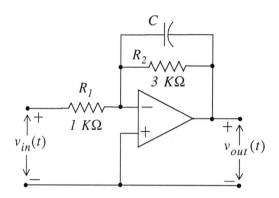

Figure 7.53. Circuit for Problem 9

7.12 Answers to Exercises

Multiple Choice

1. E $V = V_L + V_C$ where $V_L = 1\angle 0° \times j1/2 = j1/2\ V$ and V_C is found from the nodal equa-

tion $\dfrac{V_C}{1} + \dfrac{V_C}{-j} = 1 + j0$ or $(1+j)V_C = 1$ or $V_C = \dfrac{1}{1+j} \times \dfrac{1-j}{1-j} = \dfrac{1-j}{2} = \dfrac{1}{2} - j\dfrac{1}{2}\ V$. Therefore,

$V = j1/2 + 1/2 - j1/2 = 1/2 + j0\ V$

2. C Denoting the resistor in series with the voltage source as z_1, the resistor in series with the capacitor as z_2, and the resistor in series with the capacitor as z_3, the equivalent impedance is

$$Z_{eq} = z_1 + \frac{z_2 \cdot z_3}{z_2 + z_3} = 1 + \frac{(1-j1)(1+j1)}{1-j1+1+j1} = 1 + \frac{2}{2} = 2 + j0$$

and

$$I = \frac{V_S}{Z} = \frac{2+j0}{2+j0} = 1 + j0\ A$$

3. B $8\sin(2000t + 90°) = 8\cos 2000t \Leftrightarrow 8\angle 0°\ V$, $j\omega L_1 = j6$, $j\omega L_2 = j4$, $-j/\omega C_1 = -j1$,

$-j/\omega C_1 = -j1$ and the phasor equivalent circuit is shown below.

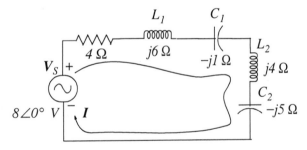

$Z = 4 + j6 - j1 + j4 - j5 = 4 + j4$, $I = \dfrac{V_S}{Z} = \dfrac{8+j0}{4+j4} = \dfrac{8+j0}{4+j4} \cdot \dfrac{4-j4}{4-j4} = \dfrac{32-j32}{32} = 1 - j1$,

and thus $V_{C_2} = -j5 \times (1-j) = 5 - j5 = \sqrt{50}\angle{-45°} \Leftrightarrow \sqrt{50}\cos(2000t - 45°)\ V$

4. D $4\cos 2000t \Leftrightarrow 4\angle 0°$, $G = 1/R = 1\ \Omega^{-1}$, $j\omega C = j1\ \Omega^{-1}$, $-j/\omega L = -j1\ \Omega^{-1}$, and the phasor equivalent circuit is shown below.

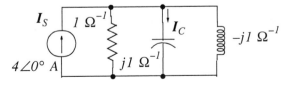

Denoting the parallel combination of the conductance and inductance as $Y_1 = 1 - j1$ and using the *current division expression for admittances* we get

$$I_C = \frac{j\omega C}{j\omega C + Y_1} \cdot I_S = \frac{j1}{j1 + 1 - j1} = j1 \times 4\angle 0° = 1\angle 90° \times 4\angle 0° = 4\angle 90° \text{ A}$$

and thus

$$i_S(t) = 4\cos(2000t + 90°) \text{ A}$$

5. B   By the voltage division expression

$$V_{TH} = V_{AB} = \frac{-j5}{4 + j2 - j5} \cdot 10\angle 0° = \frac{5\angle -90° \times 10\angle 0°}{4 - j3} = \frac{50\angle -90°}{5\angle -36.9°} = 10\angle -53.1° \text{ V}$$

6. C   We short the voltage source and looking to the left of points A and B we observe that the capacitor is in parallel with the series combination of the resistance and inductance. Thus,

$$Z_{TH} = \frac{(-j5)(4 + j2)}{4 + j2 - j5} = \frac{10 - j20}{4 - j3} = \frac{10 - j20}{4 - j3} \cdot \frac{4 + j3}{4 + j3} = \frac{100 - j50}{25} = 4 - j2$$

7. D $I_X = \dfrac{20\angle 0°}{4 + j3} = \dfrac{20\angle 0°}{5\angle 36.9°} = 4\angle -36.9°$, $4I_X = 16\angle -36.9°$

and

$$I_C = \frac{4I_X}{-j4} = \frac{16\angle -36.9°}{4\angle -90°} = 4\angle 53.1°$$

8. E $V_X = \dfrac{4}{4 + j4} \times 4\angle 0° \times j4 = \dfrac{64\angle 90°}{\sqrt{32}\angle 45°} = \dfrac{64 \times \sqrt{32}}{32}\angle 45° = 2\sqrt{32}\angle 45°$

and

$$V_{R_{5\,\Omega}} = 2V_X \times 5 = 20 \times \sqrt{32}\angle 45° = 20\sqrt{32}\left(\frac{\sqrt{2}}{2} + j\frac{\sqrt{2}}{2}\right) = 80 + j80$$

9. B $V_{OUT\,1} = -\dfrac{10}{j5} \times 1\angle 0° = j2 \times 1\angle 0° = 2\angle 90° \times 1\angle 0° = 2\angle 90°$

and

$$V_{OUT\,2} = -\frac{10}{-j5} \times V_{OUT\,1} = -j2 \times 2\angle 0° = 2\angle -90° \times 2\angle 90° = 4\angle 0° = 4 + j0$$

10. A   We write the nodal equation at Node A for V_{AB} as

$$\frac{V_{AB} - 2\angle 0°}{-j} + \frac{V_{AB}}{2} + \frac{V_{AB}}{2+j2} = 0$$

$$\left(\frac{1}{2} + j + \frac{1}{2+j2}\right) V_{AB} = 2\angle 90°$$

$$V_{AB} = \frac{2\angle 90°}{1/2 + j + 1/4 - j/4} = \frac{2\angle 90°}{3/4 + j3/4} = \frac{2\angle 90°}{1.06\angle 45°}$$

or $V_{AB} = 1.89\angle 45°$ and in the $t-domain$ $v_{AB}(t) = 1.89\cos(\omega t + 45°)$

Problems

1. We transform the current source and its parallel resistance to a voltage source series resistance, we combine the series resistors, and we draw the phasor circuit below.

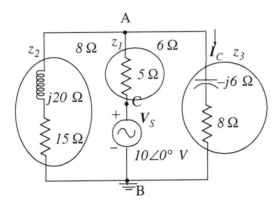

For this phasor circuit, $V_S = 2\angle 0° \times 5 = 10\angle 0°$ V, $j\omega L = j10^3 \times 20 \times 10^{-3} = j20\ \Omega$ and

$-j/\omega C = -j/(10^3 \times 10^3/6 \times 10^{-6}) = -j6$, $z_1 = 5\ \Omega$, $z_2 = (15 + j20)\ \Omega$, and $z_3 = (8 - j6)\ \Omega$

We observe that $V_A = V_{AB} = V_{AC} + V_{CB} = V_{AC} + 10\angle 0°$ V and $V_B = 0$. At Node A,

$$\frac{V_A - V_B}{z_2} + \frac{V_A - 10\angle 0°}{z_1} + \frac{V_A - V_B}{z_3} = 0$$

$$\left(\frac{1}{z_1} + \frac{1}{z_2} + \frac{1}{z_3}\right) V_A = \frac{10\angle 0°}{z_1}$$

$$\left(\frac{1}{5} + \frac{1}{15 + j20} + \frac{1}{8 - j6}\right) V_A = \frac{10\angle 0°}{5} = 2\angle 0°$$

and

$$V_A = \frac{2\angle 0°}{0.2 + \dfrac{1}{25\angle 53.1°} + \dfrac{1}{10\angle -36.9°}} = \frac{2\angle 0°}{0.2 + 0.04\angle -53.1° + 0.1\angle 36.9°}$$

$$= \frac{2\angle 0°}{0.2 + 0.04\cos 53.1° - j0.04\sin 53.1° + 0.1\cos 36.9° + j0.1\sin 36.9°}$$

$$= \frac{2\angle 0°}{0.2 + 0.04\times 0.6 - j0.04\times 0.8 + 0.1\times 0.8 + j0.1\times 0.6} = \frac{2\angle 0°}{0.304 + j0.028}$$

$$= \frac{2\angle 0°}{0.305\angle 5.26°} = 6.55\angle -5.26°$$

Then, in the $t - domain$ $v_{AB}(t) = 6.55\cos(1000 + 5.26°)$.

Also,

$$I_C = \frac{V_A}{z_3} = \frac{6.55\angle -5.26°}{10\angle -36.9°} = 0.655\angle 31.7°$$

and

$$i_C(t) = 0.655\cos(1000 + 31.7°)$$

Check with MATLAB:

```
z1=5; z2=15+20j; z3=8-6j; VA=(10+0j)/(z1*(1/z1+1/z2+1/z3)); fprintf(' \n');...
fprintf('magVA = %5.2f V \t',abs(VA));...
fprintf('phaseVA = %5.2f deg \t',angle(VA)*180/pi); fprintf(' \n'); fprintf(' \n');

magVA = 6.55 V    phaseVA = -5.26 deg
```

2. The equivalent phasor circuit is shown below where $j\omega L = j10^3 \times 5\times 10^{-3} = j5$ and
$-j/\omega C = -j/(10^3 \times 10^{-4}) = -j10$

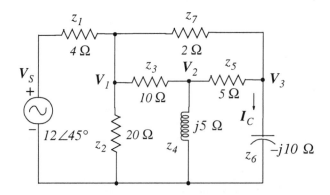

Node V_1:

$$\frac{V_1 - V_S}{z_1} + \frac{V_1 - V_2}{z_3} + \frac{V_1}{z_2} + \frac{V_1 - V_3}{z_7}$$

or

$$\left(\frac{1}{z_1}+\frac{1}{z_2}+\frac{1}{z_3}+\frac{1}{z_7}\right)V_1-\frac{1}{z_3}V_2-\frac{1}{z_7}V_3 = \frac{1}{z_1}V_S$$

Node V_2:

$$\frac{V_2-V_1}{z_3}+\frac{V_2}{z_4}+\frac{V_2-V_3}{z_5} = 0$$

or

$$-\frac{1}{z_3}V_1+\left(\frac{1}{z_3}+\frac{1}{z_4}+\frac{1}{z_5}\right)V_2-\frac{1}{z_5}V_3 = 0$$

Node V_3:

$$\frac{V_3-V_2}{z_5}+\frac{V_3-V_1}{z_7}+\frac{V_2-V_3}{z_6} = 0$$

or

$$-\frac{1}{z_7}V_1-\frac{1}{z_5}V_2+\left(\frac{1}{z_5}+\frac{1}{z_6}+\frac{1}{z_7}\right)V_3 = 0$$

and in matrix form

$$\begin{bmatrix} \left(\frac{1}{z_1}+\frac{1}{z_2}+\frac{1}{z_3}+\frac{1}{z_7}\right) & -\frac{1}{z_3} & -\frac{1}{z_7} \\ -\frac{1}{z_3} & \left(\frac{1}{z_3}+\frac{1}{z_4}+\frac{1}{z_5}\right) & -\frac{1}{z_5} \\ -\frac{1}{z_7} & -\frac{1}{z_5} & \left(\frac{1}{z_5}+\frac{1}{z_6}+\frac{1}{z_7}\right) \end{bmatrix} \cdot \begin{bmatrix} V_1 \\ V_2 \\ V_3 \end{bmatrix} = \begin{bmatrix} \frac{1}{z_1}V_S \\ 0 \\ 0 \end{bmatrix}$$

Shown below is the MATLAB code to solve this system of equations.

```
Vs=12*(cos(pi/4)+j*sin(pi/4)); % Express Vs in rectangular form
z1=4; z2=20; z3=10; z4=5j; z5=5; z6=−10j; z7=2;...
Y=[1/z1+1/z2+1/z3+1/z7 −1/z3 −1/z7;...
−1/z3 1/z3+1/z4+1/z5 −1/z5;...
−1/z7 −1/z5 1/z5+1/z6+1/z7];...
I=[Vs/z1 0 0]'; V=Y\I; Ic=V(3)/z6;...
magIc=abs(Ic); phaseIc=angle(Ic)*180/pi;...
disp('V1='); disp(V(1)); disp('V2='); disp(V(2));...
disp('V3='); disp(V(3)); disp('Ic='); disp(Ic);...
format bank % Display magnitude and angle values with two decimal places
disp('magIc='); disp(magIc); disp('phaseIc='); disp(phaseIc);...
fprintf(' \n');

V1 = 5.9950 - 4.8789i
```

```
V2 = 5.9658 - 0.5960i

V3 = 5.3552 - 4.4203i

Ic = 0.4420 + 0.5355i

magIc = 0.69

phaseIc = 50.46
```

Therefore, $I_C = 0.69\angle 50.46° \Leftrightarrow i_C(t) = 0.69\cos(1000t + 50.46°)$ A

3. The equivalent phasor circuit is shown below where $j\omega L = j10^4 \times 2 \times 10^{-3} = j20$ and $-j/\omega C = -j/(10^4 \times 10 \times 10^{-6}) = -j10$

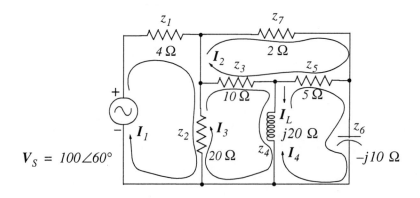

Mesh I_1:

$$(z_1 + z_2)I_1 - z_2I_3 = V_S$$

Mesh I_2:

$$(z_1 + z_2 + z_7)I_2 - z_3I_3 - z_5I_4 = 0$$

Mesh I_3:

$$-z_2I_1 - z_3I_2 + (z_2 + z_3 + z_4)I_3 - z_4I_4 = 0$$

Mesh I_4:

$$-z_5I_2 - z_4I_3 + (z_4 + z_5 + z_6)I_4 = 0$$

and in matrix form

$$\begin{bmatrix} z_1 + z_2 & 0 & -z_2 & 0 \\ 0 & z_1 + z_2 + z_7 & -z_3 & -z_5 \\ -z_2 & -z_3 & z_2 + z_3 + z_4 & -z_4 \\ 0 & -z_5 & -z_4 & z_4 + z_5 + z_6 \end{bmatrix} \cdot \begin{bmatrix} I_1 \\ I_2 \\ I_3 \\ I_4 \end{bmatrix} = \begin{bmatrix} V_S \\ 0 \\ 0 \\ 0 \end{bmatrix}$$

Shown below is the MATLAB code to solve this system of equations.

```
Vs=100*(cos(pi/3)+j*sin(pi/3)); % Express Vs in rectangular form
z1=4; z2=20; z3=10; z4=20j; z5=5; z6=-10j; z7=2;...
Z=[z1+z2 0 -z2 0;...
0 z3+z5+z7 -z3 -z5;...
-z2 -z3 z2+z3+z4 -z4;...
0 -z5 -z4 z4+z5+z6];...
V=[Vs 0 0 0]'; I=Z\V; IL=I(3)-I(4);...
magIL=abs(IL); phaseIL=angle(IL)*180/pi;...
disp('I1='); disp(I(1)); disp('I2='); disp(I(2));...
disp('I3='); disp(I(3)); disp('I4='); disp(I(4));...
disp('IL='); disp(IL);...
format bank % Display magnitude and angle values with two decimal places
disp('magIL='); disp(magIL); disp('phaseIL='); disp(phaseIL);...
fprintf(' \n');
```

```
I1 = 5.4345 - 3.4110i

I2 = 4.5527 + 0.7028i

I3 = 4.0214 + 0.2369i

I4 = 7.4364 + 1.9157i

IL= -3.4150- 1.6787i

magIL = 3.81

phaseIL = -153.82
```

Therefore, $I_L = 3.81\angle-153.82° \Leftrightarrow i_L(t) = 3.81\cos(10^4 t - -153.82°)$

4. The equivalent phasor circuit is shown below where

$$j\omega L_1 = j5 \times 10^3 \times 2 \times 10^{-3} = j10$$

$$j\omega L_2 = j5 \times 10^3 \times 5 \times 10^{-3} = j25$$

$$-j/\omega C = -j/(5 \times 10^3 \times 20 \times 10^{-6}) = -j10$$

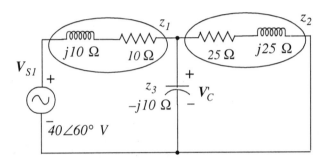

We let $V_C = V'_C + V''_C$ where V'_C is the capacitor voltage due to V_{S1} acting alone, and V''_C is the capacitor voltage due to V_{S2} acting alone. With V_{S1} acting alone the circuit reduces to that shown below.

By KCL

$$\frac{V'_C - V_{S1}}{z_1} + \frac{V'_C}{z_2} + \frac{V'_C}{z_3} = 0$$

$$\left(\frac{1}{z_1} + \frac{1}{z_2} + \frac{1}{z_3}\right)V'_C = \frac{V_{S1}}{z_1}$$

$$V'_C = \frac{V_{S1}}{z_1 \cdot \left(\frac{1}{z_1} + \frac{1}{z_2} + \frac{1}{z_3}\right)} = \frac{V_{S1}}{\left(1 + \frac{z_1}{z_2} + \frac{z_1}{z_3}\right)}$$

and with MATLAB

```
Vs1=40*(cos(pi/3)+j*sin(pi/3)); z1=10+10j; z2=−10j; z3=25+25j; V1c=Vs1/(1+z1/z2+z1/z3)
V1c = 36.7595 - 5.2962i
```

Therefore,

$$V'_C = 36.76 - j5.30 \ V$$

Next, with V_{S2} acting alone the circuit reduces to that shown below.

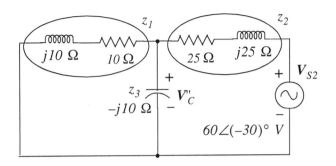

By KCL

$$\frac{V''_C}{z_1} + \frac{V''_C}{z_2} + \frac{V''_C - V_{S2}}{z_3} = 0$$

$$\left(\frac{1}{z_1} + \frac{1}{z_2} + \frac{1}{z_3}\right) V''_C = \frac{V_{S2}}{z_3}$$

$$V''_C = \frac{V_{S2}}{z_3 \cdot \left(\frac{1}{z_1} + \frac{1}{z_2} + \frac{1}{z_3}\right)} = \frac{V_{S2}}{\left(\frac{z_3}{z_1} + \frac{z_3}{z_2} + 1\right)}$$

and with MATLAB

```
Vs2=60*(cos(pi/6)–j*sin(pi/6));...
z1=10+10j; z2=–10j; z3=25+25j; V1c=36.7595–5.2962j;...
V2c=Vs2/(z3/z1+z3/z2+1); Vc=V1c+V2c; fprintf(' \n');...
disp('V1c = '); disp(V1c); disp('V2c = '); disp(V2c);...
disp('Vc=V1c+V2c'); fprintf(' \n'); disp('Vc = '); disp(Vc);...
fprintf('magVc = %4.2f V \t',abs(Vc));...
fprintf('phaseVc = %4.2f deg \t',angle(Vc)*180/pi);...
fprintf(' \n'); fprintf(' \n');
```

```
V1c = 36.7595 - 5.2962i

V2c = -3.1777 - 22.0557i

Vc = V1c+V2c

Vc = 33.5818 - 27.3519i

magVc = 43.31 V    phaseVc = -39.16 deg
```

Then,

$$V_C = V'_C + V''_C = 33.58 - j27.35 = 43.31\angle 27.35°$$

and

$$v_C(t) = 43.31\cos(5000t - 27.35°)$$

5. This circuit is excited by a DC (constant) voltage source, an AC (sinusoidal) voltage source, and an AC current source of different frequency. Therefore, we will apply the superposition principle.

Let V_C' be the capacitor voltage due to v_{S1} acting alone, V_C'' the capacitor voltage due to $v_{S2}(t)$ acting alone, and V_C''' the capacitor voltage due to $i_S(t)$ acting alone. Then, the capacitor voltage due to all three sources acting simultaneously will be $V_C = V_C' + V_C'' + V_C'''$

With the DC voltage source acting alone, after steady-state conditions have been reached the inductors behave like short circuits and the capacitor as an open circuit and thus the circuit is simplified as shown below.

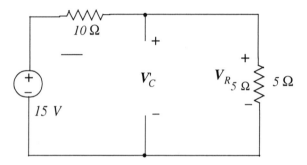

By the voltage division expression

$$V_C' = V_{R_5\Omega} = \frac{5}{10+5} \cdot 15 = 5 \ V \ DC$$

and

$$v_C'(t) = 5 \ V \ DC$$

Next, with the sinusoidal voltage source $v_{S2}(t)$ acting alone the reactances are

$$j\omega_1 L_1 = j10^3 \times 1 \times 10^{-3} = j1 \ \Omega$$

$$j\omega_1 L_2 = j10^3 \times 2 \times 10^{-3} = j2 \ \Omega$$

$$-j/\omega_1 C = -j/(10^3 \times 5 \times 10^{-4}) = -j2 \ \Omega$$

and the equivalent phasor circuit is as shown below.

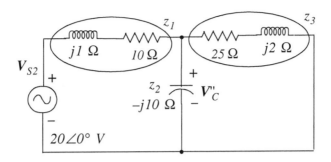

By KCL

$$\frac{V''_C - V_{S2}}{z_1} + \frac{V''_C}{z_2} + \frac{V''_C}{z_3} = 0$$

$$\left(\frac{1}{z_1} + \frac{1}{z_2} + \frac{1}{z_3}\right)V''_C = \frac{V_{S2}}{z_1}$$

$$V''_C = \frac{V_{S2}}{z_1 \cdot \left(\frac{1}{z_1} + \frac{1}{z_2} + \frac{1}{z_3}\right)} = \frac{V_{S2}}{\left(1 + \frac{z_1}{z_2} + \frac{z_1}{z_3}\right)}$$

and with MATLAB

```
Vs2=20+0j; z1=10+j; z2=-2j; z3=5+2j; V2c=Vs2/(1+z1/z2+z1/z3); fprintf(' \n');...
disp('V2c = '); disp(V2c); fprintf('magV2c = %4.2f V \t',abs(V2c));...
fprintf('phaseV2c = %4.2f deg \t',angle(V2c)*180/pi); fprintf(' \n'); fprintf(' \n');

V2c = 1.8089 - 3.5362i

magV2c = 3.97 V phaseV2c = -62.91 deg
```

Then,

$$V''_C = 1.81 - j3.54 = 3.97\angle{-62.9°}$$

and

$$v''_C(t) = 3.97\cos(1000t - 62.9°)$$

Finally, with the sinusoidal current source $i_S(t)$ acting alone the reactances are

$$j\omega_2 L_1 = j2 \times 10^3 \times 1 \times 10^{-3} = j2 \ \Omega$$

$$j\omega_2 L_2 = j2 \times 10^3 \times 2 \times 10^{-3} = j4 \ \Omega$$

$$-j/\omega_2 C = -j/(2 \times 10^3 \times 5 \times 10^{-4}) = -j1 \ \Omega$$

and the equivalent phasor circuit is as shown below where the current source and its parallel resistance have been replaced with a voltage source with a series resistor.

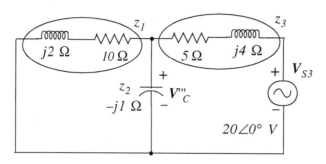

By KCL

$$\frac{V''_C}{z_1} + \frac{V''_C}{z_2} + \frac{V''_C - V_{S3}}{z_3} = 0$$

$$\left(\frac{1}{z_1} + \frac{1}{z_2} + \frac{1}{z_3} \right) V''_C = \frac{V_{S3}}{z_3}$$

$$V''_C = \frac{V_{S3}}{z_3 \cdot \left(\frac{1}{z_1} + \frac{1}{z_2} + \frac{1}{z_3} \right)} = \frac{V_{S3}}{\left(\frac{z_3}{z_1} + \frac{z_3}{z_2} + 1 \right)}$$

and with MATLAB

```
Vs3=20+0j; z1=10+2j; z2=-j; z3=5+4j; V3c=Vs3/(z3/z1+z3/z2+1); fprintf(' \n');...
disp('V3c = '); disp(V3c); fprintf('magV3c = %4.2f V \t',abs(V3c));...
fprintf('phaseV3c = %4.2f deg \t',angle(V3c)*180/pi); fprintf(' \n'); fprintf(' \n');

V3c = -1.4395 - 3.1170i

magV3c = 3.43 V phaseV3c = -114.79 deg
```

Then,

$$V''_C = -1.44 - j3.12 = 3.43\angle-114.8°$$

or

$$v'''_C(t) = 3.43\cos(2000t - 114.8°)$$

and

$$v_C(t) = v'_C + v''_C(t) + v'''_C(t) = 5 + 3.97\cos(1000t - 62.9°) + 3.43\cos(2000t - 114.8°)$$

These waveforms are plotted below using the following MATLAB code:

```
wt=linspace(0,2*2*pi); deg=wt*180/pi; V1c=5;
V2c=3.97.*cos(wt-62.9.*pi./180);
```

V3c=3.43.*cos(2.*wt−114.8.*pi./180); plot(deg,V1c,deg,V2c,deg,V3c, deg,V1c+V2c+V3c)

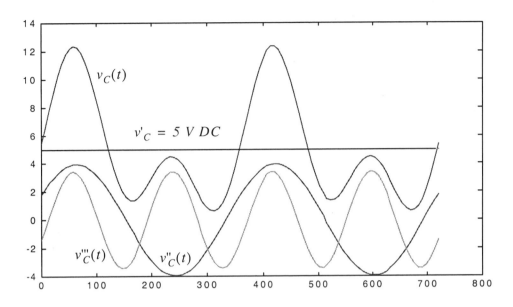

6. Since Z_S and Z_{LOAD} are complex quantities, we will express them as $Z_S = Re\{Z_S\} + jIm\{Z_S\}$ and $Z_{LOAD} = Re\{Z_{LOAD}\} + jIm\{Z_{LOAD}\}$ where Re and Im denote the real and imaginary components respectively.

We want to maximize

$$p_{LOAD} = i^2_{LOAD} \cdot Z_{LOAD} = \frac{v^2_S}{(Z_S + Z_{LOAD})^2} \cdot Z_{LOAD}$$

$$= \frac{v^2_S \cdot Z_{LOAD}}{[Re\{Z_S\} + jIm\{Z_S\} + j(Re\{Z_{LOAD}\} + jIm\{Z_{LOAD}\})]^2}$$

The only that can vary are $Re\{Z_{LOAD}\}$ and $Im\{Z_{LOAD}\}$ and we must consider them independently from each other.

From the above expression we observe that p_{LOAD} will be maximum when the denominator is minimum and this will occur when $Im\{Z_{LOAD}\} = -Im\{Z_S\}$, that is, when the imaginary parts of Z_{LOAD} and Z_S cancel each other. Under this condition, p_{LOAD} simplifies to

$$p_{LOAD} = \frac{v^2_S \cdot R_{LOAD}}{(R_S + R_{LOAD})^2}$$

and, as we found in Chapter 3, for maximum power transfer $R_{LOAD} = R_S$. Therefore, the load impedance Z_{LOAD} will receive maximum power when

$$Z_{LOAD} = Z_S{}^*$$

that is, when Z_{LOAD} is adjusted to be equal to the complex conjugate of Z_S.

7. For this, and other similar problems involving the maximum power transfer theorem, it is best to replace the circuit with its Thevenin equivalent. Moreover, we only need to compute Z_{TH}.

For this problem, to find Z_{TH} we remove Z_{LOAD} and we short the voltage source. The remaining circuit then is as shown below.

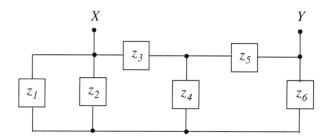

We observe that z_1 is in parallel with z_2 and this combination is shown as z_{12} in the simplified circuit below.

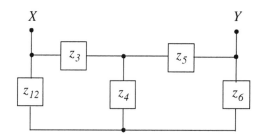

But this circuit cannot be simplified further unless we perform Wye to Delta transformation which we have not discussed. This and the Delta to Wye transformation are very useful in three-phase circuits and are discussed in *Circuit Analysis II with MATLAB Applications* by this author. Therefore, we will compute Z_{TH} using the relation $Z_{TH} = V_{OC}/I_{SC}$ where V_{OC} is the open circuit voltage, that is, V_{TH} and I_{SC} is the current that would flow between the terminals when the load is replaced by a short. Thus, we will begin our computations with the Thevenin voltage.

We disconnect Z_{LOAD} from the circuit at points X and Y as shown below.

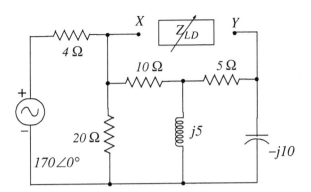

We will replace the remaining circuit with its Thevenin equivalent. Thus, with Z_{LOAD} disconnected the circuit simplifies to that shown below.

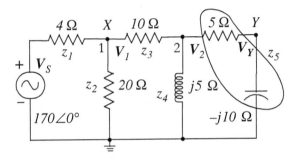

Now, we will find

$$V_{TH} = V_{XY} = V_X - V_Y = V_1 - (V_2 - V_{R_{5\,\Omega}})$$

At Node 1:

$$\frac{V_1 - V_S}{z_1} + \frac{V_1}{z_2} + \frac{V_1 - V_2}{z_3} = 0$$

$$\left(\frac{1}{z_1} + \frac{1}{z_2} + \frac{1}{z_3}\right)V_1 - \frac{1}{z_3}V_2 = \frac{V_S}{z_1}$$

At Node 2:

$$\frac{V_2 - V_1}{z_3} + \frac{V_2}{z_4} + \frac{V_2}{z_5} = 0$$

$$-\frac{1}{z_3}V_1 + \left(\frac{1}{z_3} + \frac{1}{z_4} + \frac{1}{z_5}\right)V_2$$

and with MATLAB

Vs=170; z1=4; z2=20; z3=10; z4=5j; z5=5–10j;...

```
Y=[1/z1+1/z2+1/z3 –1/z3; –1/z3 1/z3+1/z4+1/z5]; I=[Vs/z1 0]'; V=Y\I; V1=V(1); V2=V(2);...
VX=V1; VY=(5/z5)*V2; VTH=VX–VY; fprintf(' \n');...
disp('V1 = '); disp(V1); disp('V2 = '); disp(V2);...
disp('VTH = '); disp(VTH); fprintf('magVTH = %4.2f V ',abs(VTH));...
fprintf('phaseVTH = %4.2f deg ',angle(VTH)*180/pi); fprintf(' \n'); fprintf(' \n');

V1 = 1.1731e+002 + 1.1538e+001i

V2 = 44.2308+46.1538i

VTH = 1.2692e+002 - 1.5385e+001i

magVTH = 127.85 V phaseVTH = -6.91 deg
```

Thus, $V_{TH} = 127.85\angle{-6.91°}$

Next, we must find I_{SC} from the circuit shown below.

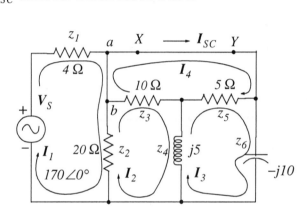

We will write four mesh equations as shown above but we only are interested in phasor current I_4. Observing that a and b are the same point the mesh equations are

$$(z_1 + z_2)\,I_1 - z_2\,I_2 = V_S$$

$$-z_2\,I_1 + (z_2 + z_3 + z_4)\,I_2 - z_4\,I_3 - z_3\,I_4 = 0$$

$$-z_4\,I_2 + (z_4 + z_5 + z_6)\,I_3 - z_5\,I_4 = 0$$

$$-z_3\,I_2 - z_5\,I_3 + (z_3 + z_5)\,I_4 = 0$$

and in matrix form

$$
\begin{bmatrix}
z_1 + z_2 & -z_2 & 0 & 0 \\
-z_2 & z_2 + z_3 + z_4 & -z_4 & -z_3 \\
0 & -z_4 & z_4 + z_5 + z_6 & -z_5 \\
& -z_3 & -z_5 & z_3 + z_5
\end{bmatrix}
\cdot
\begin{bmatrix}
I_1 \\
I_2 \\
I_3 \\
I_4
\end{bmatrix}
=
\begin{bmatrix}
V_S \\
0 \\
0 \\
0
\end{bmatrix}
$$

With MATLAB

Vs=170; VTH=126.92−15.39j; z1=4; z2=20; z3=10; z4=5j; z5=5; z6=−10j;...
Z=[z1+z2 −z2 0 0; −z2 z2+z3+z4 −z4 −z3; 0 −z4 z4+z5+z6 −z5; 0 −z3 −z5 z3+z5];...
V=[Vs 0 0 0]'; I=Z\V; I1=I(1); I2=I(2); I3=I(3); I4=I(4);...
ZTH=VTH/I4; fprintf(' \n'); disp('I1 = '); disp(I1); disp('I2 = '); disp(I2);...
disp('I3 = '); disp(I3); disp('I4 ='); disp(I4); disp('ZTH ='); disp(ZTH); fprintf(' \n');

I1 = 15.6745 - 2.6300i

I2 = 10.3094 - 3.1559i

I3 = -1.0520 + 10.7302i

I4 = 6.5223 + 1.4728i

ZTH = 18.0084 - 6.4260i

Thus, $Z_{TH} = 18.09 - j6.43 \ \Omega$ and by Problem 6, for maximum power transfer there must be
$Z_{LOAD} = Z^*_{TH}$ or

$$Z_{LOAD} = 18.09 + j6.43 \ \Omega$$

8. We assign phasor currents as shown below.

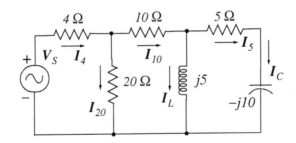

We choose I_5 as a reference, that is, we let

$$I_5 = 1\angle 0° \ A$$

Then,

$$V_5 = 5\angle 0° \ V$$

and since $I_C = I_5$

$$V_C = I_C / (-j10) = 1\angle 0° \cdot 10\angle -90° = 10\angle -90° \ V$$

Next,

$$V_L = V_5 + V_C = 5\angle 0° + 10\angle -90° = 5 + (-j10) = 5 - j10 = 11.18\angle -63.4° \ V$$

and

$$I_L = V_L/j5 = (11.18\angle-63.4°)/(5\angle90°) = 2.24\angle-153.4° = -2-j \text{ A}$$

Now

$$I_{10} = I_L + I_5 = -2-j+1 = -1-j = \sqrt{2}\angle-135° \text{ A}$$

and

$$V_{10} = 10 \times \sqrt{2}\angle-135° = 10 \times (-1-j) = -10-j10 \text{ V}$$

Continuing we find

$$V_{20} = V_{10} + V_L = -10-j10+5-j10 = -5-j20 \text{ V}$$

and

$$I_{20} = V_{20}/20 = (-5-j20)/20 = -0.25-j \text{ A}$$

Also,

$$I_4 = I_{20} + I_{10} = -0.25-j-1-j = -1.25-j2 \text{ A}$$

and

$$V_4 = 4I_4 = 4 \times (-1.25-j2) = -5-j8 \text{ V}$$

Finally,

$$V_S = V_4 + V_{20} = -5-j8-5-j20 = -10-j28 = 29.73\angle-109.7° \text{ V}$$

The magnitudes (not to scale) and the phase angles are shown below.

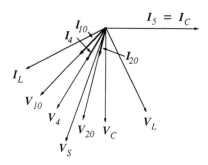

The phasor diagram above is acceptable. However, it would be more practical if we rotate it by 109.7° to show the voltage source V_S as reference at $0°$ as shown below.

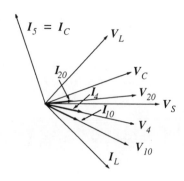

9. The equivalent phasor circuit is shown below where $z_1 = R_1 = 1\ K\Omega$, $z_2 = R_2 = 3\ K\Omega$, and

$$z_3 = -j/\omega C = -j/(10^3 \times 0.25 \times 10^{-6}) = -j4\ K\Omega$$

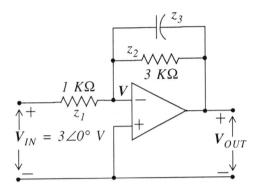

Application of KCL yields

$$\frac{V - V_{IN}}{z_1} + \frac{V - V_{OUT}}{z_2} + \frac{V - V_{OUT}}{z_3} = 0$$

and since $V = 0$ the above relation reduces to

$$\left(\frac{1}{z_2} + \frac{1}{z_3} \right) V_{OUT} = \frac{-V_{IN}}{z_1}$$

or

$$V_{OUT} = \frac{-V_{IN}}{z_1 \cdot \left(\frac{1}{z_2} + \frac{1}{z_3} \right)} = \frac{-V_{IN}}{\left(\frac{z_1}{z_2} + \frac{z_1}{z_3} \right)}$$

and with MATLAB

```
Vin=3; z1=1000; z2=3000; z3=-4000j; Vout=-Vin/(z1/z2+z1/z3);...
fprintf(' \n'); disp('Vout = '); disp(Vout); fprintf('magVout = %5.2f V \t',abs(Vout));...
fprintf('phaseVout = %5.2f deg \t',angle(Vout)*180/pi); fprintf(' \n'); fprintf(' \n');
```

```
Vout = -5.7600 + 4.3200i

magVout = 7.20 V    phaseVout = 143.13 deg
```

Thus,

$$V_{OUT} = -5.76 + j4.32 = 7.2\angle143.13°\ V$$

and

$$v_{out}(t) = 7.2\cos(1000t + 143.13°)\ V$$

NOTES

Chapter 8

Average and RMS Values, Complex Power, and Instruments

T his chapter defines average and effective values of voltages and currents, instantaneous and average power, power factor, the power triangle, and complex power. It also discusses electrical instruments that are used to measure current, voltage, resistance, power, and energy.

8.1 Periodic Time Functions

A periodic time function satisfies the expression

$$f(t) = f(t + nT) \qquad (8.1)$$

where n is a positive integer and T is the period of the periodic time function. The sinusoidal and sawtooth waveforms of Figure 8.1 are examples of periodic functions of time.

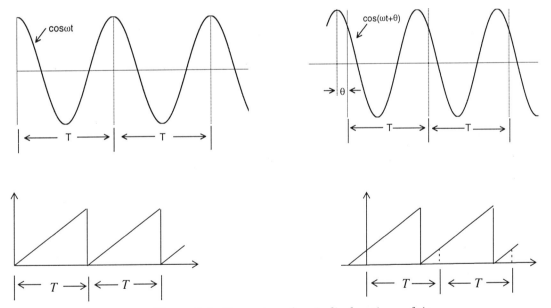

Figure 8.1. Examples of periodic functions of time

Other periodic functions of interest are the square and the triangular waveforms.

8.2 Average Values

The *average value* of any continuous function $f(t)$ such as that shown in Figure 8.2 over an interval $a \leq t \leq b$,

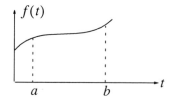

Figure 8.2. A continuous time function $f(t)$

is defined as

$$f(t)_{ave} = \frac{1}{b-a} \int_a^b f(t)dt = \frac{1}{b-a}(area\Big|_a^b) \tag{8.2}$$

The *average value of a periodic time function $f(t)$* is defined as the average of the function over one period.

Example 8.1

Compute the average value of the sinusoid shown in Figure 8.3, where V_p denotes the peak (maximum) value of the sinusoidal voltage.

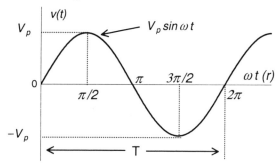

Figure 8.3. Waveform for Example 8.1

Solution:

By definition,

$$V_{ave} = \frac{1}{T}\int_0^T v(t)dt = \frac{1}{\omega T}\int_0^{\omega T} V_p \sin\omega t\, d(\omega t) = \frac{V_p}{2\pi}\int_0^{2\pi} \sin\omega t\, d(\omega t)$$

$$= \frac{V_p}{2\pi}(-\cos\omega t)\Big|_0^{2\pi} = \frac{V_p}{2\pi}(\cos\omega t)\Big|_{2\pi}^0 = \frac{V_p}{2\pi}(1-1) = 0$$

as expected since the net area of the positive and negative half cycles is zero.

Example 8.2

Compute the average value of the *half-wave rectification* waveform shown in Figure 8.4.

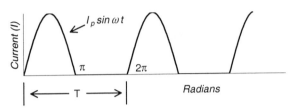

Figure 8.4. Waveform for Example 8.2

Solution:

This waveform is defined as

$$i(t) = \begin{cases} I_p sin\omega t & 0 < \omega t < \pi \\ 0 & \pi < \omega t < 2\pi \end{cases} \tag{8.3}$$

Then, its average value is found from

$$I_{ave} = \frac{1}{2\pi}\int_0^{2\pi} I_p sin\omega t d(\omega t) = \frac{1}{2\pi}\left[\int_0^{\pi} I_p sin\omega t d(\omega t) + \int_{\pi}^{2\pi} 0\, d(\omega t)\right]$$

$$= \frac{I_p}{2\pi}(-cos\omega t\Big|_0^{\pi}) = \frac{I_p}{2\pi}cos\omega t\Big|_{\pi}^0 = \frac{I_p}{2\pi}[1-(-1)] = \frac{I_p}{\pi} \tag{8.4}$$

In other words, the average value of the half-wave rectification waveform is equal to its peak value divided by π.

8.3 Effective Values

The *effective current* I_{eff} of a periodic current waveform $i(t)$ is defined as the current which produces heat in a given resistance R at the same average rate as a direct (constant) current I_{dc}, that is,

$$Average\ Power = P_{ave} = RI_{eff}^2 = RI_{dc}^2 \tag{8.5}$$

Also, in a periodic current waveform $i(t)$, the instantaneous power is

$$p(t) = R\,i^2(t) \tag{8.6}$$

and

$$P_{ave} = \frac{1}{T}\int_0^T p(t)dt = \frac{1}{T}\int_0^T Ri^2 dt = \frac{R}{T}\int_0^T i^2 dt \tag{8.7}$$

Equating (8.5) with (8.7) we get

$$RI_{eff}^2 = \frac{R}{T}\int_0^T i^2 dt$$

or

$$I_{eff}^2 = \frac{1}{T}\int_0^T i^2 dt$$

or

$$I_{eff} = \sqrt{\frac{1}{T}\int_0^T i^2 dt} = I_{Root\ Mean\ Square} = I_{RMS} = \sqrt{Ave(i^2)} \tag{8.8}$$

Caution 1: In general, $ave(i^2) \neq (i_{ave})^2$ since the expression $ave(i^2)$ implies that the function i must first be squared and the average of the squared value is then to be found. On the other hand, $(i_{ave})^2$ implies that the average value of the function must first be found and then the average must be squared. The waveform of Figure 8.5 illustrates this point.

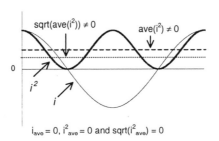

Figure 8.5. Waveforms to illustrate that $ave(i^2) \neq (i_{ave})^2$

Caution 2: In general, $P_{ave} \neq V_{ave} \times I_{ave}$. For example, if $v(t) = V_p cos\omega t$ and $i(t) = I_p cos(\omega t + \varphi)$, then $V_{ave} = 0$, and also $I_{ave} = 0$. Thus, $P_{ave} = 0$. However,

$$P_{ave} = \frac{1}{T}\int_0^T p(t)dt = \frac{1}{T}\int_0^T v(t)i(t)dt = \frac{1}{T}\int_0^T (V_p \cos\omega t)[I_p \cos(\omega t + \varphi)]dt \neq 0$$

8.4 Effective (RMS) Value of Sinusoids

Now, we will derive an expression for the *Root Mean Square (RMS)* value of a sinusoid in terms of its peak (maximum) value. We will denote the peak values of voltages and currents as V_p and I_p respectively. The value from positive to negative peak will be denoted as V_{p-p} and I_{p-p}, and the *RMS* values as V_{RMS} and I_{RMS}. Their notations and relationships are shown in Figure 8.6.

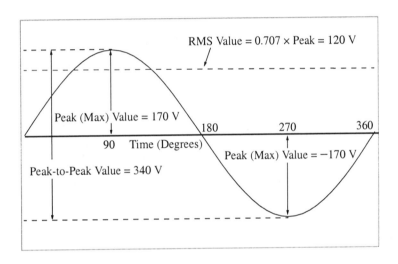

Figure 8.6. Definitions of V_{p-p}, I_{p-p}, V_{RMS}, and I_{RMS} in terms of V_p and I_p

Let

$$i = I_p \cos(\omega t - \theta)$$

then,

$$I_{RMS}^2 = \frac{1}{T}\int_0^T i^2 dt = \frac{1}{2\pi}\int_0^{2\pi} I_p^2 \cos^2(\omega t - \theta)d(\omega t)$$

and using the identity

$$\cos^2\phi = \frac{1}{2}(\cos 2\phi + 1)$$

we get

$$I_{RMS}^2 = \frac{I_p^2}{4\pi}\left[\int_0^{2\pi} cos(2\omega t - \theta)d(\omega t) + \int_0^{2\pi} d(\omega t)\right]$$

(8.9)

$$= \frac{I_p^2}{4\pi}\left[\frac{sin(2\omega t - \theta)}{2}\bigg|_0^{2\pi} + (\omega t\big|_0^{2\pi})\right] = \frac{I_p^2}{4\pi}\left[\frac{sin(4\pi - \theta) - sin(-\theta)}{2} + 2\pi\right]$$

Using the trigonometric identities

$$sin(x - y) = sinx cosy - cosx siny$$

and

$$-sin(-\alpha) = sin\alpha$$

by substitution into (8.9), we get

$$I_{RMS}^2 = \frac{I_p^2}{4\pi}\left[\frac{\overbrace{sin4\pi cos\theta}^{0} - \overbrace{cos4\pi sin\theta}^{1} + sin\theta}{2} + 2\pi\right] = \frac{I_p^2}{4\pi}(2\pi) = \frac{I_p^2}{2}$$

and therefore,

$$\boxed{\begin{array}{c} I_{RMS} = \dfrac{I_p}{\sqrt{2}} = 0.7071 I_p \\[2mm] \textbf{\textit{FOR SINUSOIDS ONLY}} \end{array}}$$

(8.10)

We observe that the *RMS* value of a sinusoid is independent of the frequency and phase angle, in other words, it is dependent on the amplitude of the sinusoid only.

Example 8.3

Compute the I_{ave} and I_{RMS} for the sawtooth waveform shown in Figure 8.7.

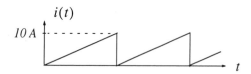

Figure 8.7. Waveform for Example 8.3

Solution:

By inspection, the period T is as shown in Figure 8.8.

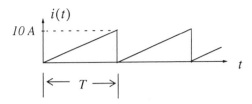

Figure 8.8. Defining the period for the waveform of Example 8.3

The average value is

$$I_{ave} = \frac{Area}{Period} = \frac{(1/2) \times 10 \times T}{T} = 5 \ A$$

To find I_{RMS} we cannot use (8.10); this is for sinusoids only. Accordingly, we must use the definition of the *RMS* value as derived in (8.8). Then,

$$I_{RMS}^2 = \frac{1}{T} \int_0^T i^2(t)dt = \frac{1}{T} \int_0^T \left(\frac{10}{T}t\right)^2 dt = \frac{1}{T}\left(\frac{100}{T^2} \cdot \frac{t^3}{3}\right)\bigg|_0^T = \frac{1}{T}\left(\frac{100}{T^2} \cdot \frac{T^3}{3}\right) = \frac{100}{3}$$

or

$$I_{RMS} = \sqrt{\frac{100}{3}} = \frac{10}{\sqrt{3}} = 5.77 \ A$$

8.5 RMS Values of Sinusoids with Different Frequencies

The *RMS* value of a waveform which consists of a sum of sinusoids of different frequencies, is equal to the square root of the sum of the squares of the *RMS* values of each sinusoid. Thus, if

$$i = I_0 + I_1 cos(\omega_1 t \pm \theta_1) + I_2 \, cos(\omega_2 t \pm \theta_2) + ... + I_N \, cos(\omega_N t \pm \theta_N) \tag{8.11}$$

where I_0 represents a constant current, and $I_1, I_2..., I_N$ represent the amplitudes of the sinusoids. Then, the *RMS* value of i is found from

$$I_{RMS} = \sqrt{I_0^2 + I_{1\,RMS}^2 + I_{2\,RMS}^2 + ... + I_{N\,RMS}^2} \tag{8.12}$$

or

$$I_{RMS} = \sqrt{I_0^2 + \frac{1}{2}I_{1\,p}^2 + \frac{1}{2}I_{2\,p}^2 + ... + \frac{1}{2}I_{N\,p}^2} \tag{8.13}$$

Example 8.4

Find the I_{RMS} value of the square waveform of Figure 8.9 by application of (8.12)

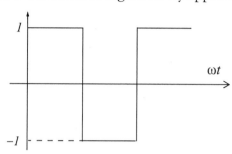

Figure 8.9. Waveform for Example 8.4

Solution:

By inspection, the period $T = 2\pi$ is as shown in Figure 8.9.

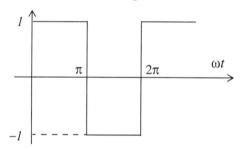

Figure 8.10. Determination of the period to the waveform of Example 8.4

Then,

a.

$$I_{RMS}^2 = \frac{1}{T}\int_0^T i^2 dt = \frac{1}{2\pi}\int_0^{2\pi} i^2 d(\omega t) = \frac{1}{2\pi}\left[\int_0^{\pi} 1^2 d(\omega t) + \int_{\pi}^{2\pi} (-1)^2 d(\omega t)\right]$$

$$= \frac{1}{2\pi}[\omega t \big|_0^{\pi} + \omega t \big|_{\pi}^{2\pi}] = \frac{1}{2\pi}[\pi + 2\pi - \pi] = 1$$

or

$$I_{RMS} = 1 \tag{8.14}$$

b. Fourier series analysis textbooks[*] show that the square waveform above can be expressed as

$$i(t) = \frac{4}{\pi}\left(\sin\omega t + \frac{1}{3}\sin 3\omega t + \frac{1}{5}\sin 5\omega t + ...\right) \tag{8.15}$$

---

[*] *Such a textbook is Signals and Systems with MATLAB Applications, ISBN 0-9709511-3-2 by this author.*

and as we know, the *RMS* value of a sinusoid is a real number independent of the frequency and the phase angle, and it is equal to *0.707* times its peak value, that is, $I_{RMS} = 0.707 \times I_p$. Then from (8.12) and (8.15),

$$I_{RMS} = \frac{4}{\pi}\sqrt{0 + \frac{1}{2}(1)^2 + \frac{1}{2}\left(\frac{1}{3}\right)^2 + \frac{1}{2}\left(\frac{1}{5}\right)^2 + \ldots} = 0.97 \qquad (8.16)$$

The numerical accuracy of (8.16) is good considering that higher harmonics have been neglected.

8.6 Average Power and Power Factor

Consider the network shown in Figure 8.11.

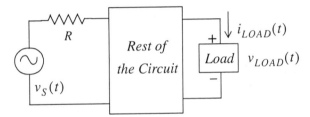

Figure 8.11. Network where it is assumed that $i_{LOAD}(t)$ and $v_{LOAD}(t)$ are out-of-phase

We will assume that the load current $i_{LOAD}(t)$ is θ degrees out-of-phase with the voltage $v_{LOAD}(t)$, i.e., if $v_{LOAD}(t) = V_p cos\omega t$, then $i_{LOAD}(t) = I_p cos(\omega t + \theta)$. We want to find an expression for the average power absorbed by the load.

As we know

$$p = vi$$

that is,

$$instantaneous\ power = instantaneous\ voltage \times instantaneous\ current$$

and the instantaneous power $p_{LOAD}(t)$ absorbed by the load is

$$p_{LOAD}(t) = v_{LOAD}(t) \times i_{LOAD}(t) = V_p I_p cos\omega t \times cos(\omega t + \theta) \qquad (8.17)$$

Using the trigonometric identity

$$cosx cosy = \frac{1}{2}[cos(x + y) + cos(x - y)]$$

we express (8.17) as

$$p_{LOAD}(t) = \frac{V_p I_p}{2}[cos(2\omega t + \theta) + cos\theta] \tag{8.18}$$

and the average power is

$$P_{ave\ LOAD} = \frac{1}{T}\int_0^T p_{LD}dt = \frac{1}{T}\int_0^T \left(\frac{V_p I_p}{2}[cos(2\omega t + \theta) + cos\theta]\right)dt$$

$$= \frac{V_p I_p}{2T}\int_0^T ([cos(2\omega t + \theta)])dt + \frac{V_p I_p}{2T}\int_0^T cos\theta dt \tag{8.19}$$

We observe that the first integral on the right side of (8.19) is zero, and the second integral, being a constant, has an average value of that constant. Then,

$$\boxed{P_{ave\ LOAD} = \frac{V_p I_p}{2}cos\theta} \tag{8.20}$$

and using the relations

$$V_{RMS} = \frac{V_p}{\sqrt{2}}$$

and

$$I_{RMS} = \frac{I_p}{\sqrt{2}}$$

we can express (8.19) as

$$\boxed{P_{ave\ LOAD} = V_{RMS\ LOAD}\ I_{RMS\ LOAD}cos\theta} \tag{8.21}$$

and it is imperative that we remember that *these relations are valid for circuits with sinusoidal excitations.*

The term $cos\theta$ in (8.20) and (8.21) is known as the *power factor* and thus

$$\boxed{Power\ Factor_{LOAD} = PF_{LOAD} = cos\theta_{LOAD} = \frac{P_{ave\ LOAD}}{V_{RMS\ LOAD}\ I_{RMS\ LOAD}}} \tag{8.22}$$

8.7 Average Power in a Resistive Load

The voltage and current in a resistive branch of a circuit are always in phase, that is, the phase angle $\theta = 0°$. Therefore, denoting that resistive branch with the subscript R we have:

$$P_{ave\ R} = V_{RMS\ R}\ I_{RMS\ R}\ cos0° = V_{RMS\ R}\ I_{RMS\ R} \qquad (8.23)$$

or

$$P_{ave\ R} = \frac{V_{RMS\ R}^2}{R} = I_{RMS\ R}^2\ R = \frac{1}{2}\frac{V_{m\ R}^2}{R} = \frac{1}{2}I_{p\ R}^2\ R \qquad (8.24)$$

8.8 Average Power in Inductive and Capacitive Loads

With inductors and capacitors there is a $90°$ phase difference between the voltage and current, that is, $\theta = 90°$ and therefore, denoting that inductive or capacitive branch with the subscript X we get:

$$P_{ave\ X} = V_{RMS\ X}\ I_{RMS\ X}\ cos90° = 0$$

Of course, the instantaneous power is zero only at specific instants.

Obviously, if the load of a circuit contains resistors, inductors and capacitors, the phase angle θ between $V_{RMS\ LOAD}$ and $I_{RMS\ LOAD}$ will be within $0 \leq \theta \leq 90°$, and the power factor $cos\theta$ will be within $0 \leq cos\theta \leq 1$.

Example 8.5

For the circuit of Figure 8.11, find the average power supplied by the voltage source, the average power absorbed by the resistor, the inductor, and the capacitor.

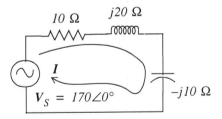

Figure 8.12. Circuit for Example 8.5

Solution:

Since this is a series circuit, we need to find the current I and its phase relation to the source voltage V_S. Then,

$$I = \frac{V_S}{Z} = \frac{170\angle0°}{10 + j20 - j10} = \frac{170\angle0°}{10 + j10} = \frac{170\angle0°}{10\sqrt{2}\angle45°} = 12\angle-45° \qquad (8.25)$$

Relation (8.25) indicates that $I_p = 12\ A$, $\theta = -45°$, and the power factor is

$$cos\theta = cos(-45°) = 0.707$$

Therefore, using (8.24) we find that the average power absorbed by the resistor is

$$P_{ave\ R} = \frac{1}{2}I_{p\ R}^2 R = \frac{1}{2}(12)^2 10 = 720\ w \qquad (8.26)$$

The average power absorbed by the inductor and the capacitor is zero since the voltages and currents in these devices are $90°$ out-of-phase with each other.

Check: The average power delivered by the voltage source is

$$P_{ave\ SOURCE} = \frac{V_p I_p}{2}cos\theta = \frac{(170)(12)}{2}0.707 = 721\ w \qquad (8.27)$$

and we observe that (8.26) and (8.27) are in close agreement.

Example 8.6

For the circuit of Figure 8.13, find the power absorbed by each resistor, and the power supplied (or absorbed) by the current sources.

Solution:

This is the same circuit as in Example 7.1 where we found that

$$V_A = -4.138 + j19.655 = 20.086\angle101.9° \qquad (8.28)$$

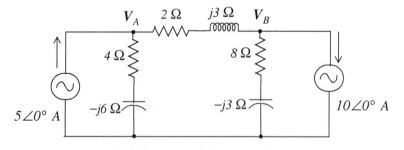

Figure 8.13. Circuit for Example 8.6

and

$$V_B = -22.414 - j1.035 = 22.440\angle-177.4° \qquad (8.29)$$

Then,

$$I_{2\ \Omega} = \frac{V_A - V_B}{2 + j3} = \frac{18.276 + j20.690}{3.61 \angle 56.3°} = \frac{32.430 \angle 145.0°}{3.61 \angle 56.3°} = 8.983 \angle 88.7°$$

and

$$P_{ave\ 2\ \Omega} = \frac{1}{2} I_{p\ 2\ \Omega}^2 (2\ \Omega) = \frac{1}{2} \times 8.983^2 \times 2 = 80.70\ w \qquad (8.30)$$

Also,

$$I_{4\ \Omega} = \frac{V_A}{4 - j6} = \frac{20.086 \angle 101.9°}{7.21 \angle -56.3°} = 2.786 \angle 158.2°$$

and

$$P_{ave\ 4\ \Omega} = \frac{1}{2} I_{p\ 4\ \Omega}^2 (4\ \Omega) = \frac{1}{2} \times 2.786^2 \times 4 = 15.52\ w \qquad (8.31)$$

Likewise,

$$I_{8\ \Omega} = \frac{V_B}{8 - j3} = \frac{22.440 \angle -177.4°}{8.54 \angle -20.6°} = 2.627 \angle -156.7°$$

and

$$P_{ave\ 8\ \Omega} = \frac{1}{2} I_{p\ 8\ \Omega}^2 (8\ \Omega) = \frac{1}{2} \times 2.627^2 \times 8 = 27.61\ w \qquad (8.32)$$

The voltages across the current sources are the same as V_A and V_B but they are $101.9°$ and $-177.4°$ out-of-phase respectively with the current sources as shown by (8.28) and (8.29). Therefore, we let $\theta_1 = 101.9°$ and $\theta_2 = -177.4°$. Then, the power absorbed by the $5\ A$ source is

$$\begin{aligned} P_{ave\ 5\ A} &= \frac{V_p I_p}{2} \cos\theta_1 = \frac{|V_A||5\ A|}{2} \cos(101.9°) \\ &= \frac{20.086 \times 5}{2} \times (-0.206) = -10.35\ w \end{aligned} \qquad (8.33)$$

and the power absorbed by the $10\ A$ source is

$$\begin{aligned} P_{ave\ 10\ A} &= \frac{V_p I_p}{2} \cos\theta_2 = \frac{|V_B||10\ A|}{2} \cos(-177.4°) \\ &= \frac{22.440 \times 10}{2} \times (-0.999) = -112.08\ w \end{aligned} \qquad (8.34)$$

The negative values in (8.33) and (8.34) indicate that both current sources supply power to the rest of the circuit.

Check: Total average power absorbed by resistors is

$$80.70 + 15.52 + 27.61 = 123.83 \ w$$

and the total average power supplied by current sources is

$$112.08 + 10.35 = 122.43 \ w$$

Thus, the total average power supplied by the current sources is equal to the total average power absorbed by the resistors. The small difference is due to rounding of fractional numbers.

8.9 Average Power in Non-Sinusoidal Waveforms

If the excitation in a circuit is non-sinusoidal, we can compute the average power absorbed by a resistor from the relations

$$P_{ave} = \frac{1}{T}\int_0^T p\,dt = \frac{1}{T}\int_0^T \frac{v^2}{R}\,dt = \frac{1}{T}\int_0^T i^2R\,dt \tag{8.35}$$

Example 8.7

Compute the average power absorbed by a $5 \ \Omega$ resistor when the voltage across it is the half-wave rectification waveform shown in Figure 8.14.

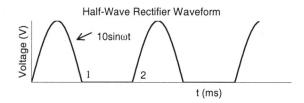

Figure 8.14. Waveform for Example 8.7

Solution:

We first need to find the numerical value of ω. It is found as follows:

$$T = 2 \ ms = 2 \times 10^{-3} \ s \qquad \omega T = 2\pi \qquad \omega = \frac{2\pi}{T} = 10^3\pi$$

and thus

$$10\sin\omega t = 10\sin 10^3\pi$$

Then,

$$P_{ave} = \frac{1}{T}\int_0^T \frac{v^2}{R}\,dt = \frac{1}{2\times 10^{-3}}\left[\int_0^{10^{-3}} \frac{10^2\sin^2 10^3\pi t}{5}\,dt + \int_{10^{-3}}^{2\times 10^{-3}} 0\right]dt$$

or

$$P_{ave} = \frac{100}{10 \times 10^{-3}} \left[\int_0^{10^{-3}} \frac{1}{2} (1 - \cos 2 \times 10^3 \pi t) dt \right] = 5 \times 10^3 \left[\int_0^{10^{-3}} dt - \int_0^{10^{-3}} \cos 2 \times 10^3 \pi t dt \right]$$

$$= \left(5 \times 10^3 \left[t - \frac{\sin 2 \times 10^3 \pi t}{2 \times 10^3 \pi} \right] \Big|_0^{10^{-3}} \right) = 5 \times 10^3 \left[10^{-3} - \frac{\sin 2 \times 10^3 \pi \times 10^{-3}}{2 \times 10^3 \pi} \right]$$

and since $sin2n\pi = 0$ for $n = integer$, the last term of the expression above reduces to

$$P_{ave} = 5 \ w$$

8.10 Lagging and Leading Power Factors

By definition an *inductive load is said to have a lagging power factor*. This refers to the phase angle of the current through the load with respect to the voltage across this load as shown in Figure 8.14.

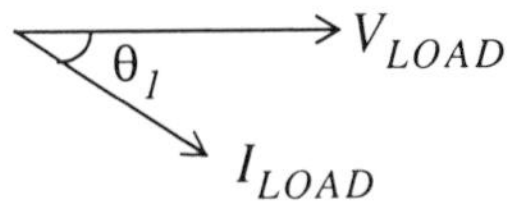

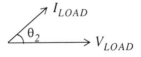

Figure 8.15. Lagging power factor

In Figure 8.14, the cosine of the angle θ_1, that is, $cos\theta_1$ is referred to as *lagging power factor* and it is denoted as *pf lag*.

The term "inductive load" means that the load is more "inductive" (with some resistance) than it is "capacitive". But in a "purely inductive load" $\theta_1 = 90°$ and thus the power factor is

$$cos\theta_1 = cos90° = 0$$

By definition a *capacitive load is said to have a leading power factor*. Again, this refers to the phase angle of the current through the load with respect to the voltage across this load as shown in Figure 8.16.

Figure 8.16. Leading power factor

In Figure 8.16, the cosine of the angle θ_2, that is, $cos\theta_2$ is referred to as *leading power factor* and it is denoted as *pf lead*.

The term "capacitive load" means that the load is more "capacitive" (with some resistance) than it is "inductive". But in a "purely capacitive load" $\theta_2 = 90°$ and thus the power factor is

$$cos\theta_2 = cos\,90° = 0$$

8.11 Complex Power - Power Triangle

We recall that

$$P_{ave} = \frac{1}{2}V_p I_p cos\theta = V_{RMS}\,I_{RMS}\,cos\theta \qquad (8.36)$$

This relation can be represented by the so-called *power triangle*. Figure 8.17 (a) shows the power triangle of an inductive load, and Figure 8.16 (b) shows the power triangle for both a capacitive load.

(a) Power Triangle for Inductive Load   (b) Power Triangle for Capacitive Load

Figure 8.17. Power triangles for inductive and capacitive loads

In a power triangle, the product $V_{RMS} \times I_{RMS}$ is referred to as the *apparent power*, and it is denoted as P_a. The apparent power is expressed in *volt – amperes* or *VA*. The product $V_{RMS} \times I_{RMS} \times sin\theta$ is referred to as the *reactive power*, and it is denoted as Q. The reactive power is expressed in *volt – amperes reactive* or *VAR*. Thus, for either triangle of Figure 8.17,

$$\boxed{P_{real} = P_{ave} = V_{RMS}\,I_{RMS}\,cos\theta = \qquad (in\ watts)} \qquad (8.37)$$

$$\boxed{Q = Reactive\ Power = V_{RMS}\,I_{RMS}\,sin\theta = \qquad (in\ VARs)} \qquad (8.38)$$

$$\boxed{P_a = Apparent\ Power = V_{RMS}\,I_{RMS} \quad (in\ VAs)} \qquad (8.39)$$

The apparent power P_a is the vector sum of the real and reactive power components, that is,

$$P_a = P_{real\ power} \pm jQ = P_{ave} \pm jQ \qquad (8.40)$$

where the (+) sign is used for inductive loads and the (−) sign for capacitive loads. For this reason, the relation of (8.41) is known as the *complex power*.

Example 8.8

For the circuit shown in Figure 8.18, find:

a. the average power delivered to the load

b. the average power absorbed by the line

c. the apparent power supplied by the voltage source

d. the power factor of the load

e. the power factor of the line plus the load

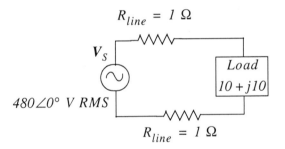

Figure 8.18. Circuit for Example 8.8

Solution:

For simplicity, we redraw the circuit as shown in Figure 8.19 where the line resistances have been combined into a single $2\ \Omega$ resistor.

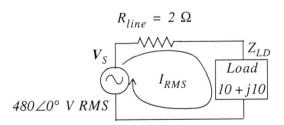

Figure 8.19. Circuit for Example 8.8 with the line resistances combined

From the circuit of Figure 8.19, we find that

$$I_{RMS} = \frac{V_{S\ RMS}}{R_{line} + Z_{LD}} = \frac{480\angle 0°}{2 + 10 + j10} = \frac{480\angle 0°}{15.62\angle 39.8°} = 30.73\angle -39.8°$$

and therefore, the current lags the voltage as shown on the phasor diagram of Figure 8.20.

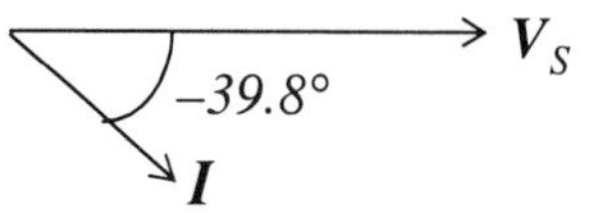

Figure 8.20. Phasor diagram for the circuit of Example 8.8

Then,

a. The average power delivered to the load is

$$P_{ave\ LOAD} = I_{RMS}^2\ Re\{Z_{LOAD}\} = (30.73)^2 \times 10 = 9443\ w = 9.443\ Kw$$

b. The average power absorbed by the line is

$$P_{ave\ line} = I_{RMS}^2\ R_{line} = (30.73)^2 \times 2 = 1889\ w = 1.889\ Kw$$

c. The apparent power supplied by the voltage source is

$$P_{a\ source} = V_{S\ RMS}\ I_{RMS} = 480 \times 30.73 = 14750\ w = 14.750\ Kw$$

d. The power factor of the load is

$$f_{LOAD} = cos\theta_{LOAD} = \frac{P_{ave\ LOAD}}{P_{a\ LOAD}} = \frac{9443}{|V_{RMS\ LOAD}||I_{RMS}|}$$

$$= \frac{9443}{|(480\angle 0° - 2(30.73\angle -39.8°))| \times 30.73} = \frac{9443}{(434.56)(30.73)} = \frac{9443}{13354} = 0.70\text{.}$$

e. The power factor of the line plus the load is

$$pf_{(line + LOAD)} = cos\theta_{(line + LOAD)} = \frac{P_{ave\ total}}{P_{a\ source}} = \frac{P_{ave\ line} + P_{ave\ LOAD}}{P_{a\ source}} = \frac{1889 + 9443}{14750} = 0.77$$

8.12 Power Factor Correction

The consumer pays the electric utility company for the average or real power, not the apparent power and, as we have seen, a low power factor (larger angle θ) demands more current. This additional current must be furnished by the utility company which must provide larger current-carrying capacity if the voltage must remain constant. Moreover, this additional current creates larger $i^2 R$

losses in the utility's transmission and distribution system. For this reason, electric utility companies impose a penalty on industrial facility customers who operate at a low power factor, typically lower than *0.85*. Accordingly, facility engineers must install the appropriate equipment to raise the power factor.

The power factor correction procedure is illustrated with the following example.

Example 8.9

In the circuit shown in Figure 8.21, the resistance of the lines between the voltage source and the load and the internal resistance of the source are considered small, and thus can be neglected.

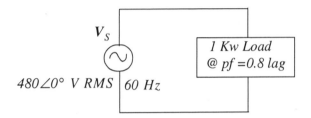

Figure 8.21. Circuit for Example 8.9

It is desired to "raise" the power factor of the load to 0.95 lagging. Compute the size and the rating of a capacitor which, when added across the load, will accomplish this.

Solution:

The power triangles for the existing and desired power factors are shown in Figure 8.22.

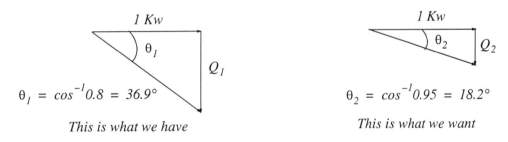

Figure 8.22. Power triangles for existing and desired power factors

Since the voltage across the given load must not change (otherwise it will affect the operation of it), it is evident that a load, say Q_3, in opposite direction of Q_1 must be added, and must be connected in parallel with the existing load. Obviously, the Q_3 load must be capacitive. Accordingly, the circuit of Figure 8.21 must be modified as shown in Figure 8.23.

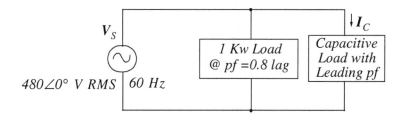

Figure 8.23. Circuit for power factor correction

For the existing load,

$$Q_1 = (1 \ Kw) \tan 36.9° = 750 \ VAR$$

and for the desired $pf = \cos\theta_2 = 0.95$, the VAR value of Q_2 must be reduced to

$$Q_2 = (1 \ Kw) \tan 18.2° = 329 \ VAR$$

Therefore, the added capacitive load must be a vector Q_3 such that

$$Q_3 = Q_1 - Q_2 = 750 - 329 = 421 \ VAR$$

The current I_C through the capacitive load is found from

$$Q_3 = I_C V_C = I_C V_S$$

Then,

$$I_C = \frac{Q_3}{V_{S \ RMS}} = \frac{421}{480} = 0.88 \ A$$

and

$$X_C = \frac{V_C}{I_C} = \frac{480}{0.88} = 547 \ \Omega$$

Therefore, the capacitive load must consist of a capacitor with the value

$$C = \frac{1}{\omega X_C} = \frac{1}{2\pi f X_C} = \frac{1}{2\pi(60)(547)} = 4.85 \ \mu F$$

However, not any $4.85 \ \mu F$ capacitor will do; the capacitor must be capable of withstanding a maximum voltage of

$$V_{C \ max} = \sqrt{2} \times 480 = 679 \ V$$

and for all practical purposes, we can choose a $5 \ \mu F$ capacitor rated at 700 volts or higher.

8.13 Instruments

Ammeters are electrical instruments used to measure current in electric circuits, *voltmeters* measure voltage, *ohmmeters* measure resistance, *wattmeters* measure power, and *watt-hour meters* measure electric energy. Voltmeters, Ohmmeters, and Milliammeters (ammeters which measure current in milliamperes) are normally combined into one instrument called VOM. Figure 8.24 shows a typical analog type VOM, and Figure 8.25 shows a typical digital type VOM. We will see how a digital VOM can be constructed from an analog VOM equivalent at the end of this section. An *oscilloscope* is an electronic instrument that produces an instantaneous trace on the screen of a cathode-ray tube corresponding to oscillations of voltage and current. A typical oscilloscope is shown in Figure 8.26.

DC ammeters and voltmeters read average values whereas AC ammeters and voltmeters read RMS values.

The basic meter movement consists of a permanent horse shoe magnet, an electromagnet which typically is a metal cylinder with very thin wire wound around it which is referred to as the coil, and a control spring. The coil is free to move on pivots, and when there is current in the coil, a torque is produced that tends to rotate the coil. Rotation of the coil is restrained by a helical spring so that the motion of the coil and the pointer which is attached to it, is proportional to the current in the coil.

Figure 8.24. A typical analog VOM

An ammeter measures current in amperes. For currents less than one ampere, a *milliammeter* or *microammeter* may be used where the former measures current in milliamperes and the latter in microamperes.

Ammeters, milliammeters, and microammeters must always be connected in series with the circuits in which they are used.

Often, the electric current to be measured, exceeds the range of the instrument. For example, we cannot directly measure a current of *5* to *10* milliamperes with a milliammeter whose range is *0* to *1* milliampere. In such a case, we can use a low range milliammeter with a *shunt* (parallel) resistor as shown in Figure 8.27, where the circle with mA represents an ideal milliammeter (a milliammeter with zero resistance). In Figure 8.27 I_T is the total current to be measured, I_M is the current through the meter, I_S is the current through the shunt resistor, R_M is the milliammeter internal resistance, and R_S is the shunt resistance.

Figure 8.25. A typical digital VOM

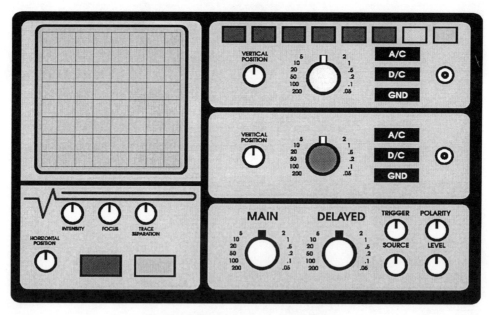

Figure 8.26. A typical oscilloscope

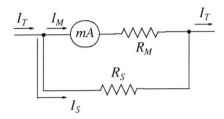

Figure 8.27. Milliammeter with shunt resistor R_S

From the circuit of Figure 8.27, we observe that the sum of the current flowing through the milliammeter I_M and the current I_S through the shunt resistor is equal to the total current I_S, that is,

$$I_T = I_M + I_S \tag{8.41}$$

Also, the shunt resistor R_S is in parallel with the milliammeter branch; therefore, the voltages across these parallel branches are equal, that is,

$$R_M I_M = R_S I_S$$

and since we normally need to calculate the shunt resistor, then

$$\boxed{R_S = \frac{I_M}{I_S} R_M} \tag{8.42}$$

Example 8.10

In the circuit of Figure 8.28, the total current entering the circuit is *5 mA* and the milliammeter range is *0* to *1* milliampere, that is, the milliammeter has a full-scale current I_{fs} of *1 mA*, and its internal resistance is *40 Ω*. Compute the value of the shunt resistor R_S.

I_M =Maximum allowable current
through the milliammeter

$I_{fs} = 1\ mA$

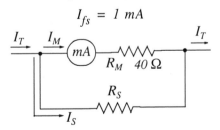

Figure 8.28. Circuit for Example 8.10

Solution:

The maximum current that the milliammeter can allow to flow through it is *1 mA* and since the total current is *5* milliamperes, the remaining *4* milliamperes must flow through the shunt resistor, that is,

$$I_S = I_T - I_M = 5 - 1 = 4 \ mA$$

The required value of the shunt resistor is found from (8.42), i.e.,

$$R_S = \frac{I_M}{I_S} R_M = \frac{1}{4} \times 40 = 10 \ \Omega$$

Check: The calculated value of the shunt resistor is *10* Ω; this is one-fourth the value of the milliammeter internal resistor of *40* Ω. Therefore, the *10* Ω resistor will allow four times as much current as the milliammeter to flow through it.

A *multi-range ammeter/milliammeter* is an instrument with two or more scales. Figure 8.29 shows the circuit of a typical multi-range ammeter/milliammeter.

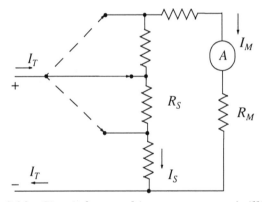

Figure 8.29. Circuit for a multi-range ammeter/milliammeter

A voltmeter, as stated earlier, measures voltage in volts. Typically, a voltmeter is a modified milliammeter where an external resistor R_V is connected in series with the milliammeter as shown in Figure 8.30 where

$$I = current \ through \ circuit$$

$$R_M = internal \ resistance \ of \ milliameter$$

$$R_V = external \ resistor \ in \ series \ with \ R_M$$

$$V_M = voltmeter \ full \ scale \ reading$$

R_V = *Voltmeter internal resistance*

V_M = *Voltmeter range*

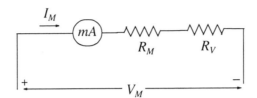

Figure 8.30. Typical voltmeter circuit

For the circuit of Figure 8.30,

$$I_M(R_M + R_V) = V_M$$

or

$$\boxed{R_V = \frac{V_M}{I_M} - R_M}$$

(8.43)

Voltmeters must always be connected in parallel with those devices of the circuit whose voltage is to be measured.

Example 8.11

Design a voltmeter which will have a *1* volt full-scale using a milliammeter with *1* milliampere full-scale and internal resistance *100* Ω.

Solution:

The voltmeter circuit consists of the milliammeter circuit and the external resistance R_V as shown in Figure 8.31.

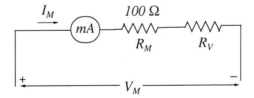

Figure 8.31. Circuit for Example 8.11

Here, we only need to compute the value of the external resistor R_V so that the voltage across the series combination will be *1 volt* full scale. Then, from (8.43),

$$R_V = \frac{V_M}{I_M} - R_M = \frac{1}{10^{-3}} - 100 = 1000 - 100 = 900 \ \Omega \tag{8.44}$$

Therefore, to convert a 1 milliampere full-scale milliammeter with an internal resistance of $100 \ \Omega$ to a *1 volt* full-scale voltmeter, we only need to attach a $900 \ \Omega$ resistor in series with that milliammeter.

Figure 8.32 shows a typical multi-range voltmeter.

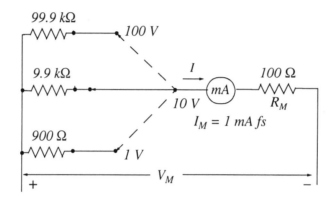

Figure 8.32. Circuit for a multi-range voltmeter

An *Ohmmeter* measures resistance in Ohms. In the series type Ohmmeter, the resistor R_X whose resistance is to be measured, is connected in series with the Ohmmeter circuit shown in Figure 8.33.

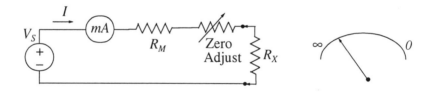

Figure 8.33. Circuit for a series type Ohmmeter

We observe from Figure 8.33 that for the series type Ohmmeter, the current I is maximum when the resistor R_X is zero (short circuit), and the current is zero when R_X is infinite (open circuit). For this reason, the 0 (zero) point appears on the right-most point of the Ohmmeter scale, and the infinity symbol appears on the left-most point of the scale.

Figure 8.34 shows the circuit of a shunt (parallel) type Ohmmeter where the resistor R_X whose value is to be measured, is in parallel with the Ohmmeter circuit.

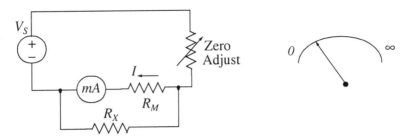

Figure 8.34. Circuit for a parallel type Ohmmeter

From Figure 8.34 we see that, for the shunt type Ohmmeter, the current through the milliammeter circuit is zero when the resistor R_X is zero (short circuit) since all current flows through that short. However, when R_X is infinite (open circuit), the current through the milliammeter branch is maximum. For this reason, the 0 (zero) point appears on the left-most point of the Ohmmeter scale, and the infinity symbol appears on the right-most point of the scale.

An instrument which can measure unknown resistance values very accurately is the *Wheatstone Bridge* shown in Figure 8.35.

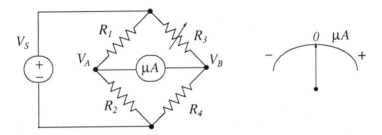

Figure 8.35. Wheatstone Bridge Circuit

One of the resistors, say R_4, is the unknown resistor whose value is to be measured, and another resistor, say R_3 is adjusted until the bridge is balanced, that is, until there is no current flow through the meter of this circuit. This balance occurs when

$$\frac{R_1}{R_2} = \frac{R_3}{R_4}$$

from which the value of the unknown resistor is found from

$$R_4 = \frac{R_2}{R_1}R_3 \tag{8.45}$$

Example 8.12

In the Wheatstone Bridge circuit of Figure 8.36, resistor R_3 is adjusted until the meter reads zero, and when this occurs, its value is $120 \ \Omega$. Compute the value of the unknown resistor R_4.

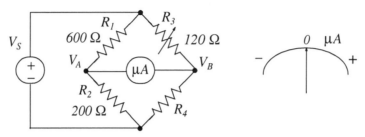

Figure 8.36. Circuit for Example 8.12

Solution:

When the bridge is balanced, that is, when the current through the meter is zero, relation (8.45) holds. Then,

$$R_4 = \frac{R_2}{R_1} R_3 = \frac{200}{600} \times 120 = 40 \ \Omega$$

When measuring resistance values, the voltage sources in the circuit to which the unknown resistance is connected must be turned off, and one end of the resistor whose value is to be measured must be disconnected from the circuit.

Because of their great accuracy, Wheatstone Bridges are also used to accept or reject resistors whose values exceed a given tolerance.

A *wattmeter* is an instrument which measures power in watts or kilowatts. It is constructed with two sets of coils, a current coil and a voltage coil where the interacting magnetic fields of these coils produce a torque which is proportional to the $V \times I$ product.

A *watt-hour meter* is an instrument which measures electric energy W, where W is the product of the average power P in watts and time t in hours, that is, $W = Pt$ in *watt-hours*. Electric utility companies use kilowatt-hour meters to bill their customers for the use of electricity.

Digital meters include an additional circuit called *analog-to-digital converter* (ADC).

There are different types of analog-to-digital converters such as the *flash converter*, the *time-window converter*, *slope converter* and *tracking converter*. We will discuss the flash converter only because of its simplicity. This and the other types are discussed in digital circuits textbooks.

As shown in Figure 8.37, the flash type ADC consists of a *resistive network*, *comparators* (denoted as triangles), and an eight-to-three *line encoder*.

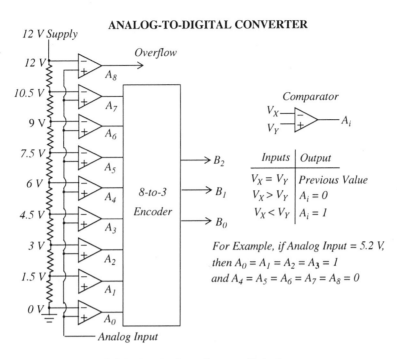

ANALOG-TO-DIGITAL CONVERTER

For Example, if Analog Input = 5.2 V,
then $A_0 = A_1 = A_2 = A_3 = 1$
and $A_4 = A_5 = A_6 = A_7 = A_8 = 0$

Figure 8.37. Typical analog-to-digital converter

Analog Input	A_8	A_7	A_6	A_5	A_4	A_3	A_2	A_1	A_0	B_2	B_1	B_0
Less than 0 V	0	0	0	0	0	0	0	0	0	x	x	x†
0 to less than 1.5 V	0	0	0	0	0	0	0	0	1	0	0	0
1.5 to less than 3.0 V	0	0	0	0	0	0	0	1	1	0	0	1
3.0 to less than 4.5 V	0	0	0	0	0	0	1	1	1	0	1	0
4.5 to less than 6.0 V	0	0	0	0	0	1	1	1	1	0	1	1
6.0 to less than 7.5 V	0	0	0	0	1	1	1	1	1	1	0	0
7.5 to less than 9.0 V	0	0	0	1	1	1	1	1	1	1	0	1
9.0 to less than 10.5 V	0	0	1	1	1	1	1	1	1	1	1	0
10.5 to 12 V	0	1	1	1	1	1	1	1	1	1	1	1
Greater than 12 V	1	1	1	1	1	1	1	1	1	x	x	x‡

† Underflow
‡ Overflow

A *digital-to-analog converter* (DAC) performs the inverse operation, that is, it converts digital values to equivalent analog values.

Figure 8.38 shows a four-bit *R-2R* ladder network and an op-amp connected to form a DAC.

DIGITAL-TO-ANALOG CONVERTER

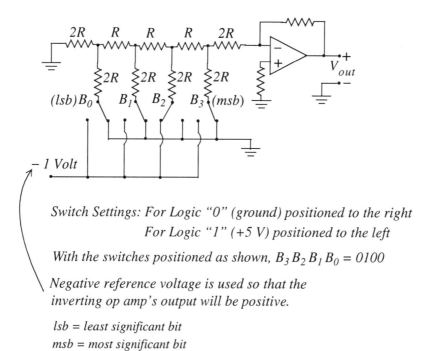

Switch Settings: For Logic "0" (ground) positioned to the right

For Logic "1" (+5 V) positioned to the left

With the switches positioned as shown, $B_3 B_2 B_1 B_0 = 0100$

Negative reference voltage is used so that the
inverting op amp's output will be positive.

lsb = least significant bit

msb = most significant bit

Figure 8.38. A typical digital-to-analog converter

8.14 Summary

- A periodic time function is one which satisfies the relation $f(t) = f(t + nT)$ where n is a positive integer and T is the period of the periodic time function.

- The average value of any continuous function $f(t)$ over an interval $a \le t \le b$, is defined as

$$f(t)_{ave} = \frac{1}{b-a} \int_a^b f(t) dt = \frac{1}{b-a} (area \big|_a^b)$$

- The average value of a periodic time function $f(t)$ is defined as the average of the function over one period.

- A half-wave rectification waveform is defined as

$$f(t) = \begin{cases} A \sin \omega t & 0 < \omega t < \pi \\ 0 & \pi < \omega t < 2\pi \end{cases}$$

- The effective current I_{eff} of a periodic current waveform $i(t)$ is defined as

$$I_{eff} = \sqrt{\frac{1}{T}\int_0^T i^2 dt} = I_{Root\ Mean\ Square} = I_{RMS} = \sqrt{Ave(i^2)}$$

- For sinusoids only, $I_{RMS} = I_p/\sqrt{2} = 0.707 I_p$

- For sinusoids of different frequencies, $I_{RMS} = \sqrt{I_0^2 + I_{1\ RMS}^2 + I_{2\ RMS}^2 + \ldots + I_{N\ RMS}^2}$

- For circuits with sinusoidal excitations the average power delivered to a load is

$$P_{ave\ LOAD} = \frac{V_p I_p}{2}\cos\theta = V_{RMS\ LOAD}\,I_{RMS\ LOAD}\cos\theta$$

where θ is the phase angle between V_{LOAD} and I_{LOAD} and it is within the range $0 \le \theta \le 90°$, and $\cos\theta$ is known as the power factor defined within the range $0 \le \cos\theta \le 1$.

- The average power in a resistive load is

$$P_{ave\ R} = \frac{V_{RMS\ R}^2}{R} = I_{RMS\ R}^2\,R$$

- The average power in inductive and capacitive loads is

$$P_{ave\ X} = V_{RMS\ X}\,I_{RMS\ X}\,\cos 90° = 0$$

- If the excitation in a circuit is non-sinusoidal, we can compute the average power absorbed by a resistor from the relations

$$P_{ave} = \frac{1}{T}\int_0^T p\,dt = \frac{1}{T}\int_0^T \frac{v^2}{R}dt = \frac{1}{T}\int_0^T i^2 R\,dt$$

- An inductive load is said to have a lagging power factor and a capacitive load is said to have a leading power factor.

- In a power triangle

$$P_{real} = P_{ave} = V_{RMS}\,I_{RMS}\,\cos\theta \quad \textit{(in watts)}$$

$$Q = Reactive\ Power = V_{RMS}\,I_{RMS}\,\sin\theta \quad \textit{(in VARs)}$$

$$P_a = Apparent\ Power = V_{RMS}\,I_{RMS} \quad \textit{(in VAs)}$$

- The apparent power P_a, also known as complex power, is the vector sum of the real and reactive power components, that is,

$$P_a = P_{real\ power} \pm jQ = P_{ave} \pm jQ$$

  where the (+) sign is used for inductive loads and the (−) sign for capacitive loads.

- A power factor can be corrected by placing a capacitive load in parallel with the load of the circuit.

- Ammeters are instruments used to measure current in electric circuits. Ammeters, milliammeters, and microammeters must always be connected in series with the circuits in which they are used.

- Voltmeters are instruments used to measure voltage. Voltmeters must always be connected in parallel with those devices of the circuit whose voltage is to be measured.

- Ohmmeters are instruments used to measure resistance. When measuring resistance values, the voltage sources in the circuit to which the unknown resistance is connected must be turned off, and one end of the resistor whose value is to be measured must be disconnected from the circuit.

- A Wheatstone Bridge is an instrument which can measure unknown resistance values very accurately.

- Voltmeters, Ohmmeters, and Milliammeters (ammeters which measure current in milliamperes) are normally combined into one instrument called VOM.

- Wattmeters are instruments used to measure power.

- Watt-Hour meters are instruments used to measure energy.

- An oscilloscope is an electronic instrument that produces an instantaneous trace on the screen of a cathode-ray tube corresponding to oscillations of voltage and current.

- DC ammeters and DC voltmeters read average values

- AC ammeters and AC voltmeters read RMS values.

- Digital meters include an additional circuit called analog-to-digital converter (ADC).

8.15 Exercises

Multiple Choice

1. The average value of a constant (DC) voltage of 12 V is

   A. *6 V*

   B. *12 V*

   C. *12/ $\sqrt{2}$ V*

   D. *12 × $\sqrt{2}$ V*

   E. none of the above

2. The average value of $i = 5 + \cos 100t$ A is

   A. *5 + $\sqrt{2}$ /2 A*

   B. *5 × $\sqrt{2}$ A*

   C. *5/ $\sqrt{2}$ A*

   D. *5 A*

   E. none of the above

3. The RMS value of a constant (DC) voltage of *12 V* is

   A. *12/ $\sqrt{2}$ V*

   B. *6 × $\sqrt{2}$/2 V*

   C. *12 V*

   D. *12 × $\sqrt{2}$ V*

   E. none of the above

4. The RMS value of $i = 5 + \cos 100t$ A is

   A. *5 + $\sqrt{2}$ /2 A*

   B. *5 × $\sqrt{2}$ A*

   C. *5/ $\sqrt{2}$ A*

   D. *5 A*

E. none of the above

5. The voltage across a load whose impedance is $Z = 75 + j38 \ \Omega$ is 115 V RMS. The average power absorbed by that load is

A. *176.33 w*

B. *157.44 w*

C. *71.3 w*

D. *352.67 w*

E. none of the above

6. The average value of the waveform of Figure 8.39 is

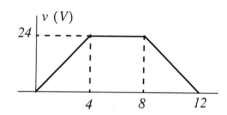

Figure 8.39. Waveform for Question 6

A. *24 V*

B. *16 V*

C. *12 V*

D. *6 V*

E. none of the above

7. The RMS value of the waveform of Figure 8.40 is

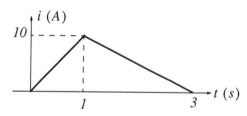

Figure 8.40. Waveform for Question 7

A. *$10/\sqrt{2}$ V*

B. $10 \times \sqrt{2} \ V$

C. $10 / \sqrt{3} \ V$

D. $10 \times \sqrt{3} \ V$

E. none of the above

8. A current with a value of $i = 5 \cos 10000t \ A$ is flowing through a load that consists of the series combination of $R = 2 \ \Omega$, $L = 1 \ mH$, and $C = 10 \ \mu F$. The average power absorbed by this load is

A. $25 \ w$

B. $10 \ w$

C. $5 \ w$

D. $0 \ w$

E. none of the above

9. If the average power absorbed by a load is $500 \ watts$ and the reactive power is $500 \ VAR$, the apparent power is

A. $0 \ VA$

B. $500 \ VA$

C. $250 \ VA$

D. $500 \times \sqrt{2} \ VA$

E. none of the above

10. A load with a leading power factor of 0.60 can be corrected to a lagging power factor of 0.85 by adding

A. a capacitor in parallel with the load

B. an inductor in parallel with the load

C. an inductor is series with the load

D. a capacitor in series with the load

E. none of the above

Problems

1. The current $i_L(t)$ through a *0.5 H* inductor is given as $i_L(t) = 5 + 10\sin t$ *A*. Compute:

   a. The average values of the current, voltage and power for this inductor.

   b. The *RMS* values of the current and voltage.

2. Compute the average and *RMS* values of the voltage waveform of Figure 8.41.

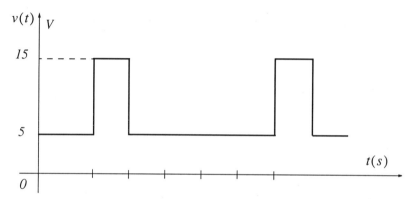

Figure 8.41. Waveform for Problem 2

3. Compute the *RMS* value of the voltage waveform of Figure 8.42.

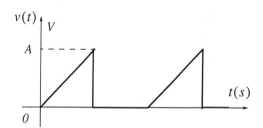

Figure 8.42. Waveform for Problem 3

4. Compute the *RMS* value of $i(t) = 10 + 2\cos 100t + 5\sin 200t$

5. A radar transmitter sends out periodic pulses. It transmits for *5 μs* and then rests. It sends out one of these pulses every *1 ms*. The average output power of this transmitter is *750 w*. Compute:

   a. The energy transmitted in each pulse.

   b. The power output during the transmission of a pulse.

6. For the circuit of Figure 8.43, $v_s(t) = 100\cos 1000t \ V$. Compute the average power delivered (or absorbed) by each device.

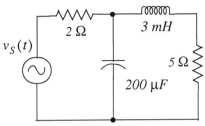

Figure 8.43. Circuit for Problem 6

7. For the circuit of Figure 8.44, the input impedance of the PCB (Printed Circuit Board) is $Z_{IN} = 100 - j100 \ \Omega$ and the board must not absorb more that *200 mw* of power; otherwise it will be damaged. Compute the largest RMS value that the variable voltage source V_S can be adjusted to.

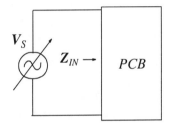

Figure 8.44. Network for Problem 7

8. For the multi-range ammeter/milliammeter shown in Figure 8.45, the meter full scale is *1 mA*. Compute the values of R_1, R_2, R_3, and R_4 so that the instrument will display the indicated values.

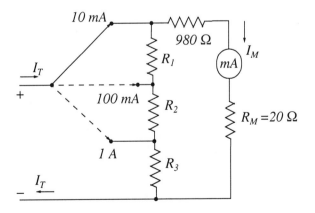

Figure 8.45. Multi-range ammeter for Problem 8

9. The circuit of Figure 8.46 is known as *full-wave rectifier*. The input and output voltage waveforms are shown in Figure 8.47. During the positive input half cycle, current flows from point A to point B, through D_2 to point C, through the resistor R to point D, through diode D_3 to point E, and returns to the other terminal point F of the input voltage source. During the negative input half cycle, current flows from point F to point E, through diode D_4 to point C, through the resistor R to point D, through the diode D_1 to point B, and returns to the other terminal point A of the input voltage source. There is a small voltage drop v_D across each diode[*] but it can be neglected if $v_{in} \gg v_D$. Compute the value indicated by the DC voltmeter.

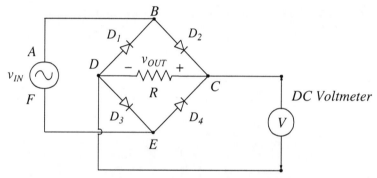

Figure 8.46. Network for Problem 9

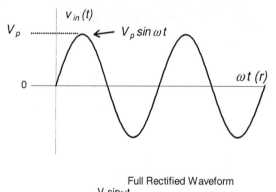

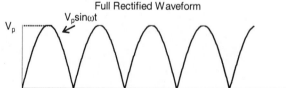

Figure 8.47. Input and output waveforms for the network of Problem 9

---

[*] *For silicon type diodes, the voltage drop is approximately 0.7 volt.*

8.16 Answers to Exercises

Multiple Choice

1. B

2. D

3. C

4. A

5. E $Z = 75 + j38 = 84.08\angle 26.87°$, $I_{RMS} = 115\angle 0° / 84.08\angle 26.87° = 1.37\angle -26.87°$ and thus

$$P_{ave} = V_{RMS} \cdot I_{RMS} \cos\theta = 115 \times 1.37 \times \cos(-26.87°) = 140.54 \ w$$

6. B

7. C

8. A

9. D

10. B

Problems

1. $i_L = 5 + 10\sin t$, $v_L = L\dfrac{di_L}{dt} = 0.5\dfrac{d}{dt}(5 + 10\sin t) = 5\cos t$

a.

$i_{L\ ave} = \dfrac{1}{T}\displaystyle\int_0^T i_L dt = \dfrac{1}{T}\displaystyle\int_0^T (5 + 10\sin t)dt$ and since $\dfrac{1}{T}\displaystyle\int_0^T 10\sin t dt = 0$, it follows that

$\dfrac{1}{T}\displaystyle\int_0^T 5dt = \dfrac{1}{T}5T = 5\ A$. Likewise, $v_{L\ ave} = \dfrac{1}{T}\displaystyle\int_0^T 5\cos t dt = 0$. Also,

$$p_{L\ ave} = \dfrac{1}{T}\int_0^T p_L dt = \dfrac{1}{T}\int_0^T v_L i_L dt = \dfrac{1}{T}\int_0^T 5\cos t(5 + 10\sin t)dt = \dfrac{1}{T}\int_0^T (25\cos t + 50\sin t\cos t)dt$$

and using $\sin 2x = 2\sin x\cos x$ it follows that $50\sin t\cos t = 25\sin 2t$ and thus

$$p_{L\ ave} = \dfrac{1}{T}\int_0^T (25\cos t + 25\sin 2t)dt = 0$$

b.

$$I^2_{L\ RMS} = \frac{1}{T}\int_0^T i^2_L dt = \frac{1}{T}\int_0^T (5 + 10\sin t)^2 dt$$

$$= \frac{1}{T}\int_0^T [5(1 + 2\sin t)]^2 dt = \frac{25}{T}\int_0^T (1 + 4\sin t + 4\sin^2 t)dt$$

Using $\sin^2 x = \dfrac{1 - \cos 2x}{2}$ and observing that $\dfrac{1}{T}\int_0^T 4\sin t\, dt = 0$ and $\dfrac{1}{T}\int_0^T \cos 2t\, dt = 0$ we get

$$I^2_{L\ RMS} = \frac{25}{T}\left(t\Big|_0^T + \frac{4}{2}t\Big|_0^T\right) = \frac{25}{T}(T + 2T) = 75 \text{ and } I_{L\ RMS} = \sqrt{75} = 8.66\ A$$

For sinusoids $V_{RMS} = V_p/(\sqrt{2}) = 0.707V_p$ and since $V_p = 5$ it follows that

$$V_{RMS} = 0.707 \times 5 = 3.54\ V$$

2. From the waveform below we observe that $Period = T = 5\tau$ and since

$$V_{ave} = Area/Period = (15\tau + 20\tau)/5\tau = 7\ V$$

Also,

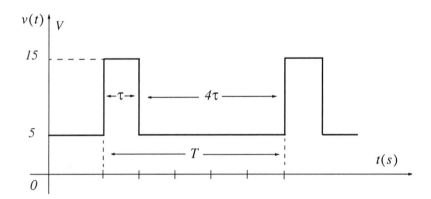

$$V^2_{RMS} = \frac{1}{T}\int_0^T v^2 dt = \frac{1}{5\tau}\left[\int_0^\tau (15)^2 dt + \int_\tau^{5\tau} (5)^2 dt\right] = \frac{1}{5\tau}(225\tau + 125\tau - 25\tau) = 65$$

and thus $V_{RMS} = \sqrt{65} = 8.06\ V$

3. We choose the period T as shown below.

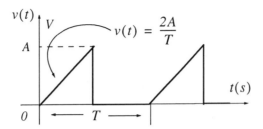

Using the straight line equation $y = mx + b$ we find that for $0 < t < T/2$, $v(t) = \dfrac{2A}{T}t$. Then,

$$V_{RMS}^2 = \frac{1}{T}\int_0^T v^2 dt = \frac{1}{T}\int_0^{T/2}\left(\frac{2A}{T}t\right)^2 dt + \frac{1}{T}\int_{T/2}^T 0 \cdot dt = \frac{4A^2}{T^3}\int_0^{T/2} t^2 dt$$

$$= \frac{4A^2}{3T^3}t^3\bigg|_0^{T/2} = \frac{4A^2}{24} = A^2/6$$

and

$$V_{RMS} = \sqrt{A^2/6} = \frac{\sqrt{6}}{6}A = 0.41A$$

4. The effective (RMS) value of a sinusoid is a real number that is independent of frequency and phase angle and for current it is equal to $I_{RMS} = I_p/\sqrt{2}$. The RMS value of sinusoids with different frequencies is given by (8.13). For this problem

$$I_{RMS} = \sqrt{10^2 + \frac{1}{2}2^2 + \frac{1}{2}5^2} = \sqrt{100 + 2 + 12.5} = 10.7 \text{ A}$$

5. The waveform representing the transmitter output pulses is shown below.

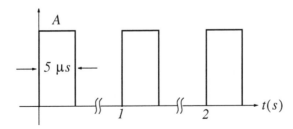

For this problem we do no know the amplitude A of each $5 \ \mu s$ pulse but we know the average power of one period $T = 1 \ s$. Since

$$P_{ave} = 750 \ w = \frac{Area}{Period} = \frac{Area}{1 \ s}$$

it follows that

a. Energy transmitted during each pulse is

$$Area\ of\ each\ pulse\ =\ 750\ w \cdot s$$

b. The power during the transmission of a pulse is

$$P\ =\ W/t\ =\ 750\ w \cdot s/5\ \mu s\ =\ 750\ w \cdot s/5 \times 10^{-6}\ =\ 150 \times 10^{6}\ w\ =\ 150\ Mw$$

6. The phasor equivalent circuit is shown below where $j\omega L\ =\ j10^{3} \times 3 \times 10^{-3}\ =\ j3\ \Omega$ and $-j/\omega C\ =\ -j/10^{3} \times 2 \times 10^{-4}\ =\ -j5\ \Omega$

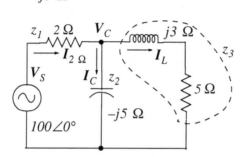

By application of KCL

$$\frac{V_C - V_S}{z_1} + \frac{V_C}{z_2} + \frac{V_C}{z_2}\ =\ 0$$

or

$$\left(\frac{1}{z_1} + \frac{1}{z_2} + \frac{1}{z_3}\right)V_C\ =\ \frac{V_S}{z_1}$$

or

$$V_C\ =\ \frac{V_S}{\left(1 + \dfrac{z_1}{z_2} + \dfrac{z_1}{z_3}\right)}$$

Also,

$$I_{2\,\Omega}\ =\ \frac{V_S - V_C}{z_1} \qquad I_C\ =\ \frac{V_C}{z_1} \qquad I_L\ =\ \frac{V_C}{z_3}$$

and with MATLAB

```
Vs=100; z1=2; z2=-5j; z3=5+3j;...
Vc=Vs/(1+z1/z2+z1/z3); I2=(Vs-Vc)/z1; Ic=Vc/z2; IL=Vc/z3; fprintf(' \n');...
disp('Vc = '); disp(Vc); disp('magVc = '); disp(abs(Vc));...
disp('phaseVc = '); disp(angle(Vc)*180/pi);...
disp('I2 = '); disp(I2); disp('magI2 = '); disp(abs(I2));...
disp('phaseI2 = '); disp(angle(I2)*180/pi);...
```

```
disp('Ic = '); disp(Ic); disp('magIc = '); disp(abs(Ic));...
disp('phaseIc = '); disp(angle(Ic)*180/pi);...
disp('IL = '); disp(IL); disp('magIL = '); disp(abs(IL));...
disp('phaseIL = '); disp(angle(IL)*180/pi);
```

Vc = 75.0341-12.9604i

magVc = 76.1452

phaseVc = -9.7998

I2 = 12.4829 + 6.4802i

magI2 = 14.0647

phaseI2 = 27.4350

Ic = 2.5921 + 15.0068i

magIc = 15.2290

phaseIc = 80.2002

IL = 9.8909 - 8.5266i
magIL = 13.0588

phaseIL = -40.7636

The average power delivered by the voltage source V_S is computed from the relation

$$P_{ave} = V_{RMS}I_{RMS}cos\theta = \frac{1}{2}V_pI_pcos\theta$$

where $\theta = 27.43°$ as shown by the phasor diagram below.

Therefore,

$$P_{S\ ave} = \frac{1}{2} \times |V_S| \times |I_{2\ \Omega}|cos\theta = 0.5 \times 100 \times 14.07 \times cos\,27.43° = 624.4\ w$$

Also,

$$P_{2\ \Omega\ ave} = \frac{1}{2}I_p^2R_{2\ \Omega} = 0.5 \times (14.07)^2 \times 2 = 197.97\ w$$

and

$$P_{5 \, \Omega \, ave} = \frac{1}{2}I_L^2 R_{5 \, \Omega} = 0.5 \times (13.06)^2 \times 5 = 426.41 \ w$$

Check:

$$P_{2 \, \Omega \, ave} + P_{5 \, \Omega \, ave} = 197.97 + 426.41 = P_{S \, ave} = 624.4 \ w$$

The average power in the capacitor and the inductor is zero since $\theta = 90°$ and $cos\theta = 0$

7. Let us consider the $t-domain$ network below.

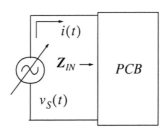

Let

$$v_S = V_p cos\omega t$$

and

$$i = I_p cos(\omega t + \theta)$$

Then,

$$p = v_S i = V_p I_p cos\omega t \cdot cos(\omega t + \theta)$$

and using

$$cosx \cdot cosy = \frac{1}{2}[cos(x+y) + cos(x-y)]$$

we get

$$p = \frac{V_p I_p}{2}[cos(2\omega t + \theta) + cos\theta]$$

We require that the power p does not exceed $200 \ mw$ or $0.2 \ w$, that is, we must satisfy the condition

$$p = \frac{V_p I_p}{2}[cos(2\omega t + \theta) + cos\theta] \le 0.2 \ w$$

and therefore we must find the phase angle θ. Since θ appears also in the $j\omega - domain$, we can find its value from the given input impedance, that is, $Z_{IN} = 100 - j100 \ \Omega$ or

$$Z_{IN} = |Z_{IN}|\angle\theta = \sqrt{100^2 + 100^2}\angle tan\frac{-1(-100)}{100} = 100\sqrt{2}\angle-45°$$

and in the $t - domain$

$$p = \frac{V_p I_p}{2}[cos(2\omega t - 45°) + cos(-45°)]$$

The maximum power p occurs when $cos(2\omega t - 45°) = 1$, that is,

$$p_{max} = \frac{V_p I_p}{2}\left(1 + \frac{\sqrt{2}}{2}\right) = 0.2 \ w$$

Then,

$$V_p I_p = 0.4/1.707$$

and now we can express I_p in terms of V_p using the relation $|Z_{IN}| = 100\sqrt{2}$ and $I_p = V_p/|Z_{IN}|$ and by substitution

$$V_p^2 = \frac{0.4 \times 100\sqrt{2}}{1.707} = 33.14$$

or

$$V_p = \sqrt{33.14} = 5.76$$

and

$$V_{RMS} = \frac{V_p}{\sqrt{2}} = \frac{5.76}{1.414} = 4.07 \ V$$

8. With the switch at the *10 mA* position, the circuit is as shown below.

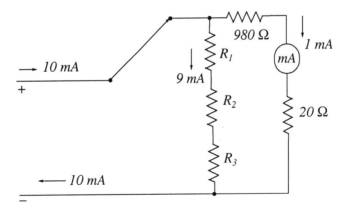

Then,

$$9 \times 10^{-3}(R_1 + R_2 + R_3) = (980 + 20) \times 10^{-3}$$

or

$$R_1 + R_2 + R_3 = \frac{1000}{9} \quad (1)$$

With the switch at the *100 mA* position, the circuit is as shown below.

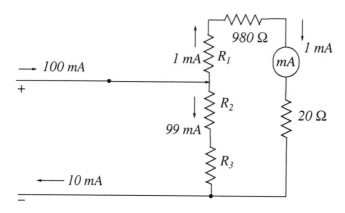

Then,

$$99 \times 10^{-3}(R_2 + R_3) = (R_1 + 980 + 20) \times 10^{-3}$$

or

$$-R_1 + 99R_2 + 99R_3 = 1000 \quad (2)$$

With the switch at the *1 A* position, the circuit is as shown below.

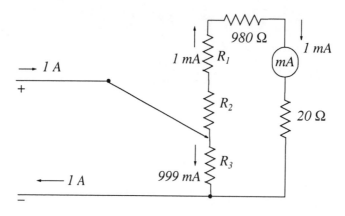

Then,

$$999 \times 10^{-3}R_3 = (R_1 + R_2 + 980 + 20) \times 10^{-3}$$

or

$$-R_1 - R_2 + 999R_3 = 1000 \quad (3)$$

Addition of (1) and (3) yields

$$1000R_3 = \frac{1000}{9} + 1000 = \frac{10000}{9}$$

or

$$R_3 = \frac{10}{9} \ \Omega \quad (4)$$

Addition of (1) and (2) yields

$$100R_2 + 100R_3 = \frac{1000}{9} + 1000 = \frac{10000}{9}$$

or

$$R_2 + R_3 = \frac{100}{9} \quad (5)$$

Substitution of (4) into (5) yields

$$R_2 = 10 \ \Omega \quad (6)$$

and substitution of (4) and (6) into (1) yields

$$R_1 = 100 \ \Omega \quad (7)$$

9. DC instruments indicate average values. Therefore, the DC voltmeter will read the average value of the voltage v_{OUT} across the resistor. The period of the full-wave rectifier waveform is taken as π.

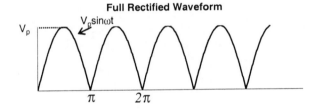

Full Rectified Waveform

Then,

$$v_{OUT\ ave} = \frac{1}{\pi} \int_0^\pi V_p \sin\omega t\, d(\omega t) = \frac{V_p}{\pi}(-\cos\omega t)\Big|_{\omega t = 0}^{\pi}$$

$$= \frac{V_p}{\pi}\cos\omega t\Big|_\pi^0 = \frac{V_p}{\pi}(1 + 1) = \frac{2V_p}{\pi}$$

As expected, this average is twice the average value of the half-wave rectifier waveform of Example 8.2.

NOTES

Chapter 9

This chapter discusses *the natural response* of electric circuits. The term natural implies that there is no excitation in the circuit, that is, the circuit is source-free, and we seek the circuit's natural response. The natural response is also referred to as the *transient response*.

9.1 The Natural Response of a Series RL circuit

Let us find the natural response of the circuit of Figure 9.1 where the desired response is the current i, and it is given that at $t = 0$, $i = I_0$, that is, the initial condition is $i(0) = I_0$.

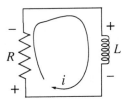

Figure 9.1. Circuit for determining the natural response of a series RL circuit

Application of KVL yields

$$v_L + v_R = 0$$

or

$$L\frac{di}{dt} + Ri = 0 \tag{9.1}$$

Here, we seek a value of i which satisfies the differential equation of (9.1), that is, we need to find the natural response which in differential equations terminology is the *complementary function*. As we know, two common methods are the separation of variables method and the assumed solution method. We will consider both.

1. Separation of Variables Method

Rearranging (9.1), so that the variables i and t are separated, we get

$$\frac{di}{i} = -\frac{R}{L}dt$$

Next, integrating both sides and using the initial condition, we get

$$\int_{I_0}^{i} \frac{1}{i} di = -\frac{R}{L} \int_{0}^{t} d\sigma$$

where σ is a dummy variable. Integration yields

$$\left. \ln i \right|_{I_0}^{i} = -\frac{R}{L} \left. \sigma \right|_{0}^{t}$$

or

$$\ln i - \ln I_0 = -\frac{R}{L} t$$

or

$$\ln \frac{i}{I_0} = -\frac{R}{L} t$$

Recalling that $x = \ln y$ implies $y = e^x$, we get

$$\boxed{i(t) = I_0 e^{-(R/L)t}} \tag{9.2}$$

Substitution of (9.2) into (9.1) yields $0 = 0$ and that at $t = 0$, $i(0) = I_0$. Thus, both the differential equation and the initial condition are satisfied.

2. Assumed Solution Method

Relation (9.1) indicates that the solution must be a function which, when added to its first derivative will become zero. An exponential function will accomplish that and therefore, we assume a solution of the form

$$i(t) = A e^{st} \tag{9.3}$$

where A and s are constants to be determined. Now, if (9.3) is a solution, it must satisfy the differential equation (9.1). Then, by substitution, we get:

$$RA e^{st} + sLA e^{st} = 0$$

or

$$\left(s + \frac{R}{L} \right) A e^{st} = 0$$

The left side of the last expression above will be zero if $A = 0$, or if $s = -\infty$, or if $s = -R/L$. But, if $A = 0$ or $s = -\infty$, then every response is zero and this represents a trivial solution. Therefore, $s = -R/L$ is the only logical solution, and by substitution into (9.3) we get

$$i(t) = A e^{-(R/L)t}$$

We must now evaluate the constant A. This is done with the use of the initial condition $i(0) = I_0$.

Thus, $I_0 = Ae^0$ or $A = I_0$ and therefore,

$$i(t) = I_0 e^{-(R/L)t}$$

as before. Next, we rewrite it as

$$\frac{i(t)}{I_0} = e^{-(R/L)t} \tag{9.4}$$

and sketch it as shown in Figure 9.2.

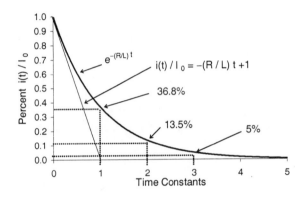

Figure 9.2. Plot for $i(t)/I_0$ in a series RL circuit

From Figure 9.2 we observe that at $t = 0$, $i/I_0 = 1$, and $i \to 0$ as $t \to \infty$.

The initial rate (slope) of decay is found from the derivative of i/I_0 evaluated at $t = 0$, that is,

$$\left.\frac{d}{dt}\left(\frac{i}{I_0}\right)\right|_{t=0} = \left.-\frac{R}{L}e^{-(R/L)t}\right|_{t=0} = -\frac{R}{L}$$

and thus the slope of the initial rate of decay is $-R/L$

Next, we define the *time constant* τ as the time required for i/I_0 to drop from unity to zero assuming that the initial rate of decay remains constant. This constant rate of decay is represented by the straight line equation

$$\frac{i(t)}{I_0} = -\frac{R}{L}t + 1$$

and at $t = \tau$, $i/I_0 = 0$. Then,

$$0 = -\frac{R}{L}\tau + 1$$

or

$$\boxed{\tau = \frac{L}{R}}$$

Time Constant for RL Circuit $\qquad\qquad$ (9.5)

Evaluating (9.4) at $t = \tau = L/R$, we get

$$\frac{i(\tau)}{I_0} = e^{-(R/L)\tau} = e^{-(R/L)(L/R)} = e^{-1} = 0.368$$

or

$$\boxed{i(\tau) = 0.368I_0} \qquad\qquad (9.6)$$

Therefore, *in one time constant, the response has dropped to approximately 36.8% of its initial value.*

If we express the rate of decay in time constant intervals as shown in Figure 9.2, we find that $i(t)/I_0 \approx 0$ after $t = 5\tau$, that is, it reaches its final value after five time constants.

Example 9.1

For the circuit shown in Figure 9.3, in how many seconds after $t = 0$ has the

a. current $i(t)$ has reached ½ of its initial value?

b. energy stored in L has reached ¼ of its initial value?

c. power dissipated in R has reached ¾ of its initial value?

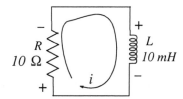

Figure 9.3. Circuit for Example 9.1

Solution:

From (9.2),

$$i(t) = I_0 e^{-(R/L)t}$$

where $I_0 = i_L(0)$. Then,

---

a. The current $i(t)$ will have reached ½ of its initial value when

$$0.5I_0 = I_0 e^{-(10/10 \times 10^{-3})t} = I_0 e^{-1000t}$$

or

$$e^{-1000t} = 0.5$$

or

$$-1000t = \ln(0.5) = -0.693$$

and therefore,

$$t = 693 \ \mu s$$

b. To find the energy stored in L which reaches ¼ of its initial value, we start with

$$W_L(t) = \frac{1}{2} L i^2(t)$$

and at $t = 0$, $I_0 = i_L(0)$. Then,

$$W_L(0) = \frac{1}{2} L I_0^2$$

and

$$\frac{1}{4} W_L(0) = \frac{1}{4} \left(\frac{1}{2} L I_0^2 \right)$$

Therefore,

$$\frac{1}{4} W_L(t) = \frac{1}{2} L i^2(t) = \frac{1}{2} L (I_0 e^{-(R/L)t})^2 = \frac{1}{4} \left(\frac{1}{2} L I_0^2 \right)$$

or

$$e^{-2(R/L)t} = 1/4$$

$$e^{-2000t} = 1/4$$

$$-2000t = \ln(0.25) = -1.386$$

and

$$t = 693 \ \mu s$$

This is the same answer as in part (a) since the energy is proportional to the square of the current.

c. To find the power dissipated in R when it reaches ¾ of its initial value, we start with the fact that the instantaneous power absorbed by the resistor is $p_R = i_R^2 R$, and since for the given circuit

$$i(t) = i_R(t) = I_0 e^{-(R/L)t}$$

then,

$$p_R = I_0^2 R e^{-2(R/L)t}$$

and the energy dissipated (in the form of heat) in the resistor is

$$W_R = \int_0^\infty p_R dt = I_0^2 R \int_0^\infty e^{-2(R/L)t} dt = I_0^2 R \left(-\frac{L}{2R}\right) e^{-2(R/L)t} \Big|_0^\infty = \frac{1}{2} L I_0^2$$

Also, from part (b) above,

$$W_L(0) = \frac{1}{2} L I_0^2$$

and thus

$$\frac{3}{4} W_R = \frac{3}{4} W_L(0) = \frac{1}{2} L i^2(t) = \frac{1}{2} L (I_0 e^{-(R/L)t})^2 = \frac{3}{4}\left(\frac{1}{2} L I_0^2\right)$$

or

$$e^{-2(R/L)t} = 3/4$$

$$e^{-2000t} = 3/4$$

$$-2000t = ln(0.75) = -0.288$$

and

$$t = 144 \ \mu s$$

In some examples and exercises that follow, the initial condition may not be given directly but it can be found from the fact that the current through an inductor cannot change instantaneously and therefore,

$$\boxed{i_L(0^-) = i_L(0) = i_L(0^+)} \qquad (9.7)$$

where $i_L(0^-)$ will be used to denote the time just before a switch is opened or closed, and $i_L(0^+)$ will be used to denote the time just after the change has occurred.

Also, in our subsequent discussion, the expression "long time" will mean that sufficient time has elapsed so that the circuit has reached its steady-state conditions. As we know from Chapter 5, when the excitations are constant, at steady state conditions the inductor behaves as a short circuit, and the capacitor behaves as an open circuit.

Example 9.2

In the circuit of Figure 9.4, the switch S has been in the closed position for a long time and opens at $t = 0$. Find $i_L(t)$ for $t > 0$, $v_R(0^-)$, and $v_R(0^+)$

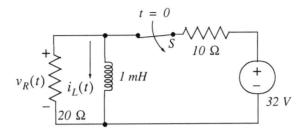

Figure 9.4. Circuit for Example 9.2

Solution:

We are not given an initial condition for this example; however, at $t = 0^-$ the inductor acts as a short thereby shorting also the *20* Ω resistor. The circuit then is as shown in Figure 9.5.

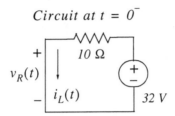

Figure 9.5. Circuit for Example 9.2 at $t = 0^-$

From the circuit of Figure 9.5, we see that

$$i_L(0^-) = i_L(0) = i_L(0^+) = 32/10 = 3.2A$$

and thus the initial condition has now been established as $I_0 = 3.2\ A$

We also observe that

$$v_R(0^-) = 0$$

At $t = 0^+$, the *32 V* source and the *10* Ω resistor are disconnected from the circuit which now is as shown in Figure 9.6.

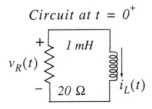

Figure 9.6. Circuit for Example 9.2 at $t = 0^+$

For the circuit of Figure 9.6,

$$i_L(t) = I_0 e^{-(R/L)t} = 3.2e^{-(20/10^{-3})t}$$

or

$$i_L(t) = 3.2e^{-20000t}$$

and

$$v_R(0^+) = 20(-I_0) = 20(-3.2)$$

or

$$v_R(0^+) = -64 \ V$$

We observe that $v_R(0^+) \ne v_R(0^-)$

Example 9.3

In the circuit shown in Figure 9.7, the switch S has been closed for a long time and opens at $t = 0$.
Find:

a. $i_L(t)$ for $t > 0$

b. $i_{60}(t)$ at $t = 100 \ \mu s$

c. $i_{48}(t)$ at $t = 200 \ \mu s$

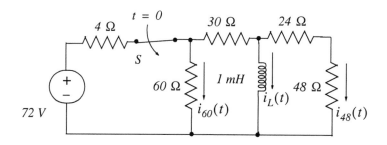

Figure 9.7. Circuit for Example 9.3

Solution:

a. At $t = 0^-$ the inductor acts as a short thereby shorting also the $24 \ \Omega$ and $48 \ \Omega$ resistors. The circuit then is as shown in Figure 9.8. Then,

$$i_T(0^-) = \frac{72 \ V}{4 + 60 \parallel 30} = \frac{72 \ V}{4 + 20} = 3 \ A$$

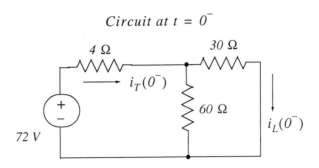

Figure 9.8. Circuit for Example 9.3 at $t = 0^-$

and by the current division expression,

$$i_L(0^-) = \frac{60}{30 + 60} \cdot i_T(0^-) = \frac{6}{9} \times 3 = 2\ A$$

and thus the initial condition has been established as $I_0 = 2\ A$

At $t = 0^+$, the *72 V* source and the *4 Ω* resistor are disconnected from the circuit which now is as shown in Figure 9.9.

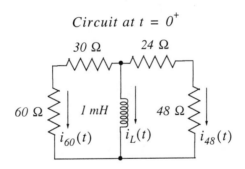

Figure 9.9. Circuit for Example 9.3 at $t = 0^+$

From (9.2),

$$i_L(t) = I_0 e^{-(R_{eq}/L)t}$$

where

$$R_{eq} = (60 + 30) \parallel (24+48) = 40\ Ω$$

and thus

$$i_L(t) = 2e^{-(40/10^{-3})t}$$

or

$$i_L(t) = 2e^{-40000t}$$

Also,

$$i_{60}(t)\Big|_{t = 100 \ \mu s} = \frac{(24 + 48)}{(30 + 60) + (24 + 48)}[-i_L(t)]\Big|_{t = 100 \ \mu s}$$

or

$$i_{60}(t)\Big|_{t = 100 \ \mu s} = \frac{12}{27}(-2e^{-40000t})\Big|_{t = 100 \ \mu s} = -\frac{8}{9}e^{-4} = -16.3 \ mA$$

and

$$i_{48}(t)\Big|_{t = 200 \ \mu s} = \frac{(30 + 60)}{(30 + 60) + (24 + 48)}[-i_L(t)]\Big|_{t = 200 \ \mu s}$$

or

$$i_{48}(t)\Big|_{t = 200 \ \mu s} = \frac{15}{27}(-2e^{-40000t})\Big|_{t = 200 \ \mu s} = -\frac{10}{9}e^{-8} = -0.373 \ mA$$

9.2 The Natural Response of a Series RC Circuit

In this section, we will find the natural response of the RC circuit shown in Figure 9.10 where the desired response is the capacitor voltage v_C, and it is given that at $t = 0$, $v_C = V_0$, that is, the initial condition is $v(0) = V_0$.

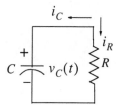

Figure 9.10. Circuit for determining the natural response of a series RC circuit

By KCL,

$$i_C + i_R = 0 \tag{9.8}$$

and with

$$i_C = C\frac{dv_C}{dt}$$

and

$$i_R = \frac{v_C}{R}$$

by substitution into (9.8), we obtain the differential equation

$$\frac{dv_C}{dt} + \frac{v_C}{RC} = 0 \tag{9.9}$$

As before, we assume a solution of the form

$$v_C(t) = Ae^{st}$$

and by substitution into (9.9)

$$Ase^{st} + \frac{Ae^{st}}{RC} = 0$$

or

$$\left(s + \frac{1}{RC}\right)Ae^{st} = 0 \tag{9.10}$$

Following the same reasoning as with the RL circuit, (9.10) will be satisfied when $s = -1/RC$ and therefore,

$$v_C(t) = Ae^{-(1/RC)t}$$

The constant A is evaluated from the initial condition, i.e., $v_C(0) = V_0 = Ae^0$ or $A = V_0$. Therefore, the natural response of the RC circuit is

$$\boxed{v_C(t) = V_0 e^{-(1/RC)t}} \tag{9.11}$$

We express (9.11) as

$$\frac{v_C(t)}{V_0} = e^{-(1/RC)t}$$

and we sketch it as shown in Figure 9.11. We observe that at $t = 0$, $v_C/V_0 = 1$, and $i \to 0$ as $t \to \infty$

The initial rate (slope) of decay is found from the derivative of $v_C(t)/V_0$ evaluated at $t = 0$, that is,

$$\left.\frac{d}{dt}\left(\frac{v_C}{V_0}\right)\right|_{t=0} = \left.-\frac{1}{RC}e^{-(1/RC)t}\right|_{t=0} = -\frac{1}{RC}$$

and thus the slope of the initial rate of decay is $-1/(RC)$

Next, we define the *time constant* τ as the time required for $v_C(t)/V_0$ to drop from unity to zero assuming that the initial rate of decay remains constant. This constant rate of decay is represented by the straight line equation

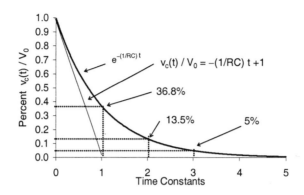

Figure 9.11. Circuit for determining the natural response of a series RC circuit

$$\frac{v_C(t)}{V_0} = -\frac{1}{RC}t + 1 \qquad (9.12)$$

and at $t = \tau$, $v_C(t)/V_0 = 0$. Then,

$$0 = -\frac{1}{RC}\tau + 1$$

or

$$\boxed{\begin{array}{c} \tau = RC \\ \textit{Time Constant for RC Circuit} \end{array}} \qquad (9.13)$$

Evaluating (9.11) at $t = \tau = RC$, we get

$$\frac{v_C(\tau)}{V_0} = e^{-\tau/RC} = e^{-RC/RC} = e^{-1} = 0.368$$

or

$$\boxed{v_C(\tau) = 0.368 V_0} \qquad (9.14)$$

Therefore, *in one time constant, the response has dropped to approximately 36.8% of its initial value.*

If we express the rate of decay in time constant intervals as shown in Figure 9.11, we find that $v_C(t)/V_0 \approx 0$ after $t = 5\tau$, that is, it reaches its final value after five time constants.

In the examples that follow, we will make use of the fact that

$$\boxed{v_C(0^-) = v_C(0) = v_C(0^+)} \qquad (9.15)$$

Example 9.4

In the circuit of Figure 9.12, the switch S has been in the closed position for a long time, and opens at $t = 0$. Find $v_C(t)$ for $t > 0$, $i(0^-)$, and $i(0^+)$.

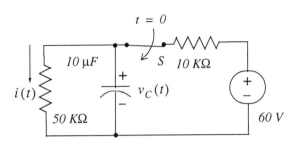

Figure 9.12. Circuit for Example 9.4

Solution:

At $t = 0^-$ the capacitor acts as an open. The circuit then is as shown in Figure 9.13.

Circuit at $t = 0^-$

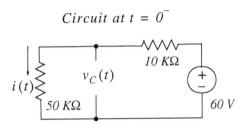

Figure 9.13. Circuit for Example 9.4 at $t = 0^-$

From the circuit of Figure 9.13 we see that

$$v_C(0^-) = v_C(0^+) = 50 \ K\Omega \times i(0^-) = 50 \times \frac{60 \ V}{10 \ K\Omega + 50 \ K\Omega} = 50 \ V$$

and thus the initial condition has been established as $V_0 = 50 \ V$. We also observe that

$$i(0^-) = \frac{60 \ V}{10 \ K\Omega + 50 \ K\Omega} = 1 \ mA$$

At $t = 0^+$ the $60 \ V$ source and the $10 \ K\Omega$ resistor are disconnected from the circuit which now is as shown in Figure 9.14.

From (9.11),

$$v_C(t) = V_0 e^{-(1/RC)t}$$

Circuit at t = 0⁺

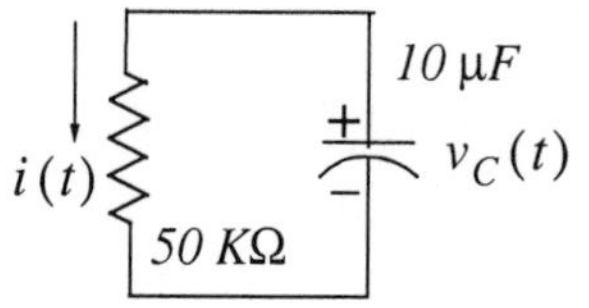

Figure 9.14. Circuit for Example 9.4 at $t = 0^+$

where

$$RC = 50 \times 10^3 \times 10 \times 10^{-6} = 0.5$$

Then,

$$v_C(t) = 50e^{-(1/0.5)t} = 50e^{-2t}$$

and

$$i(0^+) = \frac{V_0}{R} = \frac{50\ V}{50\ K\Omega} = 1\ mA$$

We observe that $i(0^+) = i(0^-)$. This is true because the voltage across the capacitor cannot change instantaneously; hence, the voltage across the resistor must be the same at $t = 0^-$ and at $t = 0^+$.

Example 9.5

In the circuit of Figure 9.15, the switch S has been in the closed position for a long time and opens at $t = 0$. Find:

a. $v_C(t)$ for $t > 0$

b. $v_{60}(t)$ at $t = 100\ \mu s$

c. $v_{10}(t)$ at $t = 200\ \mu s$

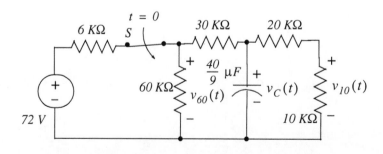

Figure 9.15. Circuit for Example 9.5

Solution:

a. At $t = 0^-$ the capacitor acts as an open and the circuit then is as shown in Figure 9.16.

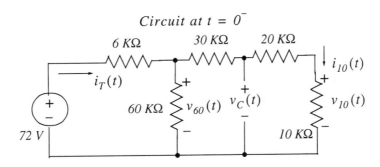

Figure 9.16. Circuit for Example 9.5 at $t = 0^-$

From the circuit of Figure 9.16,

$$i_T(0^-) = \frac{72 \ V}{6 \ K\Omega + 60 \ K\Omega \parallel 60 \ K\Omega} = \frac{72 \ V}{6 \ K\Omega + 30 \ K\Omega} = 2 \ mA$$

and using the current division expression, we get

$$i_{10}(0^-) = \frac{60 \ K\Omega}{60 \ K\Omega + 60 \ K\Omega} \cdot i_T(0^-) = \frac{1}{2} \times 2 = 1 \ mA$$

Then,

$$v_C(0^-) = (20 \ K\Omega + 10 \ K\Omega) \cdot i_{10}(0^-) = 30 \ V$$

and thus the initial condition has been established as $V_0 = 30 \ V$

At $t = 0^+$, the 72 V source and the 6 KΩ resistor are disconnected from the circuit which now is as shown in Figure 9.17.

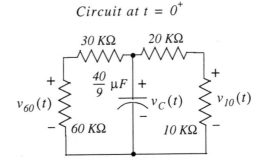

Figure 9.17. Circuit for Example 9.5 at $t = 0^+$

From (9.11),

$$v_C(t) = V_0 e^{-(1/R_{eq}C)t}$$

where

$$R_{eq} = (60\ K\Omega + 30\ K\Omega) \parallel (20\ K\Omega + 10\ K\Omega) = 22.5\ K\Omega$$

Then,

$$R_{eq}C = 22.5 \times 10^3 \times \frac{40}{9} \times 10^{-6} = 0.1$$

and

$$v_C(t) = 30 e^{-(1/0.1)t} = 30 e^{-10t}$$

b.

$$v_{60}(t)\Big|_{t = 100\ ms} = \frac{60\ K\Omega}{30\ K\Omega + 60\ K\Omega} \cdot v_C(t)\Big|_{t = 100\ ms}$$

or

$$v_{60}(t)\Big|_{t = 100\ ms} = \frac{2}{3}(30 e^{-10t})\Big|_{t = 100\ ms} = 20 e^{-1} = 7.36\ V$$

c.

$$v_{10}(t)\Big|_{t = 200\ ms} = \frac{10\ K\Omega}{10\ K\Omega + 20\ K\Omega} \cdot v_C(t)\Big|_{t = 200\ ms}$$

or

$$v_{10}(t)\Big|_{t = 200\ ms} = \frac{1}{3}(30 e^{-10t})\Big|_{t = 200\ ms} = 10 e^{-2} = 1.35\ V$$

Example 9.6

For the circuit of Figure 9.18, it is known that $v_C(0) = V_0 = 25\ V$.

a. To what value should the resistor R be adjusted so that the initial rate of change would be $-200\ V/s$?

b. What would then the energy in the capacitor be after two time constants?

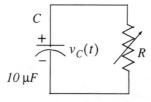

Figure 9.18. Circuit for Example 9.6

Solution:

a. The capacitor voltage decays exponentially as

$$v_C(t) = V_0 e^{-(1/RC)t}$$

and with the given values,

$$v_C(t) = 25 e^{-(100000/R)t}$$

Now, if the initial rate (slope) is to be $-200\ V/s$, then

$$\left.\frac{dv_C}{dt}\right|_{t=0} = \left(-\frac{100000}{R}\right) 25 e^{-(100000/R)t}\bigg|_{t=0} = -\frac{2.5 \times 10^6}{R} = -200$$

and solving for R we get $R = 12.5\ K\Omega$

b. After two time constants the capacitor voltage will drop to the value of

$$v_C(2\tau) = 25 e^{-(1/RC)2\tau} = 25 e^{-(2RC/RC)} = 25 e^{-2} = 3.38\ V$$

Therefore, the energy after two time constants will be

$$W_C\big|_{t=2\tau} = \frac{1}{2} C v_C^2 = 5 \times 10^{-6} \times 3.38^2 = 57.2\ \mu J$$

9.3 Summary

- The natural response of the inductor current $i_L(t)$ in a simple RL circuit has the form $i_L(t) = I_0 e^{-(R/L)t}$ where I_0 denotes the value of the current in the inductor at $t = 0$

- In a simple RL circuit the time constant τ is the time required for $i_L(t)/I_0$ to drop from unity to zero assuming that the initial rate of decay remains constant, and its value is $\tau = L/R$

- In one time constant the natural response of the inductor current in a simple RL circuit has dropped to approximately 36.8% of its initial value.

- The natural response of the inductor current in a simple RL circuit reaches its final value, that is, it decays to zero, after approximately 5 time constants.

- The initial condition I_0 can be established from the fact that the current through an inductor cannot change instantaneously and thus $i_L(0^-) = i_L(0) = i_L(0^+)$

- The natural response of the capacitor voltage $v_C(t)$ in a simple RC circuit has the form $v_C(t) = V_0 e^{-(1/RC)t}$ where V_0 denotes the value of the voltage across the capacitor at $t = 0$

- In a simple RC circuit the time constant τ is the time required for $v_C(t)/V_0$ to drop from unity to zero assuming that the initial rate of decay remains constant, and its value is $\tau = RC$

- In one time constant the natural response of the capacitor voltage in a simple RC circuit has dropped to approximately 36.8% of its initial value.

- The natural response of capacitor voltage in a simple RC circuit reaches its final value, that is, it decays to zero after approximately 5 time constants.

- The initial condition V_0 can be established from the fact that the voltage across a capacitor cannot change instantaneously and thus $v_C(0^-) = v_C(0) = v_C(0^+)$

9.4 Exercises

Multiple Choice

1. In a simple RL circuit the unit of the time constant τ is

    A. dimensionless

    B. the millisecond

    C. the microsecond

    D. the reciprocal of second, i.e., s^{-1}

    E. none of the above

2. In a simple RC circuit the unit of the term $1/RC$ is

    A. the second

    B. the reciprocal of second, i.e., s^{-1}

    C. the millisecond

    D. the microsecond

    E. none of the above

3. In the circuit of Figure 9.19 Switch S_1 has been closed for a long time while Switch S_2 has been open for a long time. At $t = 0$. Switch S_1 opens and Switch S_2 closes. The current $i_L(t)$ for all $t > 0$ is

    A. $2\ A$

    B. $2e^{-100t}\ A$

    C. $2e^{-50t}\ A$

    D. $e^{-50t}\ A$

    E. none of the above

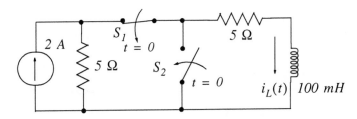

Figure 9.19. Circuit for Question 3

4. In the circuit of Figure 9.20 Switch S_1 has been closed for a long time while Switch S_2 has been open for a long time. At $t = 0$. Switch S_1 opens and Switch S_2 closes. The voltage $v_C(t)$ for all $t > 0$ is

A. *10 V*

B. $10e^{-10t}$ *V*

C. $10e^{-t}$ *V*

D. $10e^{-0.1t}$ *V*

E. none of the above

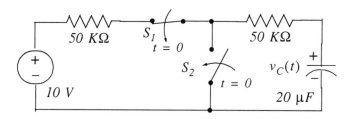

Figure 9.20. Circuit for Question 4

5. In the circuit of Figure 9.21 Switch S_1 has been closed for a long time while Switch S_2 has been open for a long time. At $t = 0$. Switch S_1 opens and Switch S_2 closes. The power absorbed by the inductor at $t = +\infty$ will be

A. *0 w*

B. *1 w*

C. *2 w*

D. *0.2 w*

E. none of the above

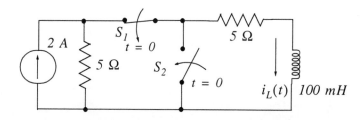

Figure 9.21. Circuit for Question 5

6. In the circuit of Figure 9.22 Switch S_1 has been closed for a long time while Switch S_2 has been open for a long time. At $t = 0$. Switch S_1 opens and Switch S_2 closes. The power absorbed by the capacitor at $t = +\infty$ will be

A. $0\ w$

B. $10\ w$

C. $5\ w$

D. $10\ mw$

E. none of the above

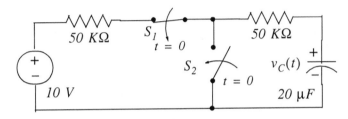

Figure 9.22. Circuit for Question 6

7. In a simple RL circuit where $R = 10\ M\Omega$ and $L = 10\ \mu H$ the time constant τ is

A. $1\ s$

B. $100\ s$

C. $10^{12}\ s$

D. $10^{-12}\ s$

E. none of the above

8. In a simple RC circuit where $R = 10\ M\Omega$ and $C = 10\ \mu F$ the time constant τ is

A. $100\ s$

B. $0.01\ s$

C. $100\ \mu s$

D. $0.01\ \mu s$

E. none of the above

9. In a simple RL circuit the condition(s) ____ are always true

A. $i_L(0^-) = i_L(0) = i_L(0^+)$ and $v_L(0^-) = v_L(0) = v_L(0^+)$

B. $i_L(0^-) = i_L(0) = i_L(0^+)$ and $i_R(0^-) = i_R(0) = i_R(0^+)$

C. $i_L(0^-) = i_L(0) = i_L(0^+)$ and $v_R(0^-) = v_R(0) = v_R(0^+)$

D. $i_L(0^-) = i_L(0) = i_L(0^+)$

E. none of the above.

10. In a simple RC circuit the condition(s) ____ are always true

A. $v_C(0^-) = v_C(0) = v_C(0^+)$ and $i_C(0^-) = i_C(0) = i_C(0^+)$

B. $v_C(0^-) = v_C(0) = v_C(0^+)$ and $v_R(0^-) = v_R(0) = v_R(0^+)$

C. $v_C(0^-) = v_C(0) = v_C(0^+)$

D. $v_C(0^-) = v_C(0) = v_C(0^+)$ and $i_R(0^-) = i_R(0) = i_R(0^+)$

E. none of the above.

Problems

1. In the circuit of Figure 9.23, switch S_1 has been closed for a long time and switch S_2 has been open for a long time. Then, at $t = 0$ switch S_1 opens while S_2 closes. Compute the current $i_{S2}(t)$ through switch S_2 for $t > 0$

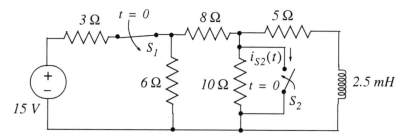

Figure 9.23. Circuit for Problem 1

2. In the circuit of Figure 9.24, both switches S_1 and S_2 have been closed for a long time and both are opened at $t = 0$. Compute and sketch the current $i_L(t)$ for the time interval $0 \le t \le 1 \ ms$

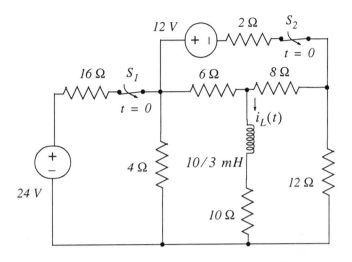

Figure 9.24. Circuit for Problem 2

3. In a series RL circuit, the voltage v_L across the inductor is $v_L = 0.2e^{-2000t}$ V and the current i_L at $t = 0$ is $i_L(0) = 10 \ mA$. Compute the values of R and L for that circuit.

4. In the circuit of Figure 9.25, both switches S_1 and S_2 have been closed for a long time, while switch S_3 has been open for a long time. At $t = 0$ S_1 and S_2 are opened and S_3 is closed. Compute the current $i_L(t)$ for $t > 0$.

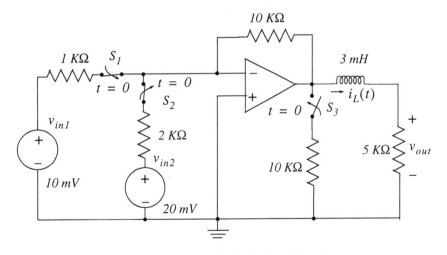

Figure 9.25. Circuit for Problem 4

5. In the circuit of Figure 9.26, switch S_1 has been closed and S_2 has been open for a long time. At $t = 0$ switch S_1 is opened and S_2 is closed. Compute the voltage $v_{C2}(t)$ for $t > 0$.

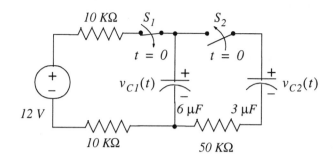

Figure 9.26. Circuit for Problem 5

6. In the circuit of Figure 9.27, switch S has been in the A position for a long time and at $t = 0$ is thrown in the B position. Compute the voltage $v_C(t)$ across the capacitor for $t > 0$, and the energy stored in the capacitor at $t = 1\ ms$.

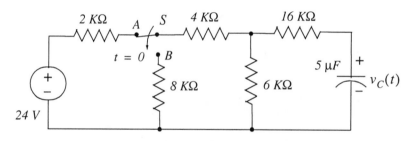

Figure 9.27. Circuit for Problem 6

7. In the circuit of Figure 9.28, switch S has been open for a long time and closes at $t = 0$. Compute $i_{SW}(t)$ for $t > 0$

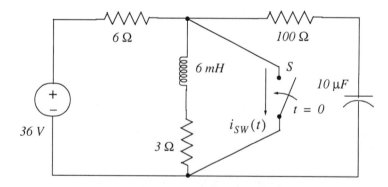

Figure 9.28. Circuit for Problem 7

9.5 Answers to Exercises

Multiple Choice

1. E $\tau = L/R = [volt/(ampere/second)]/[volt/ampere] = second\ (s)$

2. B

3. D

4. C

5. A

6. A

7. D

8. B

9. D

10. C

Problems

1. The circuit at $t = 0^-$ is as shown below.

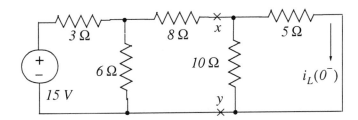

Replacing the circuit above with its Thevenin equivalent to the left of points x and y we find that $v_{TH} = \dfrac{6}{3+6} \cdot 15 = 10\ V$ and $R_{TH} = \dfrac{3 \times 6}{3+6} + 8 = 10\ \Omega$ and attaching the rest of the circuit to it we get the circuit below.

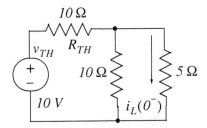

By voltage-source to current-source transformation we get the circuit below.

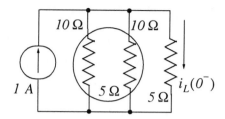

and by inspection, $i_L(0^-) = 0.5\ A$, that is, the initial condition has been established as

$$i_L(0^-) = i_L(0) = i_L(0^+) = I_0 = 0.5\ A$$

The circuit at $t = 0^+$ is as shown below.

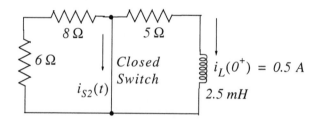

We observe that the closed shorts out the $6\ \Omega$ and $8\ \Omega$ resistors and the circuit simplifies to that shown below.

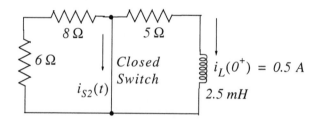

Thus for $t > 0$, $i_{S2}(t) = -i_L(t) = -I_0 e^{-(R/L)t} = -0.5 e^{-(5/2.5 \times 10^{-3})t} = -0.5 e^{-2000t}\ A$

2. The circuit at $t = 0^-$ is as shown below and the mesh equations are

$$20i_1 \qquad\quad -4i_3 \qquad\quad = 24$$
$$16i_2 - 6i_3 - 8i_4 = -12$$
$$-4i_1 - 6i_2 + 20i_3 - 10i_4 = 0$$
$$8i_2 - 10i_3 + 30i_4 = 0$$

Then, $i_L(0^-) = i_3 - i_4$

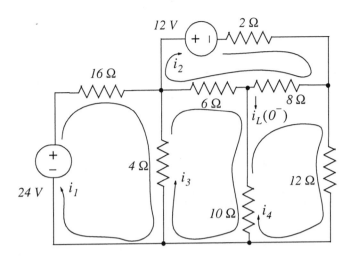

and with MATLAB

```
R=[20 0 -4 0; 0 16 -6 -8; -4 -6 20 -10; 0 -8 -10 30];...
V=[24 -12 0 0]'; I=R\V; iL0=I(3)-I(4); fprintf(' \n');...
fprintf('i1 = %4.2f A \t', I(1)); fprintf('i2 = %4.2f A \t', I(2));...
fprintf('i3 = %4.2f A \t', I(3)); fprintf('i4 = %4.2f A \t', I(4));...
fprintf('iL0 = %4.2f A \t', I(3)-I(4)); fprintf(' \n'); fprintf(' \n');

i1 = 1.15 A i2 = -1.03 A i3 = -0.26 A i4 = -0.36 A iL0 = 0.10 A
```

Therefore, $i_L(0^-) = i_L(0) = i_L(0^+) = I_0 = 0.1\ A$

Shown below is the circuit at $t = 0^+$ and the steps of simplification.

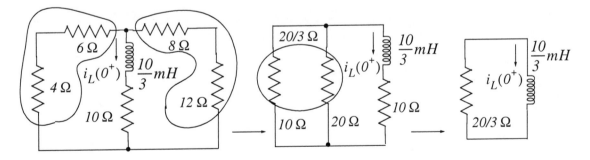

Thus for $t > 0$,

$$i_L(t) = I_0 e^{-(R/L)t} = 0.1 e^{-5000t}\ A$$

and

$$i_L\big|_{t = 0.4\ ms} = 0.1 e^{-2} = 0.0137\ A = 13.7\ mA$$

To compute and sketch the current $i_L(t)$ for the time interval $0 \le t \le 1 \ ms$ we use MATLAB as shown below.

```
t=(0: 0.01: 1)*10^(−3);...
iLt=0.1.*10.^(−3).*exp(−5000.*t);...
plot(t,iLt); grid
```

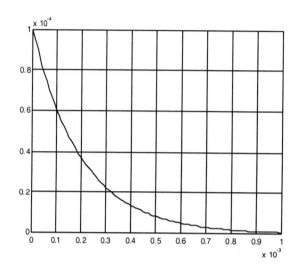

3. From the figure below $v_L = Ri_L = 0.2e^{-2000t}$ for $t > 0$

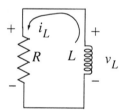

and with $i_L(0) = 10 \ mA$, by substitution $R(10 \times 10^{-3}) = 0.2e^{-0} = 0.2$

or $R = 0.2/10^{-2} = 20 \ \Omega$ Also, from $R/L = 2000$, $L = 20/2000 = 0.01 = 10 \ mH$

4. The circuit at $t = 0^-$ is as shown below and using the relation

$$v_{out} = -R_f\left(\frac{v_{in1}}{R_{in1}} + \frac{v_{in2}}{R_{in2}}\right)$$

that was developed in Example 4.11 we have

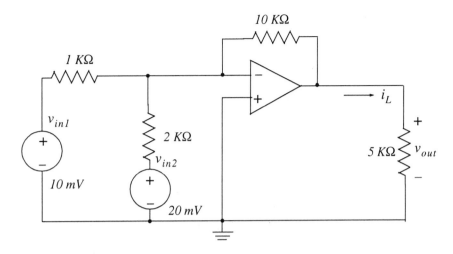

$$v_{out} = -10 \ K\Omega\left(\frac{10^{-2}}{1 \ K\Omega} + \frac{2 \times 10^{-2}}{2 \ K\Omega}\right) = -10 \times 2 \times 10^{-2} = -0.2 \ V$$

and

$$i_L(0^-) = I_0 = i_{5 \ K\Omega} = \frac{-0.2 \ V}{5 \ K\Omega} = -40 \times 10^{-6} \ A = -40 \ \mu A$$

The circuit at $t = 0^+$ is as shown below where $i_L(0^+) = 40 \ \mu A$ with the direction shown.

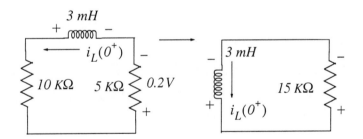

Then for $t > 0$

$$i_L(t) = I_0 e^{(R/L)t} = 40 \times 10^{-6} e^{(15 \times 10^3/3 \times 10^{-3})t} = 40e^{-5 \times 10^6 t} \ \mu A$$

with the direction shown.

5. The circuit at $t = 0^-$ is as shown below. As we've learned in Chapter 5, when a circuit is excited by a constant (DC) source, after sufficient time has elapsed the capacitor behaves as an open and thus the voltage across the capacitor C_1 is $12 \ V$ as shown.

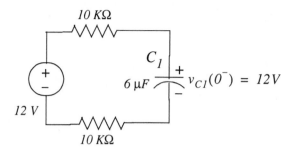

The circuit at $t = 0^+$ is as shown below where the $12\ V$ represents the voltage across capacitor C_1.

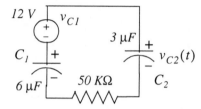

Now, $v_{C2}(t) = v_{C1}e^{-(1/RC_{eq})}$ where $v_{C1} = 12\ V$ and $C_{eq} = \dfrac{C_1 \cdot C_2}{C_1 + C_2} = \dfrac{6 \times 3}{6 + 3} = 2\ \mu F$

Then, $1/RC_{eq} = 1/(5 \times 10^4 \times 2 \times 10^{-6}) = 10$ and thus $v_{C2}(t) = 12e^{-10t}$

6. The circuit at $t = 0^-$ is as shown below.

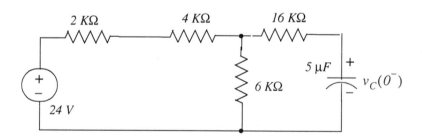

Then,

$$v_C(0^-) = V_0 = v_{6K\Omega} = \frac{6\ K\Omega}{6\ K\Omega + 6\ K\Omega} \times 24 = 12\ V$$

The circuit at $t = 0^+$ is as shown below where $v_C(0^+) = 12\ V$.

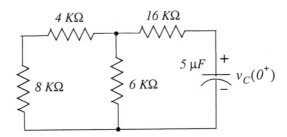

Series and parallel resistances reduction yields

$$R_{eq} = [(8\ K\Omega + 4\ K\Omega) \parallel 6\ K\Omega] + 16\ K\Omega = \frac{12 \times 6}{12 + 6} + 16 = 4 + 16 = 20\ K\Omega$$

and the circuit for $t > 0$ reduces to the one shown below.

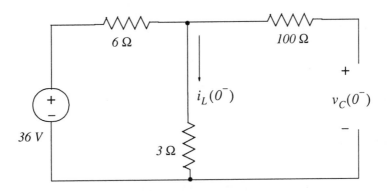

Now, $R_{eq}C = 2 \times 10^4 \times 5 \times 10^{-6} = 0.1$, $1/R_{eq}C = 10$ and $v_C(t) = V_0 e^{-10t} = 12e^{-10t}$. Also,

$$W_C \Big|_{1\ ms} = \frac{1}{2}Cv_C^2(t)\Big|_{1\ ms} = 0.5 \times 5 \times 10^{-6} \times 144 e^{-20t}\Big|_{t\ =\ 1\ ms}$$

$$= 360 \times 10^{-6} e^{-20t}\Big|_{t\ =\ 1\ ms} = 0.35\ mJ$$

7. The circuit at $t = 0^-$ is as shown below.

Then, $i_L(0^-) = \dfrac{36\ V}{(6+3)\ \Omega} = 4\ A$ and $v_C(0^-) = 3 \times i_L(0^-) = 3 \times 4 = 12\ V$

The circuit at $t = 0^+$ is as shown below and the current $i_{SW}(t)$ through the switch is the sum of the currents due to the $36\ V$ voltage source, due to $i_L(0^+) = 4\ A$, and due to $v_C(0^+) = 12\ V$.

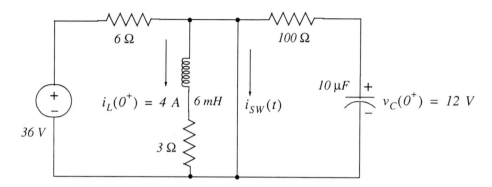

We will apply superposition three times. Thus for $t > 0$:

I. With the $36\ V$ voltage source acting alone where $i_L = 0$ (open) and $v_C = 0$ (shorted), the circuit is as shown below.

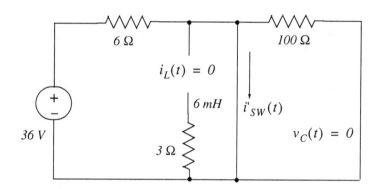

Since the $100\ \Omega$ is shorted out, we have

$$i'_{SW}(t) = 36/6 = 6\ A$$

II. With the $i_L(0^+) = 4\ A$ current source acting alone the circuit is as shown below where we observe that the $6\ \Omega$ and $100\ \Omega$ resistors are shorted out and thus $i''_{SW}(t) = -i_L(t)$ where

$i_L(t) = I_0 e^{-(R/L)t}$, $I_0 = 4\ A$, $R = 3\ \Omega$, $L = 6\ mH$, $R/L = 3/(6 \times 10^{-3}) = 500$ and thus

$$i''_{SW}(t) = -i_L(t) = -4e^{-500t}$$

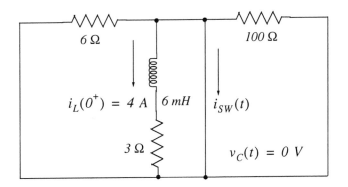

III. With the $v_C(0^+) = V_0 = 12\ V$ voltage source acting alone the circuit is as shown below where we observe that the $6\ \Omega$ resistor is shorted out.

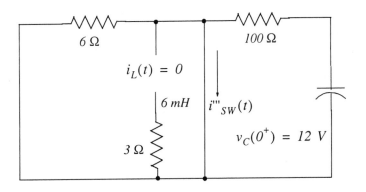

and thus

$$v_C(t) = V_0 e^{-(1/RC)t} = 12 e^{-[1/(100 \times 10^{-5})]t} = 12 e^{-1000t}.$$

Then,

$$i'''_{SW}(t) = v_C(t)/100\ \Omega = 0.12 e^{-1000t}$$

Therefore, the total current through the closed switch for $t > 0$ is

$$i_{SW}(t) = i'_{SW}(t) + i''_{SW}(t) + i'''_{SW}(t) = 6 - 4e^{-500t} + 0.12 e^{-1000t}\ A$$

NOTES

Chapter 10

Forced and Total Response in RL and RC Circuits

This chapter discusses the forced response of electric circuits. The term "forced" here implies that the circuit is excited by a voltage or current source, and its response to that excitation is analyzed. Then, the forced response is added to the natural response to form the total response.

10.1 The Unit Step Function $u_0(t)$

A function is said to be *discontinuous* if it exhibits points of discontinuity, that is, if the function jumps from one value to another without taking on any intermediate values.

A well-known discontinuous function is the *unit step function $u_0(t)$* [*] that is defined as

$$u_0(t) = \begin{cases} 0 & t < 0 \\ 1 & t > 0 \end{cases} \qquad (10.1)$$

It is also represented by the waveform of Figure 10.1.

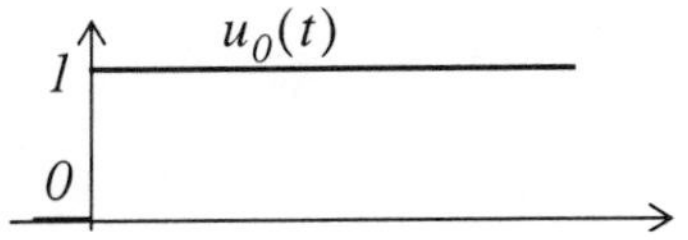

Figure 10.1. Waveform for $u_0(t)$

In the waveform of Figure 10.1, the unit step function $u_0(t)$ changes abruptly from 0 to 1 at $t = 0$. But if it changes at $t = t_0$ instead, its waveform and definition are as shown in Figure 10.2.

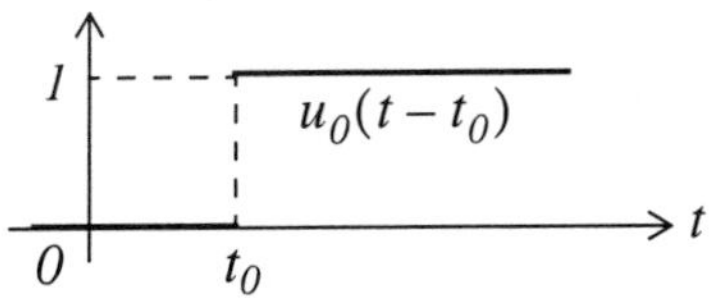

$$u_0(t - t_0) = \begin{cases} 0 & t < t_0 \\ 1 & t > t_0 \end{cases}$$

Figure 10.2. Waveform and definition of $u_0(t - t_0)$

---

[*] *In some books, the unit step function is denoted as $u(t)$, that is, without the subscript 0. In this text we will reserve this designation for any input.*

Likewise, if the unit step function changes from 0 to 1 at $t = -t_0$ as shown in Figure 10.3, it is denoted as $u_0(t + t_0)$

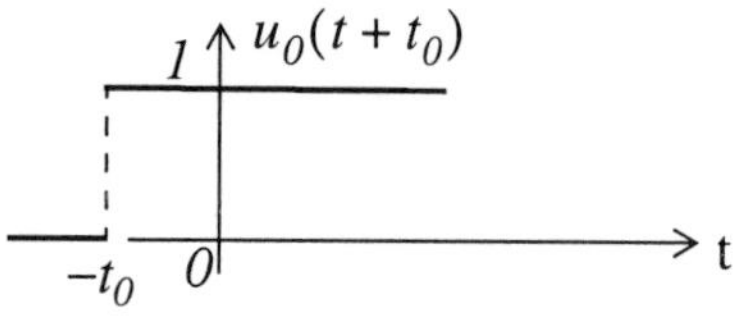

$$u_0(t + t_0) = \begin{cases} 0 & t < -t_0 \\ 1 & t > -t_0 \end{cases}$$

Figure 10.3. Waveform and definition of $u_0(t + t_0)$

Other forms of the unit step function are shown in Figure 10.4.

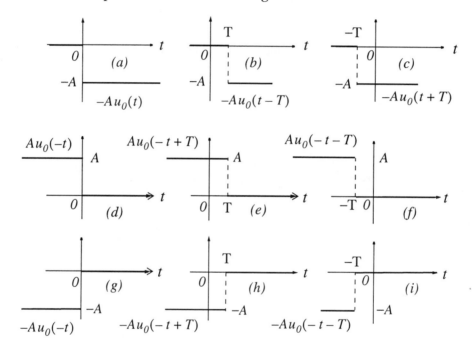

Figure 10.4. Other forms of the unit step function

Unit step functions can be used to represent other time-varying functions such as the rectangular pulse shown in Figure 10.5. This pulse is represented as $u_0(t) - u_0(t - 1)$.

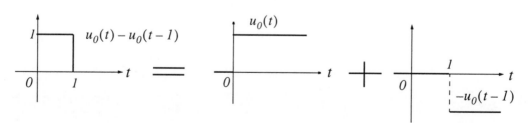

Figure 10.5. A rectangular pulse expressed as the sum of two unit step functions

The unit step function offers a convenient method of describing the sudden application of a voltage or current source. For example, a constant voltage source of $24\ V$ applied at $t = 0$, can be denoted as $24u_0(t)\ V$. Likewise, a sinusoidal voltage source $v(t) = V_m\cos\omega t\ V$ that is applied to a circuit at $t = t_0$, can be described as $v(t) = (V_m\cos\omega t)u_0(t - t_0)\ V$. Also, if the excitation in a circuit is a rectangular, or triangular, or sawtooth, or any other recurring pulse, it can be represented as a sum (difference) of unit step functions.

Example 10.1

Express the square waveform of Figure 10.6 as a sum of unit step functions. The vertical dotted lines indicate the discontinuities at $T, 2T, 3T$, and so on.

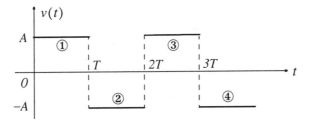

Figure 10.6. Square waveform for Example 10.1

Solution:

The line segment ① has height A, starts at $t = 0$, and terminates at $t = T$ on the time axis. Then, as in Figure 10.5, this segment can be expressed as

$$v_1(t) = A[u_0(t) - u_0(t - T)] \tag{10.2}$$

The line segment ② has height $-A$, starts at $t = T$, on the time axis, and terminates at $t = 2T$. This segment can be expressed as

$$v_2(t) = -A[u_0(t - T) - u_0(t - 2T)] \tag{10.3}$$

Line segment ③ has height A, starts at $t = 2T$, and terminates at $t = 3T$. This segment can be expressed as

$$v_3(t) = A[u_0(t - 2T) - u_0(t - 3T)] \tag{10.4}$$

Line segment ④ has height $-A$, starts at $t = 3T$, and terminates at $t = 4T$. This segment can be expressed as

$$v_4(t) = -A[u_0(t - 3T) - u_0(t - 4T)] \tag{10.5}$$

Thus, the square waveform of Figure 10.6 can be expressed as the summation of (10.2) through (10.5), that is,

$$v(t) = v_1(t) + v_2(t) + v_3(t) + v_4(t)$$
$$= A[u_0(t) - u_0(t - T)] - A[u_0(t - T) - u_0(t - 2T)] \qquad (10.6)$$
$$+ A[u_0(t - 2T) - u_0(t - 3T)] - A[u_0(t - 3T) - u_0(t - 4T)]$$

Combining like terms, we get

$$v(t) = A[u_0(t) - 2u_0(t - T) + 2u_0(t - 2T) - 2u_0(t - 3T) + \ldots] \qquad (10.7)$$

Example 10.2

Express the symmetric rectangular pulse of Figure 10.7 as a sum of unit step functions.

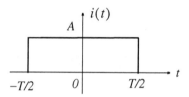

Figure 10.7. Symmetric rectangular pulse for Example 10.2

Solution:

This pulse has height A, it starts at $t = -T/2$, and terminates at $t = T/2$. Therefore, with reference to Figures 10.3 and 10.4 (b), we get

$$i(t) = Au_0\left(t + \frac{T}{2}\right) - Au_0\left(t - \frac{T}{2}\right) = A\left[u_0\left(t + \frac{T}{2}\right) - u_0\left(t - \frac{T}{2}\right)\right] \qquad (10.8)$$

Example 10.3

Express the symmetric triangular waveform shown in Figure 10.8 as a sum of unit step functions.

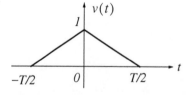

Figure 10.8. Symmetric triangular waveform for Example 10.3

Solution:

As a first step, we derive the equations of the linear segments ① and ② shown in Figure 10.9.

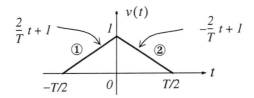

Figure 10.9. Equations for the linear segments of Figure 10.8

For line segment ①,

$$v_1(t) = \left(\frac{2}{T}t + 1\right)\left[u_0\left(t + \frac{T}{2}\right) - u_0(t)\right]$$ (10.9)

and for line segment ②,

$$v_2(t) = \left(-\frac{2}{T}t + 1\right)\left[u_0(t) - u_0\left(t - \frac{T}{2}\right)\right]$$ (10.10)

Combining (10.9) and (10.10), we get

$$v(t) = v_1(t) + v_2(t)$$
$$= \left(\frac{2}{T}t + 1\right)\left[u_0\left(t + \frac{T}{2}\right) - u_0(t)\right] + \left(-\frac{2}{T}t + 1\right)\left[u_0(t) - u_0\left(t - \frac{T}{2}\right)\right]$$ (10.11)

Example 10.4

Express the waveform shown in Figure 10.10 as a sum of unit step functions.

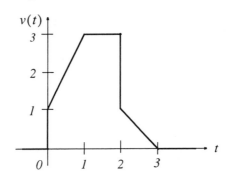

Figure 10.10. Waveform for Example 10.4

Solution:

As in the previous example, we first find the equations of the linear segments ① and ② shown in Figure 10.11.

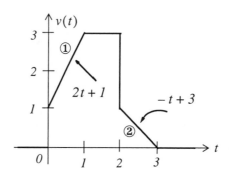

Figure 10.11. Equations for the linear segments of Figure 10.10

Following the same procedure as in the previous examples, we get

$$v(t) = (2t+1)[u_0(t) - u_0(t-1)] + 3[u_0(t-1) - u_0(t-2)]$$
$$+ (-t+3)[u_0(t-2) - u_0(t-3)]$$

Multiplying the values in parentheses by the values in the brackets, we get

$$v(t) = (2t+1)u_0(t) - (2t+1)u_0(t-1) + 3u_0(t-1)$$
$$- 3u_0(t-2) + (-t+3)u_0(t-2) - (-t+3)u_0(t-3)$$

or

$$v(t) = (2t+1)u_0(t) + [-(2t+1)+3]u_0(t-1)$$
$$+ [-3 + (-t+3)]u_0(t-2) - (-t+3)u_0(t-3)$$

and combining terms inside the brackets, we get

$$v(t) = (2t+1)u_0(t) - 2(t-1)u_0(t-1) - tu_0(t-2) + (t-3)u_0(t-3) \qquad (10.12)$$

Two other functions of interest are the *unit ramp function* and the unit impulse or delta function. We will discuss the unit ramp function first.

10.2 The Unit Ramp Function $u_1(t)$

The *unit ramp function*, denoted as $u_1(t)$, is defined as

$$u_1(t) = \int_{-\infty}^{t} u_0(\tau)d\tau \qquad (10.13)$$

where τ is a dummy variable.

We can evaluate the integral of (10.13) by considering the area under the unit step function $u_0(t)$ from $-\infty$ to t as shown in Figure 10.12.

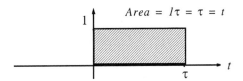

Figure 10.12. Area under the unit step function from $-\infty$ to t

Therefore,

$$u_1(t) = \begin{cases} 0 & t < 0 \\ t & t \geq 0 \end{cases} \tag{10.14}$$

and since $u_1(t)$ is the integral of $u_0(t)$, then $u_0(t)$ must be the derivative of $u_1(t)$, i.e.,

$$\frac{d}{dt}u_1(t) = u_0(t) \tag{10.15}$$

Higher order functions of t can be generated by repeated integration of the unit step function. For example, integrating $u_0(t)$ twice and multiplying by 2, we define $u_2(t)$ as

$$u_2(t) = \begin{cases} 0 & t < 0 \\ t^2 & t \geq 0 \end{cases} \qquad or \qquad u_2(t) = 2\int_{-\infty}^{t} u_1(\tau)d\tau \tag{10.16}$$

Similarly,

$$u_3(t) = \begin{cases} 0 & t < 0 \\ t^3 & t \geq 0 \end{cases} \qquad or \qquad u_3(t) = 3\int_{-\infty}^{t} u_2(\tau)d\tau \tag{10.17}$$

and in general,

$$u_n(t) = \begin{cases} 0 & t < 0 \\ t^n & t \geq 0 \end{cases} \qquad or \qquad u_n(t) = 3\int_{-\infty}^{t} u_{n-1}(\tau)d\tau \tag{10.18}$$

Also,

$$u_{n-1}(t) = \frac{1}{n}\frac{d}{dt}u_n(t) \tag{10.19}$$

10.3 The Delta Function $\delta(t)$

The *unit impulse* or *delta function*, denoted as $\delta(t)$, is the derivative of the unit step $u_0(t)$. It is generally defined as

$$\int_{-\infty}^{t} \delta(\tau)d\tau = u_0(t) \tag{10.20}$$

where

$$\delta(t) = 0 \ \ for \ all \ \ t \neq 0 \tag{10.21}$$

To better understand the delta function $\delta(t)$, let us represent the unit step $u_0(t)$ as shown in Figure 10.13 (a).

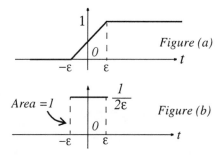

Figure 10.13. Representation of the unit step as a limit.

The function of Figure 10.13 (a) becomes the unit step as $\varepsilon \to 0$. Figure 10.13 (b) is the derivative of Figure 10.13 (a), where we see that as $\varepsilon \to 0$, $1/2\varepsilon$ becomes unbounded, but the area of the rectangle remains *1*. Therefore, in the limit, we can think of $\delta(t)$ as approaching a very large spike or impulse at the origin, with unbounded amplitude, zero width, and area equal to *1*.

Two useful properties of the delta function are the sampling property and the sifting property.

The *Sampling Property of the Delta Function* states that

$$f(t)\delta(t) = f(0)\delta(t) \tag{10.22}$$

or

$$f(t)\delta(t-a) = f(a)\delta(t) \tag{10.23}$$

that is, multiplication of any function $f(t)$ by the delta function $\delta(t)$ results in sampling the function at the time instants where the delta function is not zero. The study of discrete-time systems is based on this property.

The *Sifting Property of the Delta Function* states that

$$\int_{-\infty}^{\infty} f(t)\delta(t-\alpha)dt = f(\alpha)$$

(10.24)

that is, if we multiply any function $f(t)$ by $\delta(t-\alpha)$ and integrate from $-\infty$ to $+\infty$, we will obtain the value of $f(t)$ evaluated at $t-\alpha$.

The proofs of (10.22) through (10.24) and additional properties of the delta function are beyond the scope of this book. They are provided in *Signals and Systems with MATLAB Applications*, ISBN 0-9709511-3-2 by this author, Orchard Publications 2001.

MATLAB has two built-in functions for the unit step and the delta functions. These are designated by the names of the mathematicians who used them in their work. The unit step $u_0(t)$ is called **Heavyside(t)** and the delta function $\delta(t)$ is called **Dirac(t)**. Shown below are examples of how they are being used.

```
syms k a t
u=k*sym('Heaviside(t-a)')  % Create unit step function at t=a

u =
k*Heaviside(t-a)

d=diff(u)  % Compute the derivative of the unit step function

d =
k*Dirac(t-a)

int(d)  % Integrate the delta function

ans =
Heaviside(t-a)*k
```

Example 10.5

For the circuit shown in Figure 10.14, the inputs are applied at different times as indicated. Compute v_{out} at:

a. $t = -0.5\ s$

b. $t = 1.5\ s$

c. $t = 5\ s$

$$v_{in1} = 0.8u_0(t-3) \ V$$
$$v_{in2} = 0.5u_0(t-1) \ V$$
$$i_{in} = 0.14[u_0(t+1) + u_0(t-2)] \ mA$$

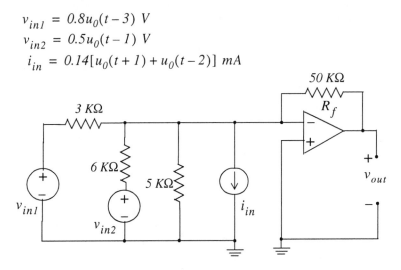

Figure 10.14. Circuit for Example 10.5

Solution:

Let us first sketch the step functions for each of the inputs.

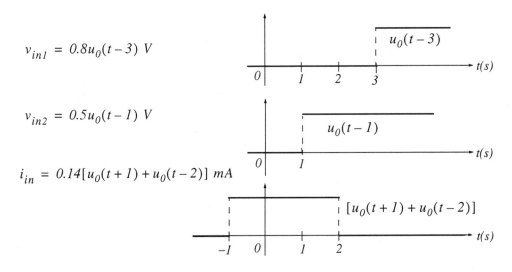

a. At $t = -0.5 \ s$ only the signal due to i_{in} is active; therefore, exchanging the current source and its parallel resistance with an equivalent voltage source with a series resistance, the input circuit becomes as shown in Figure 10.15.

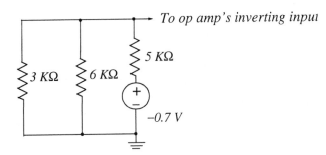

Figure 10.15. Input to the circuit of Example 10.5 when i_{in} is acting alone

Replacing the circuit of Figure 10.15 with its Thevenin equivalent, we get the network of Figure 10.16.

$3 \ K\Omega \parallel 6 \ K\Omega = 2 \ K\Omega$

$v_{TH1} = V_{2 \ K\Omega} = \dfrac{2 \ K\Omega}{2 \ K\Omega + 5 \ K\Omega}(-0.7) = -0.2 \ V$

$R_{TH1} = \dfrac{2 \ K\Omega \times 5 \ K\Omega}{7 \ K\Omega} = 10/7 \ K\Omega$

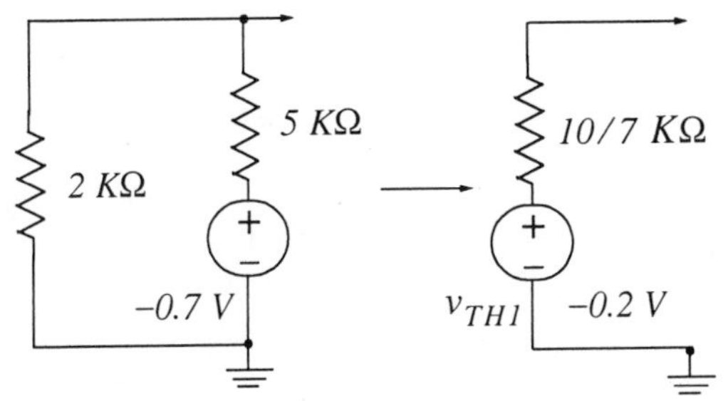

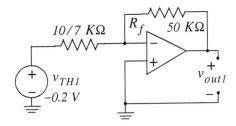

Figure 10.16. Simplified input to the circuit of Example 10.5 when i_{in} is acting alone

Now, we can compute v_{out1} with the circuit of Figure 10.17.

Figure 10.17. Circuit for computation of v_{out1}

$$v_{out1} = -\left(\dfrac{50}{10/7}\right)v_{TH1} = -35 \times (-0.2 \ mV) = 7 \ V \tag{10.25}$$

b. At $t = 1.5 \ s$ the active inputs are

$$i_{in} = 0.14[u_0(t+1) + u_0(t-2)] \ mA$$

and

$$v_{in2} = 0.5u_0(t-1) \ V$$

Since we already know the output due to i_{in} acting alone, we will find the output due to v_{in2} acting alone and then apply superposition to find the output when both of these inputs are present. Thus, with the input v_{in2} acting alone, the input circuit is as shown in Figure 10.18.

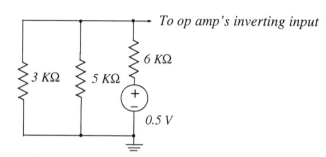

Figure 10.18. Input to the circuit of Example 10.5 when v_{in2} is acting alone

Replacing this circuit of Figure 10.18 with its Thevenin equivalent, we get the network of Figure 10.19.

$$3 \ K\Omega \ || \ 5 \ K\Omega \ = 15/8 \ K\Omega$$

$$v_{TH2} = v_{(15/8) \ K\Omega} = \frac{15/8}{15/8+6}(0.5) = \frac{5}{42} \ V$$

$$R_{TH2} = R_{TH1} = 10/7 \ K\Omega$$

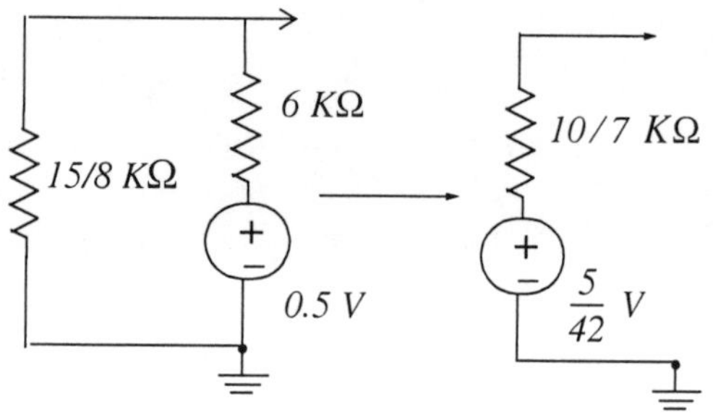

Figure 10.19. Simplified input to the circuit of Example 10.5 when v_{in2} is acting alone

Now, we can compute v_{out2} with the circuit of Figure 10.20.

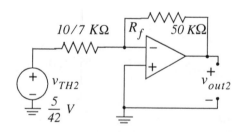

Figure 10.20. Circuit for computation of v_{out2}

$$v_{out2} = -\left(\frac{50}{10/7}\right)v_{TH2} = -35 \times \left(\frac{5}{42}\right) = -\frac{25}{6} \ V \tag{10.26}$$

Therefore, from (10.25) and (10.26) the op amp's output voltage at $t = 1.5 \ s$ is

$$v_{out1} + v_{out2} = 7 - \frac{25}{6} = \frac{17}{6} \ V \tag{10.27}$$

c. At $t = 5 \ s$ the active inputs are

$$v_{in1} = 0.8u_0(t-3) \ V$$

and

$$v_{in2} = 0.5u_0(t-1) \ V$$

Since we already know the output due to v_{in2} acting alone, we will find the output due to v_{in1} acting alone and then apply superposition to find the output when both of these inputs are present. Thus, with the input v_{in1} acting alone, the input circuit is as shown in Figure 10.21.

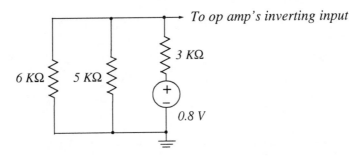

Figure 10.21. Input to the circuit of Example 10.5 when v_{in1} is acting alone

Replacing this circuit of Figure 10.21 with its Thevenin equivalent, we get the network of Figure 10.22.

$6 \ K\Omega \parallel 5 \ K\Omega = 30/11 \ k\Omega$

$$v_{TH3} = v_{(30/11) \ K\Omega} = \frac{30/11}{30/11+3}(0.8) = \frac{10}{21} \ V$$

$$R_{TH3} = R_{TH2} = R_{TH1} = 10/7 \ K\Omega$$

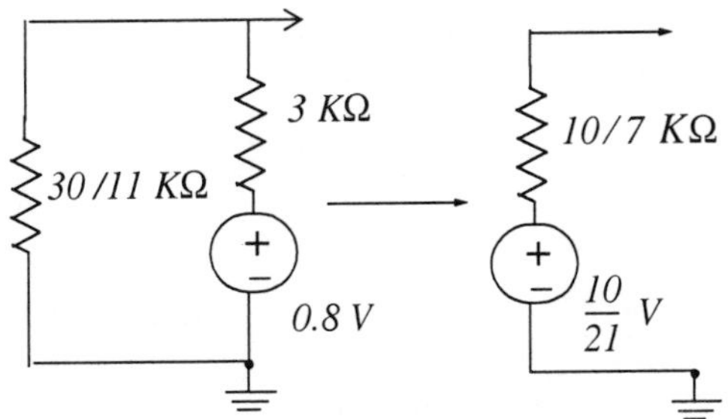

Figure 10.22. Simplified input to the circuit of Example 10.5 when v_{in1} is acting alone

Now, we can compute v_{out3} with the circuit of Figure 10.23.

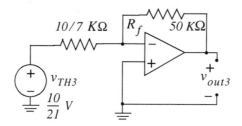

Figure 10.23. Circuit for computation of v_{out3}

$$v_{out3} = -\left(\frac{50}{10/7}\right)v_{TH3} = -35 \times \left(\frac{10}{21}\right) = -\frac{50}{3} \ V \tag{10.28}$$

Therefore, from (10.26) and (10.28) the op amp's output voltage at $t = 5 \ s$ is

$$v_{out2} + v_{out3} = -\frac{25}{6} - \frac{50}{3} = -\frac{125}{6} \ V \tag{10.29}$$

10.4 The Forced and Total Response in an RL Circuit

For the circuit shown in Figure 10.24 (a), V_S is constant. We will derive an expression for the inductor current $i_L(t) = i(t)$ for $t > 0$ given that the initial condition is $i_L(0^-) = 0$. Here, the inductor current $i_L(t)$ will be referred to as the *total response*.

The switch in Figure 10.24 (a) can be omitted if we multiply the excitation V_S by the unit step function $u_0(t)$ as shown in Figure 10.24 (b).

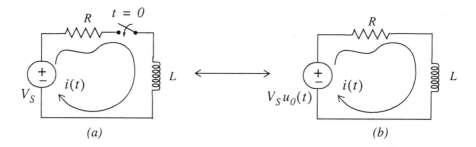

Figure 10.24. Circuits for derivation of the total response $i_L(t) = i(t)$

We start by applying KVL, that is,

$$L\frac{di}{dt} + Ri = V_S u_0(t) \tag{10.30}$$

The initial condition states that $i_L(0^-) = 0$; thus for $t < 0$, $i(t) = 0$

For $t > 0$, we must solve the differential equation

$$L\frac{di}{dt} + Ri = V_S \qquad (10.31)$$

It is shown in differential equations textbooks that a differential equation such as the above, can be solved by the method of separation of the variables. Thus, rearranging (10.32), separating the variables, and integrating we get:

$$L\frac{di}{dt} = V_S - Ri$$

or

$$\frac{Ldi}{V_S - Ri} = dt$$

or

$$\int\frac{Ldi}{V_S - Ri} = \int dt$$

and referring to a table of integrals, we get

$$-\frac{L}{R}ln(V_S - Ri) = t + k \qquad (10.32)$$

The constant k in (10.32) represents the constant of integration of both sides and it can be evaluated from the initial condition, and as we stated in the previous chapter

$$i_L(0^-) = i_L(0) = i_L(0^+) \qquad (10.33)$$

Therefore, at $t = 0^+$

$$-\frac{L}{R}ln(V_S - 0) = 0 + k$$

or

$$k = -\frac{L}{R}ln V_S$$

and by substitution into (10.32), we get

$$-\frac{L}{R}ln(V_S - Ri) = t - \frac{L}{R}ln V_S$$

or

$$-\frac{L}{R}[ln(V_S - Ri) - ln V_S] = t$$

or

$$-\frac{L}{R}\left[ln\frac{V_S - Ri}{V_S}\right] = t$$

or

$$ln\frac{V_S - Ri}{V_S} = -\frac{R}{L}t$$

or

$$\frac{V_S - Ri}{V_S} = e^{-(R/L)t}$$

or

$$Ri = V_S - V_S e^{-(R/L)t}$$

or

$$i(t) = \frac{V_S}{R} - \frac{V_S}{R}e^{-(R/L)t}$$

The general expression for all t is

$$i(t) = \left(\frac{V_S}{R} - \frac{V_S}{R}e^{-(R/L)t}\right)u_0(t) \qquad (10.34)$$

We observe that the right side of (10.34) consists of two terms, V_S/R which is constant called the *forced response*, and the exponential term $-\frac{V_S}{R}e^{-(R/L)t}$ that has the same form as that of the previous chapter which we call the *natural response*.

The forced response V_S/R is a result of the application of the excitation (forcing) function $V_S u_0(t)$ applied to the *RL* circuit. This value represents the steady-state condition reached as $t \to \infty$ since the inductor L at this state behaves as a short circuit.

The amplitude of the natural response is $-V_S/R$ and depends on the values of V_S and R.

The summation of the forced response and the natural response constitutes the *total response* or *complete response*, that is,

$$i(t)_{total} = i(t)_{forced\ response} + i(t)_{natural\ response}$$

or

$$i_{total} = i_f + i_n \qquad (10.35)$$

Now, let us return to the *RL* circuit of Figure 10.24 to find the complete (total) response i_{total} by the summation of the forced and the natural responses as indicated in (10.35).

The forced response i_f is found from the circuit of Figure 10.25 where we let $t \to \infty$

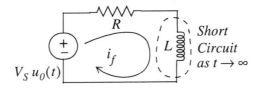

Figure 10.25. Circuit for derivation of the forced response i_f

Then, from the circuit of Figure 10.25,

$$i_f = \frac{V_S}{R} \tag{10.36}$$

Next, we need to find the natural response. This is found by letting the excitation (forcing function) $V_S u_0(t)$ go to zero as shown in the circuit of Figure 10.26.

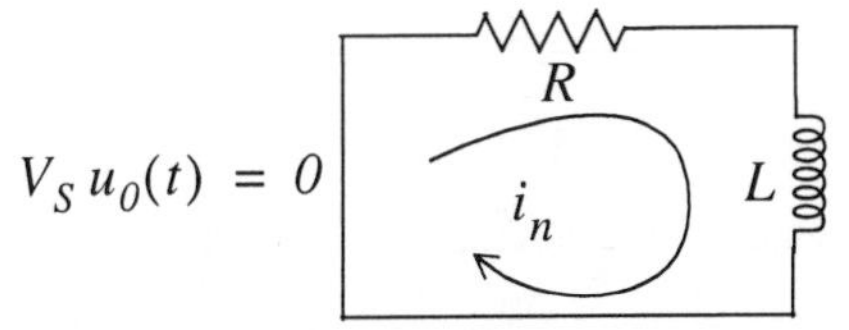

Figure 10.26. Circuit for derivation of the natural response i_n

We found in Chapter 9 that the natural response i_n has the exponential form

$$i_n = A e^{-(R/L)t} \tag{10.37}$$

Therefore, the total response is

$$i_{total} = i_f + i_n = \frac{V_S}{R} + A e^{-(R/L)t} \tag{10.38}$$

where the constant A is evaluated from the initial condition $i_L(0^-) = i_L(0) = i_L(0^+)$

Substitution of the initial condition into (10.38) yields

$$i(0) = 0 = \frac{V_S}{R} + A e^0$$

or

$$A = -\frac{V_S}{R}$$

and with this substitution (10.38) is rewritten as

$$i_{total} = \left(\frac{V_S}{R} - \frac{V_S}{R} e^{-(R/L)t} \right) u_0(t) \tag{10.39}$$

and this is the same as (10.34).

We can sketch i_{total} easily if we sketch $\dfrac{V_S}{R}$ and $-\dfrac{V_S}{R}e^{-(R/L)t}$ separately and then add these. This is done with Excel and the plots are shown in Figure 10.27.

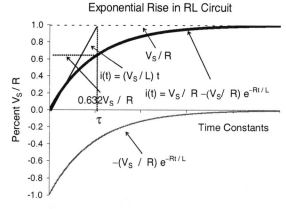

Figure 10.27. Curves for forced, natural, and total responses in a series RL circuit

The time constant τ is defined as before, and its numerical value can be found from the circuit constants R and L as follows:

The equation of the straight line with $slope = V_S/L$ is found from

$$\frac{d}{dt}(i_{total})\bigg|_{t=0} = \frac{R}{L} \cdot \frac{V_S}{R} e^{-(R/L)t}\bigg|_{t=0} = \frac{V_S}{L}$$

Assuming constant rate of change as shown in Figure 10.27, at $t = \tau$,

$$i(t) = \frac{V_S}{R}$$

and thus

$$\frac{V_S}{R} = \frac{V_S}{L}\tau$$

or

$$\tau = \frac{L}{R}$$

as before. Also, from (10.39)

$$i(\tau) = \frac{V_S}{R} - \frac{V_S}{R}e^{-(R/L)(L/R)} = \frac{V_S}{R}(1 - e^{-1}) = \frac{V_S}{R}(1 - 0.368)$$

or

$$\boxed{i(\tau) = 0.632\frac{V_S}{R}} \tag{10.40}$$

Therefore, the current in a series RL circuit which has been excited by a constant source, in one time constant has reached 63.2% of its final value.

Example 10.6

For the circuit of Figure 10.28, compute the energy stored in the $10\ mH$ inductor at $t = 100\ ms$.

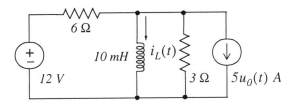

Figure 10.28. Circuit for Example 10.6

Solution:

For $t < 0$, the circuit is as shown in Figure 10.29 where the $3\ \Omega$ resistor is shorted out by the inductor.

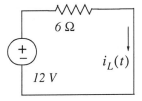

Figure 10.29. Circuit of Example 10.6 for $t < 0$

From the circuit of Figure 10.29,

$$i_L(0^-) = \frac{12}{6} = 2\ A$$

and this value establishes our initial condition as

$$i_L(0^+) = 2\ A \tag{10.41}$$

For $t > 0$, the circuit is as shown in Figure 10.30.

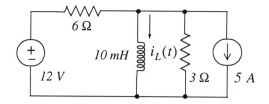

Figure 10.30. Circuit of Example 10.6 for $t > 0$

We will find $i_L(t)$ from the relation

$$i_L(t) = i_f + i_n$$

The forced component i_f is found from the circuit at steady state conditions. It is shown in Figure 10.31 where the voltage source and its series resistance have been exchanged for an equivalent current source with a parallel resistor. The resistors have been shorted out by the inductor.

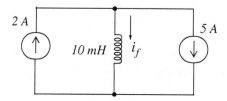

Figure 10.31. Circuit of Example 10.6 under steady-state conditions

By inspection, $i_f = 2 - 5$ or

$$i_f = -3 \ A \tag{10.42}$$

To find i_n we short the voltage source and open the current source. The circuit then reduces to that shown in Figure 10.32.

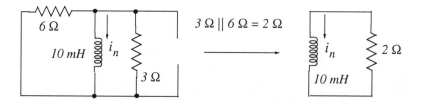

Figure 10.32. Circuit of Example 10.6 for determining the natural response

The natural response of RL circuit of Figure 10.32 is

$$i_n = Ae^{-(R/L)t} = Ae^{-(2/10 \times 10^{-3})t}$$

or

$$i_n = Ae^{-200t} \tag{10.43}$$

The total response is the summation of (10.42) and (10.43), that is,

$$i_{total} = i_f + i_n = -3 + Ae^{-200t} \tag{10.44}$$

Using the initial condition of (10.42), we get

$$i_L(0^+) = 2 = -3 + Ae^{-0}$$

or

$$A = 5$$

Finally, by substitution into (10.44) we get

$$i_{total} = (-3 + 5e^{-200t})u_0(t) \tag{10.45}$$

and the energy stored in the inductor at $t = 100 \ ms$ is

$$W_L\Big|_{t = 100 \ ms} = \frac{1}{2}Li_L^2\Big|_{t = 100 \ ms} = \frac{1}{2}10 \times 10^{-3}\left(-3 + 5e^{-200 \times 100 \times 10^{-3}}\right)^2 \tag{10.46}$$

$$= 5 \times 10^{-3}(-3 + 5e^{-20})^2 = 45 \ mJ$$

10.5 The Forced and Total Response in an RC Circuit

For the circuit shown in Figure 10.33 (a), V_S is constant. We will derive an expression for the capacitor voltage $v_C(t)$ for $t > 0$ given that the initial condition is $v_C(0^-) = 0$. Here, the capacitor voltage $v_C(t)$ will be referred to as the *total response*.

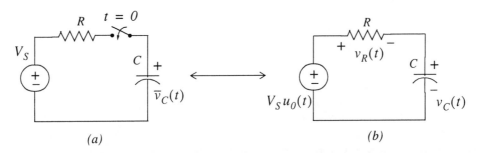

Figure 10.33. Circuits for derivation of the total response $v_C(t)$

The switch in Figure 10.33 (a) can be omitted if we multiply the excitation V_S by the unit step function $u_0(t)$ as shown in Figure 10.33 (b).

We start by applying KVL, that is,

$$v_R + v_C = V_S u_0(t) \tag{10.47}$$

and since

$$i = i_C = C\frac{dv_C}{dt}$$

we can express v_R as

$$v_R = Ri = RC\frac{dv_C}{dt}$$

By substitution into (10.47), we get

$$RC\frac{dv_C}{dt} + v_C = V_S u_0(t) \tag{10.48}$$

The initial condition states that $v_C(0^-) = 0$; thus for $t < 0$, $v_C(t) = 0$

For $t > 0$, we must solve the differential equation

$$RC\frac{dv_C}{dt} + v_C = V_S \tag{10.49}$$

Rearranging, separating variables and integrating, we get:

$$RCdv_C = (V_S - v_C)dt$$

or

$$\frac{dv_C}{v_C - V_S} = -\frac{1}{RC}dt$$

or

$$\int\frac{dv_C}{v_C - V_S}dt = -\frac{1}{RC}\int dt \tag{10.50}$$

or

$$ln(v_C - V_S) = -\frac{1}{RC}t + k$$

where k represents the constant of integration of both sides of (10.51). Then,

$$v_C - V_S = e^{-(1/RC)t + k} = e^k e^{-(1/(RC))t} = k_1 e^{-(1/RC)t}$$

or

$$v_C = V_S - k_1 e^{-(1/RC)t} \tag{10.51}$$

The constant k_1 can be evaluated from the initial condition $v_C(0^+) = v_C(0^-) = 0$ where by substitution into (10.51) we get

$$v_C(0^+) = 0 = V_S - k_1 e^0$$

or

$$k_1 = V_S$$

Therefore, the solution of (10.49) is

$$\boxed{v_C(t) = (V_S - V_S e^{-(1/RC)t})u_0(t)} \tag{10.52}$$

As with the *RL* circuit of the previous section, we observe that the solution consists of a *forced response* and a *natural response*. The constant term V_S is the voltage attained across the capacitor as $t \to \infty$ and represents the steady-state condition since the capacitor C at this state behaves as an open circuit.

The amplitude of the exponential term natural response is $-V_S$.

The summation of the forced response and the natural response constitutes the total response, i.e.,

$$v_C(t)_{complete\ response} = v_C(t)_{forced\ response} + v_C(t)_{natural\ response}$$

or

$$v_{Ctotal} = v_{Cf} + v_{Cn} \tag{10.53}$$

Now, let us return to the *RC* circuit of Figure 10.33 to find the complete (total) response by summing the forced and the natural responses indicated in (10.53).

The forced response v_{Cf} is found from the circuit of Figure 10.34 where we let $t \to \infty$

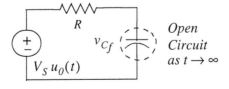

Figure 10.34. Circuit for derivation of the forced response v_{Cf}

Then, from the circuit of Figure 10.34,

$$\boxed{v_{C_f} = V_S}$$ (10.54)

Next, we need to find the natural response and this is found by letting the excitation (forcing function) $V_S u_0(t)$ go to zero as shown in Figure 10.35.

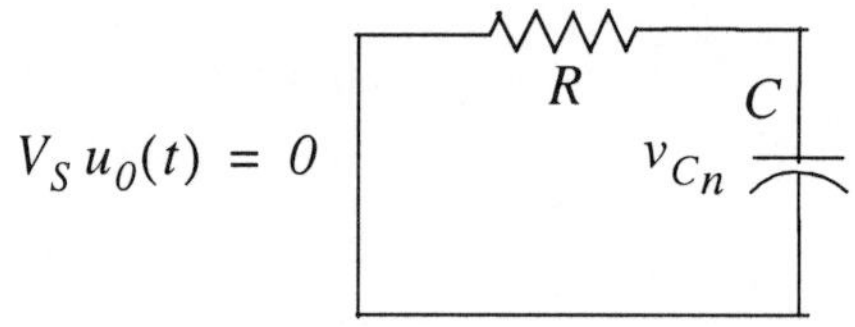

Figure 10.35. Circuit for derivation of the natural response v_{C_n}

We found in Chapter 9 that the natural response v_{C_n} has the exponential form

$$v_{C_n} = Ae^{-(1/RC)t}$$

and thus the total response is

$$v_C(t) = v_{C_f} + v_{C_n} = V_S + Ae^{-(1/RC)t}$$ (10.55)

where the constant A is evaluated from the initial condition $v_C(0^-) = v_C(0) = v_C(0^+) = 0$

Substitution of the initial condition into (10.55) yields

$$v_C(0^+) = 0 = V_S - Ae^0$$

or

$$A = -V_S$$

With this substitution (10.55) is rewritten as

$$v_C(t) = (V_S - V_S e^{-(1/RC)t})u_0(t)$$ (10.56)

and this is the same as (10.52).

We can sketch $v_{C_{total}}$ easily if we sketch V_S and $-V_S e^{-(1/RC)t}$ separately and then add these. This is done with Excel and the plots are shown in Figure 10.36.

The time constant τ is defined as before, and its numerical value can be found from the circuit constants R and C as follows:

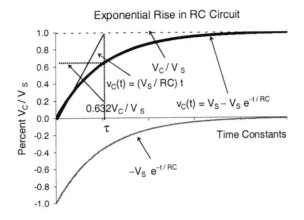

Figure 10.36. Curves for forced, natural, and total responses in a series RC circuit

The equation of the straight line with $slope = V_S/RC$ is found from

$$\frac{d}{dt}v_C(t)\Big|_{t=0} = \frac{1}{RC}\cdot V_S e^{-(1/RC)t}\Big|_{t=0} = \frac{V_S}{RC}$$

Assuming constant rate of change as shown in Figure 10.36, at $t = \tau$,

$$v_C(t) = V_S$$

and thus

$$V_S = \frac{V_S}{RC}\tau$$

or

$$\tau = RC$$

as before. Also, from (10.56)

$$v_C(\tau) = V_S - V_S e^{-(1/RC)RC} = V_S(1 - e^{-1}) = V_S(1 - 0.368)$$

or

$$\boxed{v_C(\tau) = 0.632V_S} \tag{10.57}$$

Therefore, the voltage across a capacitor in a series RC circuit which has been excited by a constant source, in one time constant has reached *63.2%* of its final value.

Example 10.7

For the circuit shown in Figure 10.37 find:

a. $v_C(1^-)$ and $i_C(1^-)$

b. $v_C(1^+)$ and $i_C(1^+)$

c. $v_C(t = 10 \ min.)$ and $i_C(t = 10 \ min.)$

d. $i_C(t)$ for $t > 1$

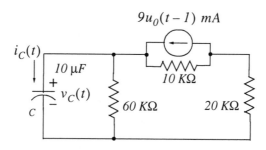

Figure 10.37. Circuit for Example 10.7

Solution:

a. No initial condition is given so we must assume that sufficient time has elapsed for steady-state conditions to exist for all $t < 1 \ s$. We assume time is in seconds since we are not told otherwise. Then, since there is no voltage or current source present to cause current to flow, we get

$$v_C(1^-) = 0$$

and

$$i_C(1^-) = 0$$

b. Exchanging the current source and the $10 \ K\Omega$ resistor with a voltage source with a $10 \ K\Omega$ series resistor, the circuit at $t = 1^+$ is as shown in Figure 10. 38.

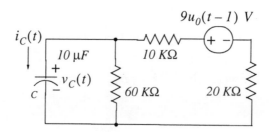

Figure 10.38. Circuit for Example 10.7 at $t = 1^+$

Now, since $v_C(1^+) = v_C(1^-)$, no current flows through the $60\ k\Omega$ resistor at $t = 1^+$; if it did, the voltage across the capacitor would change instantaneously, and as we know, this is a physical impossibility. Instead, the current path is through the capacitor which at exactly $t = 1^+$ acts as a short circuit since $v_C(1^+) = v_C(1^-) = 0$. Therefore,

$$i_C(1^+) = \frac{90\ V}{(20+10)\ K\Omega} = 3\ mA \tag{10.58}$$

c. The time $t = 10\ min$ is the essentially the same as $t = \infty$, and at this time the capacitor voltage $v_C(t = 10\ min)$ is constant and equal to the voltage across the $60\ K\Omega$ resistor, i.e.,

$$v_C(t = 10\ min) = v_C(\infty) = v_{60\ K\Omega} = \frac{90\ V}{(20+10+60)\ K\Omega} \cdot 60\ K\Omega = 60\ V$$

Also,

$$i_C(t)\big|_{t\,=\,\infty} = C\frac{dv_C}{dt} = 0$$

d. For $t > 1$

$$i_C(t)\big|_{t\,>\,1} = i_{Cf} + i_{Cn}$$

where from part (c)

$$i_{Cf}(\infty) = 0$$

and

$$i_{Cn} = Ae^{-(1/R_{eq}C)t}$$

With the voltage source shorted in the circuit of Figure 10.38, the equivalent resistance is

$$R_{eq} = (10\ K\Omega + 20\ K\Omega) \parallel 60\ K\Omega = 20\ K\Omega$$

or

$$R_{eq}C = 20 \times 10^3 \times 10 \times 10^{-6} = 0.2\ s$$

Therefore,

$$i_{Cn} = Ae^{-(1/0.2)t} = Ae^{-5t} \tag{10.59}$$

We can evaluate the constant A using (10.59) where

$$i_C(1^+) = 3\ mA = Ae^{-5}$$

or

$$A = \frac{3 \times 10^{-3}}{e^{-5}} = 0.445$$

and by substitution into (10.59),

$$i_C(t)\big|_{t>1} = i_{Cf} + i_{Cn} = i_{Cn} = 0.445e^{-5t}u_0(t-1) \ mA \qquad (10.60)$$

Example 10.8

In the circuit shown in Figure 10.39, the switch is actually an electronic switch and it is open for
$15 \ \mu s$ and closed for $15 \ \mu s$. Initially, the capacitor is discharged, i.e., $v_C(0) = 0$. Compute and
sketch the voltage across the capacitor for two repetitive cycles.

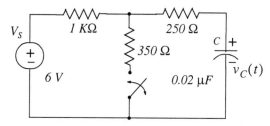

Figure 10.39. Circuit for Example 10.8

Solution:

With the switch in the open position the circuit is as shown in Figure 10.40.

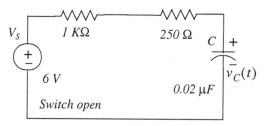

Figure 10.40. Circuit for Example 10.8 with the switch in the open position

For the time period $0 < t_{open} < 15 \ \mu s$ the time constant for the circuit of Figure 10.40 is

$$\tau_{open} = R_{eq}C = (1 \ K\Omega + 0.25 \ K\Omega) \times 0.02 \times 10^{-6} = 25 \ \mu s$$

Thus, at the end of the first period when the switch is open, the voltage across the capacitor is

$$v_C(t)\big|_{t=15 \ \mu s} = v_{Cf} + v_{Cn} = V_S - V_S e^{-t/RC} = 6 - 6e^{-4 \times 10^4 t} = 6 - 6e^{-0.6} = 2.71 \ V \qquad (10.61)$$

Next, with the switch closed for $15 < t_{closed} < 30 \ \mu s$ the circuit is as shown in Figure 10.41.

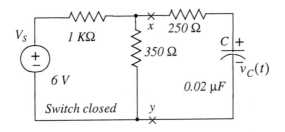

Figure 10.41. Circuit for Example 10.8 with the switch in the closed position

Replacing the circuit to the left of points x and y by its Thevenin equivalent, we get the circuit shown in Figure 10.42.

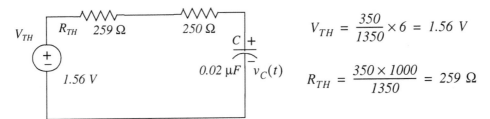

$$V_{TH} = \frac{350}{1350} \times 6 = 1.56 \text{ V}$$

$$R_{TH} = \frac{350 \times 1000}{1350} = 259 \text{ }\Omega$$

Figure 10.42. Thevenin equivalent circuit for the circuit of Figure 10.41

The time constant for the circuit of Figure 10.42 where the switch is closed, is

$$\tau_{closed} = R_{eq}C = (259 \text{ }\Omega + 250 \text{ }\Omega) \times 0.02 \times 10^{-6} = 10.2 \text{ }\mu s$$

The capacitor voltage $v_C(t)$ for the circuit of Figure 10.42 is

$$v_C(t) = v_{Cf} + v_{Cn} = V_{TH} + A_1 e^{-(1/RC)(t-15)} = 1.56 + A_1 e^{-(1/10.2)(t-15)} \qquad (10.62)$$

and the constant A_1 is evaluated from initial condition at $t = 15 \text{ }\mu s$ which by (10.62) is

$$v_C(t)\big|_{t = 15 \text{ }\mu s} = 2.71 \text{ V}$$

Then,

$$v_C(t)\big|_{15 < t < 30 \text{ }\mu s} = 2.71 = 1.56 + A_1 e^{-(1/10.2)(15-15)}$$

or

$$A_1 = 1.15$$

and by substitution into (10.62)

$$v_C(t)\big|_{15 < t < 30 \text{ }\mu s} = 1.56 + 1.15 e^{-(1/10.2)(t-15)} \qquad (10.63)$$

At the end of the first period when the switch is closed, the voltage across the capacitor is

$$v_C(t)\big|_{t = 30 \ \mu s} = 1.56 + 1.15e^{-(1/10.2)(30-15)} = 1.82 \ V \tag{10.64}$$

For the next cycle, that is, for $30 < t_{open} < 45 \ \mu s$ when the switch is open, the time constant τ_{open} is the same as before, i.e., $\tau_{open} = 25 \ \mu s$ and the capacitor voltage is

$$v_C(t) = v_{Cf} + v_{Cn} = 6 + A_2 e^{-(1/25)(t-30)} \tag{10.65}$$

The constant A_2 is computed with (10.65) as

$$v_C(t)\big|_{t = 30 \ \mu s} = 1.82 = 6 + A_2 e^{-(1/25)(30-30)}$$

or

$$A_2 = -4.18$$

and by substitution into (10.65)

$$v_C(t)\big|_{30 < t < 45 \ \mu s} = 6 - 4.18 e^{-(1/25)(t-30)} \tag{10.66}$$

At the end of the second period when the switch is open, the voltage across the capacitor is

$$v_C(t)\big|_{t = 45 \ \mu s} = 6 - 4.18 e^{-(1/25)(45-30)} = 3.71 \ V \tag{10.67}$$

The second period when the switch is closed is $45 < t_{closed} < 60 \ \mu s$

Then,

$$v_C(t)\big|_{45 < t < 60 \ \mu s} = v_{Cf} + v_{Cn} = V_{TH} + A_3 e^{-(1/RC)(t-45)} = 1.56 + A_3 e^{-(1/10.2)(t-45)} \tag{10.68}$$

and with (10.67) we get

$$A_3 = 2.15$$

Therefore,

$$v_C(t)\big|_{45 < t < 60 \ \mu s} = 1.56 + 2.15 e^{-(1/10.2)(t-45)} \tag{10.69}$$

and

$$v_C(t)\big|_{t = 60 \ \mu s} = 1.56 + 2.15 e^{-(1/10.2)(60-45)} = 2.05 \ V \tag{10.70}$$

Repeating the above steps for the third open and closed switch periods, we get

$$v_C(t)\big|_{60 < t < 75 \ \mu s} = 6 - 3.95 e^{-(1/25)(t-60)} \tag{10.71}$$

and

$$v_C(t)\big|_{t = 75\ \mu s} = 3.83\ V \tag{10.72}$$

Likewise,

$$v_C(t)\big|_{75 < t < 90\ \mu s} = 1.56 + 2.27e^{-(1/10.2)(t - 75)} \tag{10.73}$$

and

$$v_C(t)\big|_{t = 90\ \mu s} = 2.08\ V \tag{10.74}$$

and using Excel we get the waveform shown in Figure 10.43.

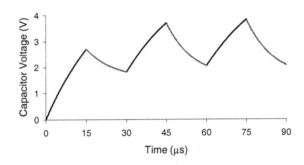

Figure 10.43. Voltage across the capacitor for the circuit of Example 10.8

10.6 Summary

- The unit step function $u_0(t)$ is defined as

$$u_0(t) = \begin{cases} 0 & t < 0 \\ 1 & t > 0 \end{cases}$$

and it is represented by the waveform below.

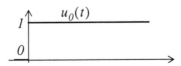

- Unit step functions can be used to represent other time-varying functions.

- The unit step function offers a convenient method of describing the sudden application of a voltage or current source.

- The unit ramp function $u_1(t)$, is defined as the integral of the unit step function, that is,

$$u_1(t) = \int_{-\infty}^{t} u_0(\tau)d\tau$$

where τ is a dummy variable. It is also expressed as

$$u_1(t) = \begin{cases} 0 & t < 0 \\ t & t \geq 0 \end{cases}$$

- The unit impulse or delta function, denoted as $\delta(t)$, is the derivative of the unit step $u_0(t)$. It is defined as

$$\delta(\tau) = \frac{d}{dt}u_0(t)$$

or

$$\int_{-\infty}^{t} \delta(\tau)d\tau = u_0(t)$$

and

$$\delta(t) = 0 \ \ for \ all \ \ t \neq 0$$

- In a simple RL circuit that is excited by a voltage source $V_S u_0(t)$ the current is

$$i(t) = i_f + i_n = \left(\frac{V_S}{R} - \frac{V_S}{R}e^{-(R/L)t}\right)u_0(t)$$

where the forced response i_f represents the steady-state condition reached as $t \to \infty$. Since the inductor L at this state behaves as a short circuit, $i_f = V_S/R$. The natural response i_n is the second term in the parenthesis of the above expression, that is, $i_n = (-V_S/R)e^{-(R/L)t}$

- In a simple RC circuit that is excited by a voltage source $V_S u_0(t)$ the voltage across the capacitor is

$$v_C(t) = v_{Cf} + v_{Cn} = (V_S + Ae^{-(1/RC)t})u_0(t)$$

where the forced response v_{Cf} represents the steady-state condition reached as $t \to \infty$. Since the capacitor C at this state behaves as an open circuit, $v_{Cf} = V_S$. The natural response v_{Cn} is the second term in the parenthesis of the above expression, that is, $v_{Cn} = Ae^{-(1/RC)t}$. The constant A must be evaluated from the total response.

10.7 Exercises

Multiple Choice

1. For the circuit of Figure 10.44 the time constant is

   A. *0.5 ms*

   B. *71.43 μs*

   C. *2,000 s*

   D. *0.2 ms*

   E. none of the above

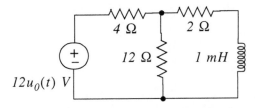

Figure 10.44. Circuit for Question 1

2. For the circuit of Figure 10.45 the time constant is

   A. *50 ms*

   B. *100 ms*

   C. *190 ms*

   D. *78.6 ms*

   E. none of the above

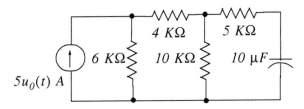

Figure 10.45. Circuit for Question 2

3. The forced response component i_{Lf} of the inductor current for the circuit of Figure 10.46 is

   A. *16 A*

   B. *10 A*

C. *6 A*

D. *2 A*

E. none of the above

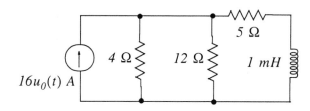

Figure 10.46. Circuit for Question 3

4. The forced response component v_{Cf} of the capacitor voltage for the circuit of Figure 10.47 is

A. *10 V*

B. *2 V*

C. *32/3 V*

D. *8 V*

E. none of the above

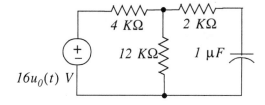

Figure 10.47. Circuit for Question 4

5. For the circuit of Figure 10.48 $i_L(0^-) = 2\ A$. For $t > 0$ the total response of $i_L(t)$ is

A. *6 A*

B. $6e^{-5000t}\ A$

C. $6 + 6e^{-5000t}\ A$

D. $6 - 4e^{-5000t}\ A$

E. none of the above

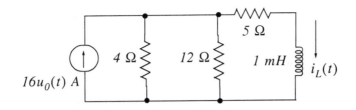

Figure 10.48. Circuit for Question 5

6. For the circuit of Figure 10.49 $v_C(0^-) = 5\ V$. For $t > 0$ the total response of $v_C(t)$ is

A. $12\ V$

B. $10 - 5e^{-500t}\ V$

C. $12 - 7e^{-200t}\ V$

D. $12 + 7e^{-200t}\ V$

E. none of the above

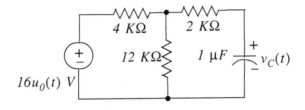

Figure 10.49. Circuit for Question 6

7. For the circuit of Figure 10.50 $i_L(0^-) = 2\ A$. For $t > 0$ the total response of $v_L(t)$ is

A. $20e^{-5000t}\ V$

B. $20e^{-5000t}\ V$

C. $-32e^{-8000t}\ V$

D. $32e^{-8000t}\ V$

E. none of the above

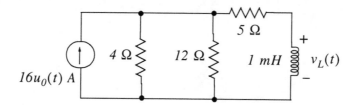

Figure 10.50. Circuit for Question 7

8. For the circuit of Figure 10.51 $v_C(0^-) = 5\ V$. For $t > 0$ the total response $i_C(t)$ is

A. $1400e^{-200t}\ A$

B. $1.4e^{-200t}\ A$

C. $3500e^{-500t}\ A$

D. $3.5e^{-500t}\ A$

E. none of the above

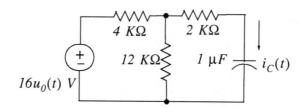

Figure 10.51. Circuit for Question 8

9. The waveform of Figure 10.52 can be expressed as

A. $3tu_0(t)\ A$

B. $3[u_0(t)] - 3[u_0(t-3)]\ A$

C. $3t[u_0(t) - u_0(t-1)] + (-1.5t + 4.5)[u_0(t-1) - u_0(t-3)]\ A$

D. $3t[u_0(t) - u_0(t-3)] + (-1.5t + 4.5)[u_0(t) - u_0(t-3)]\ A$

E. none of the above

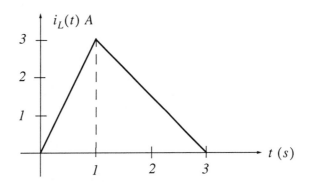

Figure 10.52. Waveform for Question 9

10. The waveform of Figure 10.53 can be expressed as

A. $2(1 - e^{-\alpha t} - e^{-\beta t})u_0(t)$ V

B. $(2 - 2e^{-\alpha t})[u_0(t) - u_0(t-2)] + (2e^{-\beta t})[u_0(t-2) - u_0(t-3)]$ V

C. $(2 - 2e^{-\alpha t})[u_0(t) - u_0(t-2)] - (2e^{-\beta t})[u_0(t-2) - u_0(t-3)]$ V

D. $(2 - 2e^{-\alpha t})[u_0(t)] - (2e^{-\beta t})[u_0(t-3)]$ V

E. none of the above

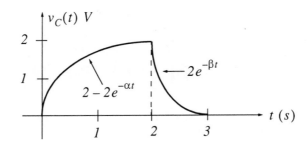

Figure 10.53. Waveform for Question 10

Problems

1. In the circuit of Figure 10.54, the voltage source $v_S(t)$ varies with time as shown by the waveform of Figure 10.55. Compute, sketch, and express $v_{LOAD}(t)$ as a sum of unit step functions for $0 \le t \le 5$ s.

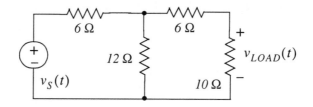

Figure 10.54. Circuit for Problem 1

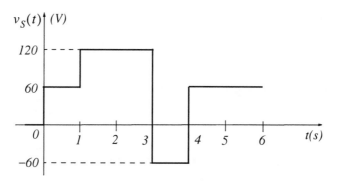

Figure 10.55. Waveform for Problem 1

2. In the circuit of Figure 10.56, $v_S(t) = 15u_0(t) - 30u_0(t-2)$ V. Compute $i_L(t)$ for $t > 0$.

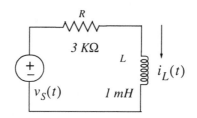

Figure 10.56. Circuit for Problem 2

3. In the circuit of Figure 10.57 (a), the excitation $v_S(t)$ is a pulse as shown in Figure 10.57 (b).

   a. Compute $i_L(t)$ for $0 < t < 0.3$ ms

   b. Compute and sketch $i_L(t)$ for all $t > 0$

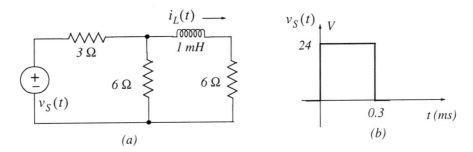

Figure 10.57. Circuit and waveform for Problem 3

4. In the circuit of Figure 10.58, switch S has been open for a very long time and closes at $t = 0$.
Compute and sketch $i_L(t)$ and $i_{SW}(t)$ for $t > 0$.

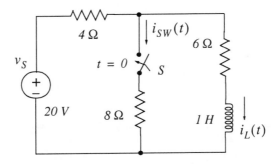

Figure 10.58. Circuit for Problem 4

5. For the circuit of Figure 10.59, compute $v_C(t)$ for $t > 0$.

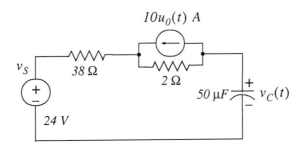

Figure 10.59. Circuit for Problem 5

6. For the circuit of Figure 10.60, compute $v_{out}(t)$ for $t > 0$ in terms of R, C, and $v_{in}u_0(t)$ given

that $v_C(0^-) = 0$

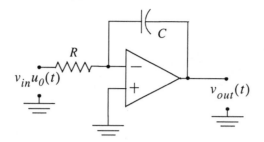

Figure 10.60. Circuit for Problem 6

7. In the circuit of Figure 10.61, switch S has been open for a very long time and closes at $t = 0$. Compute and sketch $v_C(t)$ and $v_{R3}(t)$ for $t > 0$.

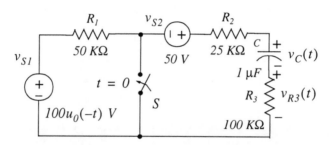

Figure 10.61. Circuit for Problem 7

8. For the circuit of Figure 10.62, it is given that $v_C(0^-) = 5\ V$. Compute $i_C(t)$ for $t > 0$. Hint: Be careful in deriving the time constant for this circuit.

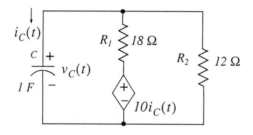

Figure 10.62. Circuit for Problem 8

10.8 Answers to Exercises

Multiple Choice

1. D

2. B

3. C

4. E *12 V*

5. E $6 - 4e^{-8000t}$ *A*

6. C

7. D

8. A

9. C

10. B

Problems

1. We replace the given circuit shown below with its Thevenin equivalent.

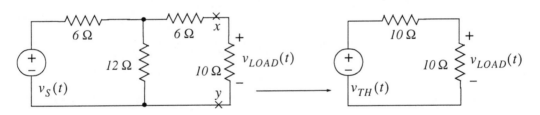

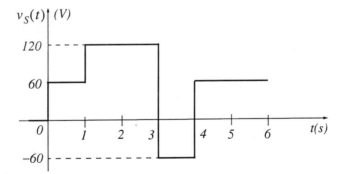

For the Thevenin equivalent voltage at different time intervals is as shown below.

$$v_{TH}(t) = \begin{cases} \dfrac{12}{18}v_s(t) = \dfrac{2}{3} \times 60 = 40 \ V & 0 < t < 1 \ s \\[2mm] \dfrac{12}{18}v_s(t) = \dfrac{2}{3} \times 120 = 80 \ V & 1 < t < 3 \ s \\[2mm] \dfrac{12}{18}v_s(t) = \dfrac{2}{3} \times (-60) = -40 \ V & 3 < t < 4 \ s \\[2mm] \dfrac{12}{18}v_s(t) = \dfrac{2}{3} \times 60 = 40 \ V & t > 4 \ s \end{cases}$$

and

$$v_{LOAD}(t) = \frac{10}{20}v_{TH}(t) = 0.5 v_{TH}(t) = \begin{cases} 0.5 \times 40 = 20 \ V & 0 < t < 1 \ s \\[1mm] 0.5 \times 80 = 40 \ V & 1 < t < 3 \ s \\[1mm] 0.5 \times (-40) = -20 \ V & 3 < t < 4 \ s \\[1mm] 0.5 \times 40 = 20 \ V & t > 4 \ s \end{cases}$$

The waveform of the voltage across the load is as shown below.

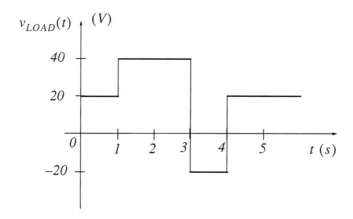

The waveform above can now be expressed as a sum of unit step functions as follows:

$$v_{LOAD}(t) = 20u_0(t) - 20u_0(t-1) + 40u_0(t-1) - 40u_0(t-3) + 20u_0(t-4)$$
$$-20u_0(t-3) + 20u_0(t-4) + 20u_0(t-4)$$
$$= 20u_0(t) + 20u_0(t-1) - 60u_0(t-3) + 40u_0(t-4)$$

2. The circuit at $t = 0^-$ is as shown below and since we are not told otherwise, we will assume that

$$i_L(0^-) = 0$$

For $t > 0$ we let $i_{L1}(t)$ be the inductor current when the $15u_0(t)$ voltage source acts alone and $i_{L2}(t)$ when the $30u_0(t-2)$ voltage source acts alone. Then, $i_{L\ TOTAL}(t) = i_{L1}(t) + i_{L2}(t)$

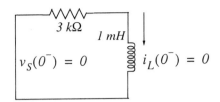

For $0 < t < 2\ s$ the circuit is as shown below.

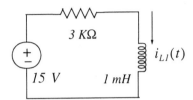

Then, $i_{L1}(t) = i_{L1f} + i_{L1n}$ where $i_{L1f} = \dfrac{15}{3\ K\Omega} = 5\ mA$ and $i_{L1n} = A_1 e^{-(R/L)t} = Ae^{-3 \times 10^6 t}$

Thus, $i_{L1}(t) = 5 + A_1 e^{-3 \times 10^6 t}\ mA$ and using the initial condition $i_L(0^-) = i_L(0^+) = 0$, we get

$i_{L1}(0) = 5 + A_1 e^0\ mA$ or $A_1 = -5$. Therefore,

$$i_{L1}(t) = 5 - 5e^{-3 \times 10^6 t} \quad (1)$$

Next, with the $30u_0(t-2)$ voltage source acts alone the circuit is as shown below.

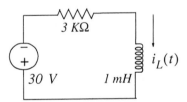

Then, $i_{L2}(t) = i_{L2f} + i_{L2n}$, $i_{L2f} = \dfrac{-30}{3\ K\Omega} = -10\ mA$ and $i_{L2n} = A_2 e^{-(R/L)(t-2)} = Be^{-3 \times 10^6 (t-2)}$

Thus, $i_{L2}(t) = -10 + A_2 e^{-3 \times 10^6 (t-2)}\ mA$ and the initial condition at $t = 2$ is found from (1)

above as $i_{L1}\big|_{t=2\ s} = 5 - 5e^{-6 \times 10^6 t} \approx 5\ mA$. Therefore,

$i_{L2}\big|_{t=2\ s} = i_{L1}\big|_{t=2\ s} = 5 = -10 + A_2 e^{-3 \times 10^6 (2-2)}\ mA$ or $A_2 = 15$ and

$$i_{L2}(t) = -10 + 15e^{-3 \times 10^6 (t-2)}\ mA \quad (2)$$

Therefore, the total current when both voltage sources are present is the summation of (1) and (2), that is,

$$i_{L\ TOTAL}(t) = i_{L1}(t) + i_{L2}(t) = 5 - 5e^{-3 \times 10^6 t} - 10 + 15e^{-3 \times 10^6 (t-2)} \ mA$$

$$= -5 - 5e^{-3 \times 10^6 t} + 15e^{-3 \times 10^6 (t-2)} \ mA$$

3.

a. For this circuit $v_S(t) = 24[u_0(t) - u_0(t - 0.3)]$ and since we are not told otherwise, we will assume that $i_L(0^-) = 0$. For $0 < t < 0.3 \ ms$ the circuit and its Thevenin equivalent are as shown below.

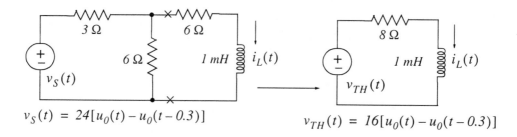

$$v_S(t) = 24[u_0(t) - u_0(t - 0.3)] \qquad\qquad v_{TH}(t) = 16[u_0(t) - u_0(t - 0.3)]$$

Then,

$$i_L(t) = i_{Lf} + i_{Ln} = 16/8 + A_1 e^{-(R/L)t} = 2 + Ae^{-8000t}$$

and at $t = 0$

$$i_L(0) = i_L(0^-) = 0 = 2 + A_1 e^0$$

or $A_1 = -2$ and thus for $0 < t < 0.3 \ ms$

$$i_L(t) = 2 - 2e^{-8000t} \quad (1)$$

b. For $t > 0.3 \ ms$ the circuit is as shown below. For this circuit

$$i_L(t) = A_2 e^{-(R/L)(t - 0.3)} = A_2 e^{-8000(t - 0.3)} \quad (2)$$

and A_2 is found from the initial condition at $t = 0.3 \ ms$, that is, with (1) above we get

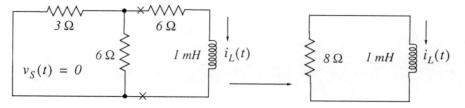

$$i_L\big|_{t\,=\,0.3\ ms} = 2 - 2e^{-8\times 10^3 \times 0.3 \times 10^{-3}} = 2 - 2e^{2.4} = 1.82\ A$$

and by substitution into (2) above

$$i_L\big|_{t\,=\,0.3\ ms} = 1.82 = A_2 e^{-8000(0.3-0.3)}$$

or $A_2 = 1.82$. Therefore for $t > 0.3\ ms$

$$i_L(t) = 1.82 e^{-(t-0.3\ ms)}$$

The waveform for the inductor current $i_L(t)$ for all $t > 0$ is shown below.

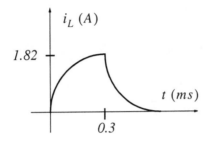

4. At $t = 0^-$ the circuit is as shown below where $i_L(0^-) = 20/(4+6) = 2\ A$ and thus the initial condition has been established.

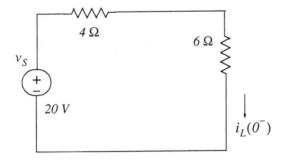

For $t > 0$ the circuit and its Thevenin equivalent are as shown below where

$$v_{TH} = \frac{8}{4+8} \times 20 = 40/3\ V$$

and

$$R_{TH} = \frac{8\times 4}{8+4} + 6 = 26/3\ \Omega$$

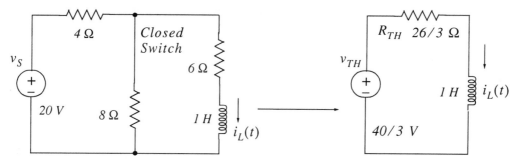

Then,

$$i_L(t) = i_{Lf} + i_{Ln} = \frac{40/3}{26/3} + Ae^{-(R/L)t} = 20/13 + Ae^{-(26/3)t}$$

and A is evaluated from the initial condition, i.e.,

$$i_L(0^-) = i_L(0^+) = 2 = 20/13 + Ae^0$$

from which $A = 6/13$ and thus for $t > 0$

$$i_L(t) = \frac{20}{13} + \frac{6}{13}e^{-(26/3)t} = 1.54 + 0.46e^{-8.67t} \quad (1)$$

Next, to find $i_{SW}(t)$ we observe that this current flows also through the $8\ \Omega$ resistor and this can be found from $v_{8\ \Omega}$ shown on the circuit below.

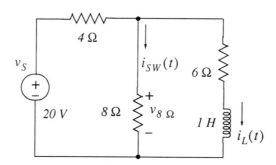

Now,

$$v_{8\ \Omega} = v_{6\ \Omega} + v_L(t) = 6i_L(t) + L\frac{di_L}{dt}$$

$$= 6(1.54 + 0.46e^{-8.67t}) + 1 \times \frac{d}{dt}(1.54 + 0.46e^{-8.67t})$$

$$= 9.24 + 2.76e^{-8.67t} - 8.67 \times 0.46e^{-8.67t}$$

$$= 9.24 - 1.23e^{-8.67t}$$

and

$$i_{SW}(t) = i_{8\,\Omega} = \frac{v_{8\,\Omega}}{8} = \frac{9.24 - 1.23e^{-8.67t}}{8} = 1.16 - 0.15e^{-8.67t} \quad (2)$$

Therefore, from the initial condition, (1) and (2) above we have

$$i_L(0^+) = 2 \qquad i_L(\infty) = 1.54 \qquad i_{SW}(0^+) = 1.16 - 0.15 = 1.01 \qquad i_{SW}(\infty) = 1.16$$

and with these values we sketch $i_L(t)$ and $i_{SW}(t)$ as shown below.

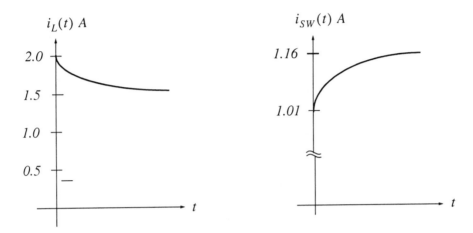

5. At $t = 0^-$ the circuit is as shown below where $v_C(0^-) = 24\ V$ and thus the initial condition has been established.

The circuit for $t > 0$ is shown below where the current source has been replaced with a voltage source.

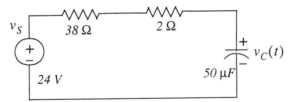

Now,

$$v_C(t) = v_{Cf} + v_{Cn} = 4 + Ae^{-1/(RC)t} = 4 + Ae^{-500t}$$

and with the initial condition $v_C(0^-) = v_C(0^+) = 24\ V = 4 + Ae^0$ from which $A = 20$ we get

$$v_C(t) = 4 + 20e^{-500t}$$

6. For $t > 0$ the op amp circuit is as shown below.

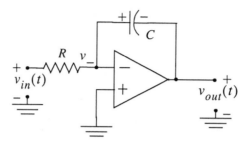

Application of KCL at the minus (−) input yields

$$\frac{v_- - v_{in}}{R} + C\frac{dv_C}{dt} = 0$$

and since $v_- = 0$

$$C\frac{dv_C}{dt} = \frac{v_{in}}{R}$$

or

$$\frac{dv_C}{dt} = \frac{v_{in}}{RC}$$

Integrating both sides and observing that $v_{out}(t) = -v_C(t)$ we get

$$v_{out}(t) = -\frac{v_{in}}{RC}t + k$$

where k is the constant of integration of both sides and it is evaluated from the given initial condition. Then,

$$v_C(0^-) = v_C(0^+) = 0 = 0 + k$$

or $k = 0$. Therefore,

$$v_{out}(t) = -\left(\frac{v_{in}}{RC}t\right)u_0(t)$$

and v_{in}/RC is the slope as shown below.

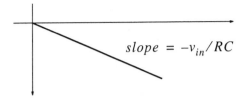

7. At $t = 0^-$ the circuit is as shown below where $v_C(0^-) = 150\ V$ and thus the initial condition has been established.

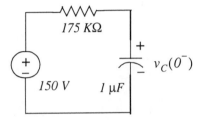

The circuit for $t > 0$ is shown below where the voltage source v_{S1} is absent for all positive time and the $50\ K\Omega$ is shorted out by the closed switch.

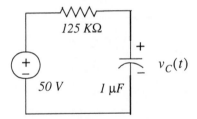

For the circuit above

$$v_C(t) = v_{Cf} + v_{Cn} = 50 + Ae^{-t/(RC)} = 50 + Ae^{-8t}$$

and with the initial condition

$$v_C(0^-) = v_C(0^+) = 150 = 50 + Ae^0$$

from which $A = 100$ and thus for $t > 0$

$$v_C(t) = 50 + 100e^{-8t}\ V$$

To find $v_{R3}(t)$ we will first find $i_C(t)$ from the circuit below where

$$i_C(t) = C\frac{dv_C}{dt} = 10^{-6} \times (-8 \times 10^{-4}e^{-8t})$$

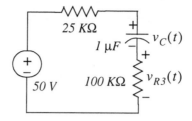

Then,

$$v_{R3}(t) = (100 \ K\Omega)i_C = 10^5(-8 \times 10^{-4} e^{-8t}) = -80e^{-8t} \ V$$

The sketches below show $v_C(t)$ and $v_{R3}(t)$ as they approach their final values.

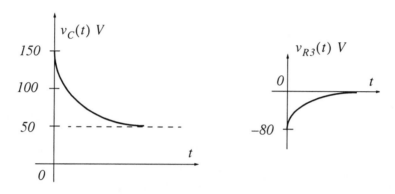

8. For this circuit we cannot short the dependent source and therefore we cannot find R_{eq} by combining the resistances R_1 and R_2 in parallel combination in order to find the time constant $\tau = RC$. Instead, we will derive the time constant from the differential equation of (9.9) of the previous chapter, that is,

$$\frac{dv_C}{dt} + \frac{v_C}{RC} = 0$$

From the given circuit shown below

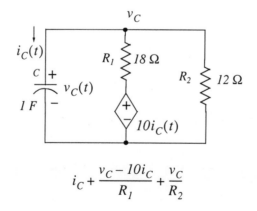

$$i_C + \frac{v_C - 10i_C}{R_1} + \frac{v_C}{R_2}$$

or

$$C\frac{dv_C}{dt} + \left(\frac{1}{R_1} + \frac{1}{R_2}\right)v_C - \frac{10}{R_1}C\frac{dv_C}{dt} = 0$$

or

$$\left(1 - \frac{10}{R_1}\right)\left(C\frac{dv_C}{dt}\right) + \left(\frac{R_1 + R_2}{R_1 \cdot R_2}\right)v_C = 0$$

or

$$\frac{dv_C}{dt} + \frac{\left(\dfrac{R_1 + R_2}{R_1 \cdot R_2}\right)}{\left(1 - \dfrac{10}{R_1}\right) \cdot C}v_C = 0$$

and from this differential equation we see that the coefficient of v_C is

$$\frac{1}{R_{eq}C} = \frac{1}{\tau} = \frac{30/216}{\left(1 - \dfrac{10}{18}\right) \cdot 1} = \frac{30/216}{8/18} = \frac{15 \times 9}{4 \times 108} = \frac{135}{432} = \frac{5}{16} = 0.3125$$

and thus

$$v_C(t) = Ae^{-0.3125t}$$

and with the given initial condition $v_C(0^-) = V_0 = A = 5\ V$ we get

$$v_C(t) = 5e^{-0.3125t}$$

Then, using

$$i_C = C\frac{dv_C}{dt}$$

we find that for $t > 0$

$$i_C(t) = (1)(-0.3125 \times 5e^{-0.3125t}) = -1.5625e^{-0.3125t}$$

and the minus (–) sign indicates that the $i_C(t)$ direction is opposite to that shown.

NOTES

Appendix A

Introduction to MATLAB ®

T his appendix serves as an introduction to the basic MATLAB commands and functions, procedures for naming and saving the user generated files, comment lines, access to MATLAB's Editor/Debugger, finding the roots of a polynomial, and making plots. Several examples are provided with detailed explanations.

A.1 MATLAB® and Simulink®

MATLAB and Simulink are products of The MathWorks, Inc. These are two outstanding software packages for scientific and engineering computations and are used in educational institutions and in industries including automotive, aerospace, electronics, telecommunications, and environmental applications. MATLAB enables us to solve many advanced numerical problems fast and efficiently. Simulink is a block diagram tool used for modeling and simulating dynamic systems such as controls, signal processing, and communications. In this appendix we will discuss MATLAB only.

A.2 Command Window

To distinguish the screen displays from the user commands, important terms, and MATLAB functions, we will use the following conventions:

Click: Click the left button of the mouse

`Courier Font`: Screen displays

Helvetica Font: User inputs at MATLAB's command window prompt >> or EDU>>*

Helvetica Bold: MATLAB functions

Times Bold Italic: Important terms and facts, notes and file names

When we first start MATLAB, we see the toolbar on top of the *command screen* and the prompt EDU>>. This prompt is displayed also after execution of a command; MATLAB now waits for a new command from the user. It is highly recommended that we use the *Editor/Debugger* to write our program, save it, and return to the command screen to execute the program as explained below.

To use the Editor/Debugger:

1. From the *File* menu on the toolbar, we choose *New* and click on *M-File*. This takes us to the *Editor*

---

* EDU>> *is the MATLAB prompt in the Student Version*

Window where we can type our *code* (list of statements) for a new file, or open a previously saved file. We must save our program with a file name which starts with a letter. *Important!* MATLAB is *case sensitive*, that is, it distinguishes between upper- and lower-case letters. Thus, *t* and *T* are two different letters in MATLAB language. The files that we create are saved with the file name we use and the extension *.m*; for example, *myfile01.m*. It is a good practice to save the code in a file name that is descriptive of our code content. For instance, if the code performs some matrix operations, we ought to name and save that file as *matrices01.m* or any other similar name. We should also use a floppy disk to backup our files.

2. Once the code is written and saved as an *m-file*, we may exit the *Editor/Debugger* window by clicking on *Exit Editor/Debugger* of the *File* menu. MATLAB then returns to the command window.

3. To execute a program, we type the file name *without* the *.m* extension at the >> prompt; then, we press <enter> and observe the execution and the values obtained from it. If we have saved our file in drive *a* or any other drive, we must make sure that it is added it to the desired directory in MATLAB's search path. The MATLAB User's Guide provides more information on this topic.

Henceforth, it will be understood that each input command is typed after the >> prompt and followed by the <enter> key.

The command **help matlab\iofun** will display input/output information. To get help with other MATLAB topics, we can type help followed by any topic from the displayed menu. For example, to get information on graphics, we type help matlab\graphics. The MATLAB User's Guide contains numerous help topics.

To appreciate MATLAB's capabilities, we type demo and we see the MATLAB Demos menu. We can do this periodically to become familiar with them. Whenever we want to return to the command window, we click on the Close button.

When we are done and want to leave MATLAB, we type **quit** or **exit**. But if we want to clear all previous values, variables, and equations without exiting, we should use the command **clear**. This command erases everything; it is like exiting MATLAB and starting it again. The command **clc** clears the screen but MATLAB still remembers all values, variables and equations that we have already used. In other words, if we want to clear all previously entered commands, leaving only the >> prompt on the upper left of the screen, we use the **clc** command.

All text after the **%** (percent) symbol is interpreted as a *comment line* by MATLAB, and thus it is ignored during the execution of a program. A comment can be typed on the same line as the function or command or as a separate line. For instance,

conv(p,q)   % performs multiplication of polynomials p and q.

% The next statement performs partial fraction expansion of p(x) / q(x)

are both correct.

One of the most powerful features of MATLAB is the ability to do computations involving *complex numbers*. We can use either i, or j to denote the imaginary part of a complex number, such as $3-4i$ or $3-4j$. For example, the statement

z=3−4j

displays

```
z = 3.0000-4.0000i
```

In the above example, a multiplication (*) sign between 4 and j was not necessary because the complex number consists of numerical constants. However, if the imaginary part is a function, or variable such as $cos(x)$, we must use the multiplication sign, that is, we must type cos(x)*j or j*cos(x) for the imaginary part of the complex number.

A.3 Roots of Polynomials

In MATLAB, a polynomial is expressed as a *row vector* of the form $[a_n \ a_{n-1} \ ... \ a_2 \ a_1 \ a_0]$. These are the coefficients of the polynomial in descending order. *We must include terms whose coefficients are zero.*

We find the roots of any polynomial with the **roots(p)** function; **p** is a row vector containing the polynomial coefficients in descending order.

Example A.1

Find the roots of the polynomial

$$p_1(x) = x^4 - 10x^3 + 35x^2 - 50x + 24$$

Solution:

The roots are found with the following two statements where we have denoted the polynomial as **p1**, and the roots as **roots_ p1**.

p1=[1 −10 35 −50 24]   % Specify and display the coefficients of p1(x)

```
p1 =

     1    -10    35    -50    24
```

roots_ p1=roots(p1)   % Find the roots of p1(x)

```
roots_p1 =

   4.0000

   3.0000
```

2.0000

1.0000

We observe that MATLAB displays the polynomial coefficients as a row vector, and the roots as a column vector.

Example A.2

Find the roots of the polynomial

$$p_2(x) = x^5 - 7x^4 + 16x^2 + 25x + 52$$

Solution:

There is no cube term; therefore, we must enter zero as its coefficient. The roots are found with the statements below, where we have defined the polynomial as **p2**, and the roots of this polynomial as **roots_ p2**. The result indicates that this polynomial has three real roots, and two complex roots. Of course, complex roots always occur in *complex conjugate** pairs.

p2=[1 –7 0 16 25 52]

p2 =

　　　　1　　　-7　　　0　　　16　　　25　　　52

roots_ p2=roots(p2)

roots_ p2 =

　　6.5014

　　2.7428

　-1.5711

　-0.3366+ 1.3202i

　-0.3366- 1.3202i

A.4 Polynomial Construction from Known Roots

We can compute the coefficients of a polynomial, from a given set of roots, with the **poly(r)** function where **r** is a row vector containing the roots.

---

* *By definition, the conjugate of a complex number $A = a + jb$ is $A^* = a - jb$*

Example A.3

It is known that the roots of a polynomial are *1, 2, 3,* and *4*. Compute the coefficients of this polynomial.

Solution:

We first define a row vector, say *r3*, with the given roots as elements of this vector; then, we find the coefficients with the **poly(r)** function as shown below.

```
r3=[1  2  3  4]          % Specify the roots of the polynomial

r3 =

       1      2      3      4

poly_r3=poly(r3)         % Find the polynomial coefficients

poly_r3 =

       1    -10     35    -50     24
```

We observe that these are the coefficients of the polynomial $p_1(x)$ of Example A.1.

Example A.4

It is known that the roots of a polynomial are -1, -2, -3, $4+j5$ and $4-j5$. Find the coefficients of this polynomial.

Solution:

We form a row vector, say *r4*, with the given roots, and we find the polynomial coefficients with the **poly(r)** function as shown below.

```
r4=[−1  −2  −3  −4+5j  −4−5j ]

r4 =

  Columns 1 through 4

   -1.0000    -2.0000    -3.0000    -4.0000+ 5.0000i

  Column 5

   -4.0000- 5.0000i

poly_r4=poly(r4)

poly_r4 =

       1     14    100    340    499    246
```

Therefore, the polynomial is

$$p_4(x) = x^5 + 14x^4 + 100x^3 + 340x^2 + 499x + 246$$

A.5 Evaluation of a Polynomial at Specified Values

The **polyval(p,x)** function evaluates a polynomial $p(x)$ at some specified value of the independent variable x.

Example A.5

Evaluate the polynomial

$$p_5(x) = x^6 - 3x^5 + 5x^3 - 4x^2 + 3x + 2 \qquad \text{(A.1)}$$

at $x = -3$.

Solution:

```
p5=[1 -3 0 5 -4 3 2];   % These are the coefficients
% The semicolon (;) after the right bracket suppresses the display of the row vector
%  that contains the coefficients of p5.
%
val_minus3=polyval(p5, -3)   % Evaluate p5 at x=-3; no semicolon is used here
                             % because we want the answer to be displayed
val_minus3 =

        1280
```

Other MATLAB functions used with polynomials are the following:

conv(a,b) – multiplies two polynomials **a** and **b**

[q,r]=deconv(c,d) –divides polynomial **c** by polynomial **d** and displays the quotient **q** and remainder **r**.

polyder(p) – produces the coefficients of the derivative of a polynomial **p**.

Example A.6

Let

$$p_1 = x^5 - 3x^4 + 5x^2 + 7x + 9$$

and

$$p_2 = 2x^6 - 8x^4 + 4x^2 + 10x + 12$$

Compute the product $p_1 \cdot p_2$ using the **conv(a,b)** function.

Solution:

p1=[1 −3 0 5 7 9];   % The coefficients of p1

p2=[2 0 −8 0 4 10 12];   % The coefficients of p2

p1p2=conv(p1,p2)   % Multiply p1 by p2 to compute coefficients of the product p1p2

p1p2 =

2   −6   −8   34   18   −24   −74   −88   78   166   174   108

Therefore,

$$p_1 \cdot p_2 = 2x^{11} - 6x^{10} - 8x^9 + 34x^8 + 18x^7 - 24x^6$$
$$-74x^5 - 88x^4 + 78x^3 + 166x^2 + 174x + 108$$

Example A.7

Let

$$p_3 = x^7 - 3x^5 + 5x^3 + 7x + 9$$

and

$$p_4 = 2x^6 - 8x^5 + 4x^2 + 10x + 12$$

Compute the quotient p_3/p_4 using the **[q,r]=deconv(c,d)** function.

Solution:

% It is permissible to write two or more statements in one line separated by semicolons

p3=[1 0 −3 0 5 7 9]; p4=[2 −8 0 0 4 10 12]; [q,r]=deconv(p3,p4)

q =

    0.5000

r =

    0   4   −3   0   3   2   3

Therefore,

$$q = 0.5 \qquad r = 4x^5 - 3x^4 + 3x^2 + 2x + 3$$

Example A.8

Let

$$p_5 = 2x^6 - 8x^4 + 4x^2 + 10x + 12$$

Compute the derivative $\dfrac{d}{dx}p_5$ using the **polyder(p)** function.

Solution:

```
p5=[2  0  -8  0  4  10  12];        % The coefficients of p5
der_p5=polyder(p5)                  % Compute the coefficients of the derivative of p5
der_p5 =
    12      0    -32      0      8      10
```

Therefore,

$$\frac{d}{dx}p_5 = 12x^5 - 32x^3 + 4x^2 + 8x + 10$$

A.6 Rational Polynomials

Rational Polynomials are those which can be expressed in ratio form, that is, as

$$R(x) = \frac{Num(x)}{Den(x)} = \frac{b_n x^n + b_{n-1} x^{n-1} + b_{n-2} x^{n-2} + \dots + b_1 x + b_0}{a_m x^m + a_{m-1} x^{m-1} + a_{m-2} x^{m-2} + \dots + a_1 x + a_0} \qquad (A.2)$$

where some of the terms in the numerator and/or denominator may be zero. We can find the roots of the numerator and denominator with the **roots(p)** function as before.

As noted in the comment line of Example A.7, we can write MATLAB statements in one line, if we separate them by commas or semicolons. *Commas will display the results whereas semicolons will suppress the display.*

Example A.9

Let

$$R(x) = \frac{P_{num}}{P_{den}} = \frac{x^5 - 3x^4 + 5x^2 + 7x + 9}{x^6 - 4x^4 + 2x^2 + 5x + 6}$$

Express the numerator and denominator in factored form, using the **roots(p)** function.

Solution:

```
num=[1 -3 0 5 7 9]; den=[1 0 -4 0 2 5 6];    % Do not display num and den coefficients
roots_num=roots(num), roots_den=roots(den)    % Display num and den roots
```

```
roots_num =

   2.4186+ 1.0712i    2.4186- 1.0712i    -1.1633

  -0.3370+ 0.9961i   -0.3370- 0.9961i

roots_den =

   1.6760+0.4922i     1.6760-0.4922i    -1.9304

  -0.2108+0.9870i    -0.2108-0.9870i    -1.0000
```

As expected, the complex roots occur in complex conjugate pairs.

For the numerator, we have the factored form

$$P_{num} = (x - 2.4186 - j1.0712)(x - 2.4186 + j1.0712)(x + 1.1633)$$
$$(x + 0.3370 - j0.9961)(x + 0.3370 + j0.9961)$$

and for the denominator, we have

$$P_{den} = (x - 1.6760 - j0.4922)(x - 1.6760 + j0.4922)(x + 1.9304)$$
$$(x + 0.2108 - j0.9870)(x + 0.2108 + j0.9870)(x + 1.0000)$$

We can also express the numerator and denominator of this rational function as a combination of *linear* and *quadratic* factors. We recall that, in a quadratic equation of the form $x^2 + bx + c = 0$ whose roots are x_1 and x_2, the negative sum of the roots is equal to the coefficient b of the x term, that is, $-(x_1 + x_2) = b$, while the product of the roots is equal to the constant term c, that is, $x_1 \cdot x_2 = c$. Accordingly, we form the coefficient b by addition of the complex conjugate roots and this is done by inspection; then we multiply the complex conjugate roots to obtain the constant term c using MATLAB as follows:

(2.4186 + 1.0712i)*(2.4186 −1.0712i)

```
ans = 6.9971
```

(−0.3370+ 0.9961i)*(−0.3370−0.9961i)

```
ans = 1.1058
```

(1.6760+ 0.4922i)*(1.6760−0.4922i)

```
ans = 3.0512
```

(−0.2108+ 0.9870i)*(−0.2108−0.9870i)

```
ans = 1.0186
```

Thus,

$$R(x) = \frac{p_{num}}{p_{den}} = \frac{(x^2 - 4.8372x + 6.9971)(x^2 + 0.6740x + 1.1058)(x + 1.1633)}{(x^2 - 3.3520x + 3.0512)(x^2 + 0.4216x + 1.0186)(x + 1.0000)(x + 1.9304)}$$

We can check this result with MATLAB's *Symbolic Math Toolbox* which is a collection of tools (functions) used in solving symbolic expressions. They are discussed in detail in MATLAB's Users Manual. For the present, our interest is in using the **collect(s)** function that is used to multiply two or more symbolic expressions to obtain the result in polynomial form. We must remember that the **conv(p,q)** function is used with numeric expressions only, that is, polynomial coefficients.

Before using a symbolic expression, we must create one or more symbolic variables such as x, y, t, and so on. For our example, we use the following code:

```
syms x  % Define a symbolic variable and use collect(s) to express numerator in polynomial form
collect((x^2–4.8372*x+6.9971)*(x^2+0.6740*x+1.1058)*(x+1.1633))

ans =

 x^5-29999/10000*x^4-1323/3125000*x^3+7813277909/
 1562500000*x^2+1750276323053/250000000000*x+4500454743147/
 500000000000
```

and if we simplify this, we find that is the same as the numerator of the given rational expression in polynomial form. We can use the same procedure to verify the denominator.

A.7 Using MATLAB to Make Plots

Quite often, we want to plot a set of ordered pairs. This is a very easy task with the MATLAB **plot(x,y)** command that plots y versus x. Here, x is the horizontal axis (abscissa) and y is the vertical axis (ordinate).

Example A.10

Consider the electric circuit of Figure A.1, where the radian frequency ω (radians/second) of the applied voltage was varied from 300 to 3000 in steps of 100 radians/second, while the amplitude was held constant. The ammeter readings were then recorded for each frequency. The magnitude of the impedance $|Z|$ was computed as $|Z| = |V/A|$ and the data were tabulated on Table A.1.

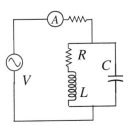

Figure A.1. Electric circuit for Example A.10

TABLE A.1 Table for Example A.10

ω (rads/s)	|Z| Ohms	ω (rads/s)	|Z| Ohms
300	39.339	1700	90.603
400	52.589	1800	81.088
500	71.184	1900	73.588
600	97.665	2000	67.513
700	140.437	2100	62.481
800	222.182	2200	58.240
900	436.056	2300	54.611
1000	1014.938	2400	51.428
1100	469.83	2500	48.717
1200	266.032	2600	46.286
1300	187.052	2700	44.122
1400	145.751	2800	42.182
1500	120.353	2900	40.432
1600	103.111	3000	38.845

Plot the magnitude of the impedance, that is, $|Z|$ versus radian frequency ω.

Solution:

We cannot type ω (omega) in the MATLAB command window, so we will use the English letter w instead.

If a statement, or a row vector is too long to fit in one line, it can be continued to the next line by typing three or more periods, then pressing *<enter>* to start a new line, and continue to enter data. This is illustrated below for the data of w and z. Also, as mentioned before, we use the semicolon (;) to suppress the display of numbers that we do not care to see on the screen.

The data are entered as follows:

w=[300 400 500 600 700 800 900 1000 1100 1200 1300 1400 1500 1600 1700 1800 1900....

2000 2100 2200 2300 2400 2500 2600 2700 2800 2900 3000];

%

z=[39.339 52.789 71.104 97.665 140.437 222.182 436.056....

1014.938 469.830 266.032 187.052 145.751 120.353 103.111....

90.603 81.088 73.588 67.513 62.481 58.240 54.611 51.468....

48.717 46.286 44.122 42.182 40.432 38.845];

Of course, if we want to see the values of *w* or *z* or both, we simply type w or z, and we press *<enter>*. To plot *z* (y-axis) versus *w* (x-axis), we use the **plot(x,y)** command. For this example, we use **plot(w,z)**. When this command is executed, MATLAB displays the plot on MATLAB's *graph screen*. This plot is shown in Figure A.2.

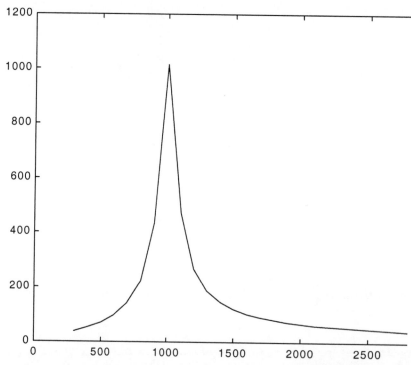

Figure A.2. Plot of impedance |z| versus frequency ω for Example A.10

This plot is referred to as the *amplitude frequency response* of the circuit.

To return to the command window, we press any key, or from the *Window* pull-down menu, we select *MATLAB Command Window*. To see the graph again, we click on the Window pull-down menu, and we select *Figure*.

We can make the above, or any plot, more presentable with the following commands:

grid on: This command adds grid lines to the plot. The **grid off** command removes the grid. The command **grid** toggles them, that is, changes from off to on or vice versa. The default[*] is off.

box off: This command removes the box (the solid lines which enclose the plot), and **box on** restores the box. The command **box** toggles them. The default is on.

title('string'): This command adds a line of the text **string** (label) at the top of the plot.

xlabel('string') and **ylabel('string')** are used to label the x- and y-axis respectively.

The amplitude frequency response is usually represented with the x-axis in a logarithmic scale. We can use the **semilogx(x,y)** command that is similar to the **plot(x,y)** command, except that the x-axis is represented as a log scale, and the y-axis as a linear scale. Likewise, the **semilogy(x,y)** command is similar to the **plot(x,y)** command, except that the y-axis is represented as a log scale, and the x-axis as a linear scale. The **loglog(x,y)** command uses logarithmic scales for both axes.

Throughout this text it will be understood that log is the common (base 10) logarithm, and ln is the natural (base e) logarithm. We must remember, however, the function **log(x)** in MATLAB is the natural logarithm, whereas the common logarithm is expressed as **log10(x)**, and the logarithm to the base 2 as **log2(x).**

Let us now redraw the plot with the above options by adding the following statements:

semilogx(w,z); grid;   % Replaces the plot(w,z) command

title('Magnitude of Impedance vs. Radian Frequency');

xlabel('w in rads/sec'); ylabel('|Z| in Ohms')

After execution of these commands, our plot is as shown in Figure A.3.

If the y-axis represents power, voltage or current, the x-axis of the frequency response is more often shown in a logarithmic scale, and the y-axis in dB (decibels). The decibel unit is defined in Chapter 4.

---

[*] *A default is a particular value for a variable that is assigned automatically by an operating system and remains in effect unless canceled or overridden by the operator.*

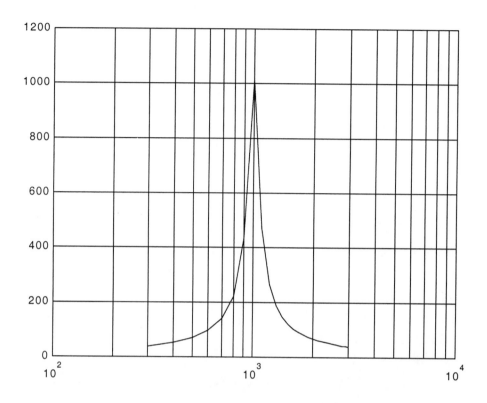

Figure A.3. Modified frequency response plot of Figure A.2.

To display the voltage v in a dB scale on the y-axis, we add the relation **dB=20*log10(v)**, and we replace the **semilogx(w,z)** command with semilogx(w,dB).

The command **gtext('string')**[*] switches to the current *Figure Window,* and displays a cross-hair that can be moved around with the mouse. For instance, we can use the command gtext('Impedance |Z| versus Frequency'), and this will place a cross-hair in the *Figure* window. Then, using the mouse, we can move the cross-hair to the position where we want our label to begin, and we press <enter>.

The command **text(x,y,'string')** is similar to **gtext('string').** It places a label on a plot in some specific location specified by **x** and **y**, and **string** is the label which we want to place at that location. We will illustrate its use with the following example that plots a *3-phase* sinusoidal waveform.

The first line of the code below has the form

linspace(first_value, last_value, number_of_values)

---

[*] *With MATLAB Versions 6 and 7 we can add text, lines and arrows directly into the graph using the tools provided on the Figure Window.*

This function specifies *the number of data points* but not the increments between data points. An alternate function is

x=first: increment: last

and this specifies *the increments between points* but not the number of data points.

The code for the 3-phase plot is as follows:

```
x=linspace(0, 2*pi, 60);        %  pi is a built-in function in MATLAB;
%  we could have used x=0:0.02*pi:2*pi or x = (0: 0.02: 2)*pi instead;
y=sin(x); u=sin(x+2*pi/3); v=sin(x+4*pi/3);
plot(x,y,x,u,x,v);              %  The x-axis must be specified for each function
grid on, box on,                %  turn grid and axes box on
text(0.75, 0.65, 'sin(x)');  text(2.85, 0.65, 'sin(x+2*pi/3)');  text(4.95, 0.65, 'sin(x+4*pi/3)')
```

These three waveforms are shown on the same plot of Figure A.4.

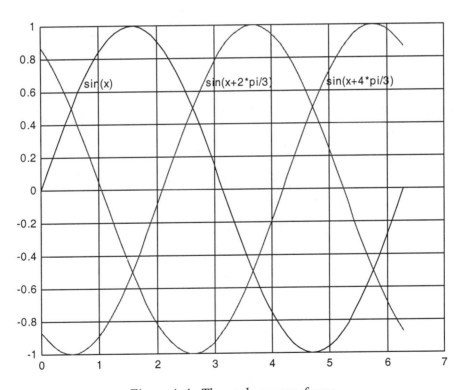

Figure A.4. Three-phase waveforms

In our previous examples, we did not specify line styles, markers, and colors for our plots. However, MATLAB allows us to specify various line types, plot symbols, and colors. These, or a combination of these, can be added with the **plot(x,y,s)** command, where **s** is a character string containing one or more characters shown on the three columns of Table A.2. MATLAB has no default color; it starts with blue and cycles through the first seven colors listed in Table A.2 for each additional line in the plot. Also, there is no default marker; no markers are drawn unless they are selected. The default line is the solid line.

TABLE A.2 Styles, colors, and markets used in MATLAB

Symbol	Color	Symbol	Marker	Symbol	Line Style
b	blue	.	point	−	solid line
g	green	o	circle	:	dotted line
r	red	x	x-mark	−.	dash-dot line
c	cyan	+	plus	−−	dashed line
m	magenta	*	star		
y	yellow	s	square		
k	black	d	diamond		
w	white	∨	triangle down		
		∧	triangle up		
		<	triangle left		
		>	triangle right		
		p	pentagram		
		h	hexagram		

For example, **plot(x,y,'m*:')** plots a magenta dotted line with a star at each data point, and **plot(x,y,'rs')** plots a red square at each data point, but does not draw any line because no line was selected. If we want to connect the data points with a solid line, we must type **plot(x,y,'rs−')**. For additional information we can type **help plot** in MATLAB's command screen.

The plots we have discussed thus far are two-dimensional, that is, they are drawn on two axes. MATLAB has also a three-dimensional (three-axes) capability and this is discussed next.

The **plot3(x,y,z)** command plots a line in *3-space* through the points whose coordinates are the elements of x, y and z, where x, y and z are three vectors of the same length.

The general format is **plot3($x_1,y_1,z_1,s_1,x_2,y_2,z_2,s_2,x_3,y_3,z_3,s_3,$...)** where x_n, y_n and z_n are vectors or matrices, and s_n are strings specifying color, marker symbol, or line style. These strings are the same as those of the two-dimensional plots.

Example A.11

Plot the function

$$z = -2x^3 + x + 3y^2 - 1 \qquad (A.3)$$

Solution:

We arbitrarily choose the interval (length) shown on the code below.

```
x= -10: 0.5: 10;          % Length of vector x
y= x;                     % Length of vector y must be same as x
z= -2.*x.^3+x+3.*y.^2-1;  % Vector z is function of both x and y*
plot3(x,y,z); grid
```

The three-dimensional plot is shown in Figure A.5.

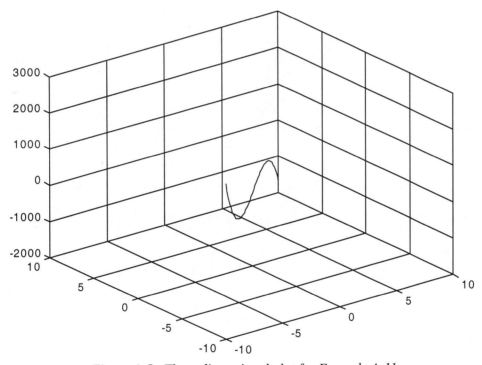

Figure A.5. Three dimensional plot for Example A.11

In a two-dimensional plot, we can set the limits of the *x*- and *y-axes* with the **axis([xmin xmax ymin ymax])** command. Likewise, in a three-dimensional plot we can set the limits of all three axes with

---

* *This statement uses the so called dot multiplication, dot division, and dot exponentiation where the multiplication, division, and exponential operators are preceded by a dot. These operations will be explained in Section A.8.*

the **axis([xmin xmax ymin ymax zmin zmax])** command. It must be placed after the **plot(x,y)** or **plot3(x,y,z)** commands, or on the same line without first executing the **plot** command. This must be done for each plot. The three-dimensional **text(x,y,z,'string')** command will place **string** beginning at the co-ordinate (x,y,z) on the plot.

For three-dimensional plots, **grid on** and **box off** are the default states.

We can also use the **mesh(x,y,z)** command with two vector arguments. These must be defined as $length(x) = n$ and $length(y) = m$ where $[m, n] = size(Z)$. In this case, the vertices of the mesh lines are the triples $\{x(j), y(i), Z(i, j)\}$. We observe that **x** corresponds to the columns of Z, and **y** corresponds to the rows.

To produce a mesh plot of a function of two variables, say $z = f(x, y)$, we must first generate the X and Y matrices that consist of repeated rows and columns over the range of the variables x and y. We can generate the matrices X and Y with the **[X,Y]=meshgrid(x,y)** function that creates the matrix X whose rows are copies of the vector **x**, and the matrix Y whose columns are copies of the vector **y**.

Example A.12

The volume V of a right circular cone of radius r and height h is given by

$$V = \frac{1}{3}\pi r^2 h \tag{A.4}$$

Plot the volume of the cone as r and h vary on the intervals $0 \leq r \leq 4$ and $0 \leq h \leq 6$ meters.

Solution:

The volume of the cone is a function of both the radius r and the height h, that is,

$$V = f(r, h)$$

The three-dimensional plot is created with the following MATLAB code where, as in the previous example, in the second line we have used the dot multiplication, dot division, and dot exponentiation. This will be explained in Section A.8.

```
[R,H]=meshgrid(0: 4, 0: 6);        % Creates R and H matrices from vectors r and h

V=(pi .* R .^ 2 .* H) ./ 3;  mesh(R, H, V)

xlabel('x-axis, radius r (meters)'); ylabel('y-axis, altitude h (meters)');

zlabel('z-axis, volume (cubic meters)'); title('Volume of Right Circular Cone'); box on
```

The three-dimensional plot of Figure A.6, shows how the volume of the cone increases as the radius and height are increased.

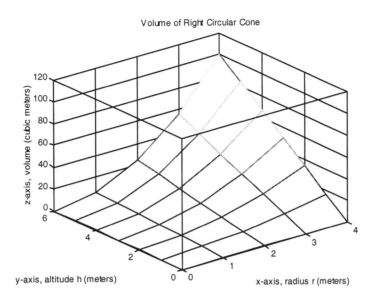

Figure A.6. Volume of a right circular cone.

This, and the plot of Figure A.5, are rudimentary; MATLAB can generate very sophisticated three-dimensional plots. The MATLAB User's manual contains more examples.

A.8 Subplots

MATLAB can display up to four windows of different plots on the *Figure* window using the command **subplot(m,n,p)**. This command divides the window into an $m \times n$ matrix of plotting areas and chooses the *pth* area to be active. No spaces or commas are required between the three integers *m, n* and *p*. The possible combinations are shown in Figure A.7.

We will illustrate the use of the **subplot(m,n,p)** command following the discussion on multiplication, division and exponentiation that follows.

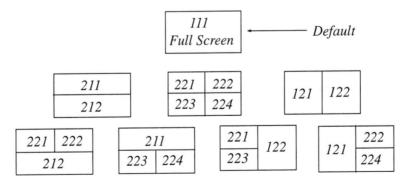

Figure A.7. Possible subplot arrangements in MATLAB

A.9 Multiplication, Division and Exponentiation

MATLAB recognizes two types of multiplication, division, and exponentiation. These are the *matrix* multiplication, division, and exponentiation, and the *element-by-element* multiplication, division, and exponentiation. They are explained in the following paragraphs.

In Section A.2, the arrays $[a \quad b \quad c \quad ...]$, such a those that contained the coefficients of polynomials, consisted of one row and multiple columns, and thus are called *row vectors*. If an array has one column and multiple rows, it is called a *column vector*. We recall that the elements of a row vector are separated by spaces. To distinguish between row and column vectors, the elements of a column vector must be separated by semicolons. An easier way to construct a column vector, is to write it first as a row vector, and then transpose it into a column vector. MATLAB uses the single quotation character (') to transpose a vector. Thus, a column vector can be written either as b=[−1; 3; 6; 11] or as b=[−1 3 6 11]'. MATLAB produces the same display with either format as shown below.

b=[−1; 3; 6; 11]

b =

        −1

         3

         6

        11

b=[−1 3 6 11]'

b =

        −1

         3

         6

        11

We will now define Matrix Multiplication and Element-by-Element multiplication.

1. **Matrix Multiplication** (multiplication of row by column vectors)

Let

$$A = [a_1 \quad a_2 \quad a_3 \quad ... \quad a_n]$$

and

$$B = [b_1 \quad b_2 \quad b_3 \quad ... \quad b_n]'$$

be two vectors. We observe that A is defined as a row vector whereas B is defined as a column vector, as indicated by the transpose operator ('). Here, multiplication of the row vector **A** by the column vector **B**, is performed with the matrix multiplication operator (*). Then,

$$A*B = [a_1b_1 + a_2b_2 + a_3b_3 + \ldots + a_nb_n] = single\ value \qquad (A.5)$$

For example, if

$$A = [1\quad 2\quad 3\quad 4\quad 5]$$

and

$$B = [-2\quad 6\quad -3\quad 8\quad 7]'$$

the matrix multiplication $A*B$ produces the single value 68, that is,

$$A*B = 1 \times (-2) + 2 \times 6 + 3 \times (-3) + 4 \times 8 + 5 \times 7 = 68$$

and this is verified with MATLAB as

```
A=[1  2  3 4 5]; B=[−2  6 −3  8  7]';
A*B

ans =

    68
```

Now, let us suppose that both A and B are row vectors, and we attempt to perform a row-by-row multiplication with the following MATLAB statements.

```
A=[1 2  3 4 5]; B=[−2 6 −3 8 7];
A*B
```

When these statements are executed, MATLAB displays the following message:

```
??? Error using ==> *

Inner matrix dimensions must agree.
```

Here, because we have used the matrix multiplication operator (*) in **A*B**, MATLAB expects vector B to be a column vector, not a row vector. It recognizes that B is a row vector, and warns us that we cannot perform this multiplication using the matrix multiplication operator (*). Accordingly, we must perform this type of multiplication with a different operator. This operator is defined below.

2.Element-by-Element Multiplication (multiplication of a row vector by another row vector)

Let

$$C = [c_1\quad c_2\quad c_3\quad \ldots\quad c_n]$$

and

$$D = [d_1\quad d_2\quad d_3\quad \ldots\quad d_n]$$

be two row vectors. Here, multiplication of the row vector C by the row vector D is performed with the *dot multiplication operator* (.*). There is no space between the dot and the multiplication symbol. Thus,

$$C.*D = [c_1 d_1 \quad c_2 d_2 \quad c_3 d_3 \quad \dots \quad c_n d_n] \tag{A.6}$$

This product is another row vector with the same number of elements, as the elements of C and D.

As an example, let

$$C = [1 \quad 2 \quad 3 \quad 4 \quad 5]$$

and

$$D = [-2 \quad 6 \quad -3 \quad 8 \quad 7]$$

Dot multiplication of these two row vectors produce the following result.

$$C.*D = 1 \times (-2) \quad 2 \times 6 \quad 3 \times (-3) \quad 4 \times 8 \quad 5 \times 7 = -2 \quad 12 \quad -9 \quad 32 \quad 35$$

Check with MATLAB:

```
C=[1 2 3 4 5];        %  Vectors C and D must have
D=[-2 6 -3 8 7];      %  same number of elements
C.*D                  % We observe that this is a dot multiplication

ans =
    -2    12    -9    32    35
```

Similarly, the division (/) and exponentiation (^) operators, are used for matrix division and exponentiation, whereas dot division (./) and dot exponentiation (.^) are used for element-by-element division and exponentiation.

We must remember that *no space is allowed between the dot (.) and the multiplication, division, and exponentiation operators.*

Note: A dot (.) is never required with the plus (+) and minus (–) operators.

Example A.13

Write the MATLAB code that produces a simple plot for the waveform defined as

$$y = f(t) = 3e^{-4t} \cos 5t - 2e^{-3t} \sin 2t + \frac{t^2}{t+1} \tag{A.7}$$

in the $0 \le t \le 5$ seconds interval.

Solution:

The MATLAB code for this example is as follows:

```
t=0: 0.01: 5            %  Define t-axis in 0.01 increments
y=3 .* exp(−4 .* t) .* cos(5 .* t)−2 .* exp(−3 .* t) .* sin(2 .* t) + t .^2 ./ (t+1);
plot(t,y); grid; xlabel('t'); ylabel('y=f(t)'); title('Plot for Example A.13')
```

Figure A.8 shows the plot for this example.

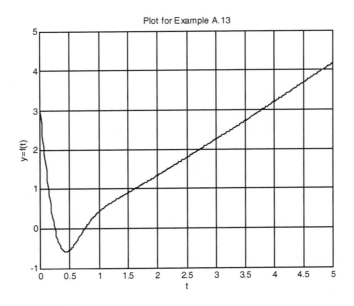

Figure A.8. Plot for Example A.13

Had we, in this example, defined the time interval starting with a negative value equal to or less than -1, say as $-3 \le t \le 3$, MATLAB would have displayed the following message:

```
Warning: Divide by zero.
```

This is because the last term (the rational fraction) of the given expression, is divided by zero when $t = -1$. To avoid division by zero, we use the special MATLAB function **eps,** which is a number approximately equal to 2.2×10^{-16}. It will be used with the next example.

The command **axis([xmin xmax ymin ymax])** scales the current plot to the values specified by the arguments **xmin, xmax, ymin and ymax.** There are no commas between these four arguments. This command must be placed *after* the plot command and must be repeated for each plot.

The following example illustrates the use of the dot multiplication, division, and exponentiation, the **eps** number, the **axis([xmin xmax ymin ymax])** command, and also MATLAB's capability of displaying up to four windows of different plots.

Example A.14

Plot the functions

$$y = sin^2x, \quad z = cos^2x, \quad w = sin^2x \cdot cos^2x, \quad v = sin^2x / cos^2x$$

in the interval $0 \leq x \leq 2\pi$ using 100 data points. Use the **subplot** command to display these functions on four windows on the same graph.

Solution:

The MATLAB code to produce the four subplots is as follows:

```
x=linspace(0,2*pi,100);          % Interval with 100 data points
y=(sin(x).^ 2);  z=(cos(x).^ 2);
w=y.* z;
v=y./ (z+eps);                    %  add eps to avoid division by zero
subplot(221);% upper left of four subplots
plot(x,y);  axis([0 2*pi 0 1]);
title('y=(sinx)^2');
subplot(222);                     % upper right of four subplots
plot(x,z);  axis([0 2*pi 0 1]);
title('z=(cosx)^2');
subplot(223);                     % lower left of four subplots
plot(x,w);  axis([0 2*pi 0 0.3]);
title('w=(sinx)^2*(cosx)^2');
subplot(224);                     % lower right of four subplots
plot(x,v);  axis([0 2*pi 0 400]);
title('v=(sinx)^2/(cosx)^2');
```

These subplots are shown in Figure A.9.

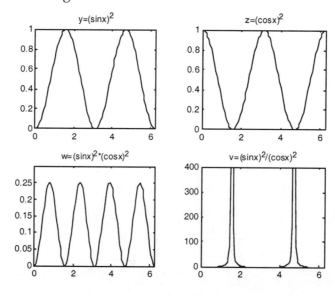

Figure A.9. Subplots for the functions of Example A.14

The next example illustrates MATLAB's capabilities with imaginary numbers. We will introduce the **real(z)** and **imag(z)** functions that display the real and imaginary parts of the complex quantity $z = x + iy$, the **abs(z)**, and the **angle(z)** functions that compute the absolute value (magnitude) and phase angle of the complex quantity $z = x + iy = r\angle\theta$. We will also use the **polar(theta,r)** function that produces a plot in polar coordinates, where **r** is the magnitude, **theta** is the angle in radians, and the **round(n)** function that rounds a number to its nearest integer.

Example A.15

Consider the electric circuit of Figure A.10.

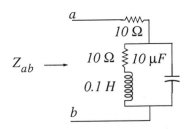

Figure A.10. Electric circuit for Example A.15

With the given values of resistance, inductance, and capacitance, the impedance Z_{ab} as a function of the radian frequency ω can be computed from the following expression:

$$Z_{ab} = Z = 10 + \frac{10^4 - j(10^6/\omega)}{10 + j(0.1\omega - 10^5/\omega)} \tag{A.8}$$

a. Plot $Re\{Z\}$ (the real part of the impedance Z) versus frequency ω

b. Plot $Im\{Z\}$ (the imaginary part of the impedance Z) versus frequency ω

c. Plot the impedance Z versus frequency ω in polar coordinates.

Solution:

The MATLAB code below computes the real and imaginary parts of Z_{ab} that is, for simplicity, denoted as z, and plots these as two separate graphs (parts a & b). It also produces a polar plot (part c).

```
w=0: 1: 2000;  % Define interval with one radian interval
z=(10+(10 .^ 4 –j .* 10 .^ 6 ./ (w+eps)) ./ (10 + j .* (0.1 .* w –10.^5./ (w+eps))));
%
%  The first five statements (next two lines) compute and plot Re{z}
real_part=real(z);  plot(w,real_part);  grid;
xlabel('radian frequency w');  ylabel('Real part of Z');
%
```

```
%  The next five statements (next two lines) compute and plot Im{z}
imag_part=imag(z);  plot(w,imag_part);  grid;
xlabel('radian frequency w');  ylabel('Imaginary part of Z');
%  The last six statements (next six lines) below produce the polar plot of z
mag=abs(z);          %  Computes |Z|
rndz=round(abs(z));  %  Rounds |Z| to read polar plot easier
theta=angle(z);      %  Computes the phase angle of impedance Z
polar(theta,rndz);   %  Angle is the first argument
grid;
ylabel('Polar Plot of Z');
```

The real, imaginary, and polar plots are shown in Figures A.11, A.12, and A.13 respectively.

Example A.15 clearly illustrates how powerful, fast, accurate, and flexible MATLAB is.

A.10 Script and Function Files

MATLAB recognizes two types of files: *script files* and *function files*. Both types are referred to as *m-files* since both require the *.m* extension.

A *script file* consists of two or more built-in functions such as those we have discussed thus far. Thus, the code for each of the examples we discussed earlier, make up a script file. Generally, a script file is one which was generated and saved as an m-file with an editor such as the MATLAB's Editor/ Debugger.

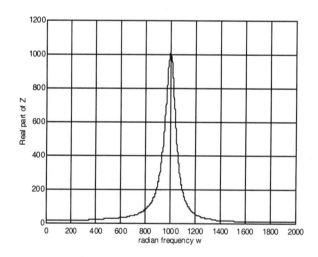

Figure A.11. Plot for the real part of the impedance in Example A.15

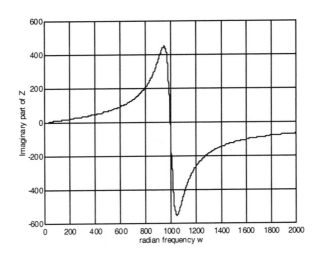

Figure A.12. Plot for the imaginary part of the impedance in Example A.15

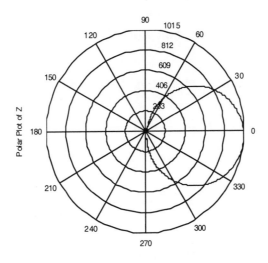

Figure A.13. Polar plot of the impedance in Example A.15

A *function file* is a user-defined function using MATLAB. We use function files for repetitive tasks. The first line of a function file must contain the word *function*, followed by the output argument, the equal sign (=), and the input argument enclosed in parentheses. The function name and file name must be the same, but the file name must have the extension *.m*. For example, the function file consisting of the two lines below

function y = myfunction(x)

y=x.^ 3 + cos(3.* x)

is a function file and must be saved as *myfunction.m*

For the next example, we will use the following MATLAB functions.

fzero(f,x) tries to find a zero of a function of one variable, where **f** is a string containing the name of a real-valued function of a single real variable. MATLAB searches for a value near a point where the function **f** changes sign, and returns that value, or returns NaN if the search fails.

Important: We must remember that we use **roots(p)** to find the roots of polynomials only, such as those in Examples A.1 and A.2.

fmin(f,x1,x2) minimizes a function of one variable. It attempts to return a value of x where $f(x)$ is minimum in the interval $x_1 < x < x_2$. The string **f** contains the name of the function to be minimized.

Note: MATLAB does not have a function to maximize a function of one variable, that is, there is no **fmax(f,x1,x2)** function in MATLAB; but since a maximum of $f(x)$ is equal to a minimum of $-f(x)$, we can use **fmin(f,x1,x2)** to find both minimum and maximum values of a function.

fplot(fcn,lims) plots the function specified by the string **fcn** between the x-axis limits specified by **lims = [xmin xmax]**. Using **lims = [xmin xmax ymin ymax]** also controls the y-axis limits. The string **fcn** must be the name of an *m-file* function or a string with variable x.

Note: **NaN** (Not-a-Number) is not a function; it is MATLAB's response to an undefined expression such as $0/0$, ∞/∞, or inability to produce a result as described on the next paragraph. We can avoid division by zero using the **eps** number, that we mentioned earlier.

Example A.16

Find the zeros, maxima and minima of the function

$$f(x) = \frac{1}{(x-0.1)^2 + 0.01} + \frac{1}{(x-1.2)^2 + 0.04} - 10$$

Solution:

We first plot this function to observe the approximate zeros, maxima, and minima using the following code.

```
x=−1.5: 0.01: 1.5;
y=1./ ((x−0.1).^ 2 + 0.01) −1./ ((x−1.2).^ 2 + 0.04) −10;
plot(x,y); grid
```

The plot is shown in Figure A.14.

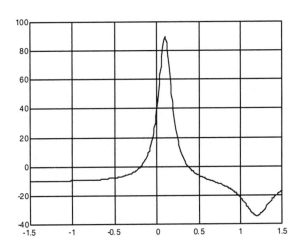

Figure A.14. Plot for Example A.16 using the plot command

The roots (zeros) of this function appear to be in the neighborhood of $x = -0.2$ and $x = 0.3$. The maximum occurs at approximately $x = 0.1$ where, approximately, $y_{max} = 90$, and the minimum occurs at approximately $x = 1.2$ where, approximately, $y_{min} = -34$.

Next, we define and save $f(x)$ as the **funczero01.m** function m-file with the following code:

```
function y=funczero01(x)
```

```
% Finding the zeros of the function shown below
```

```
y=1/((x−0.1)^2+0.01)−1/((x−1.2)^2+0.04)-10;
```

Now, we can use the **fplot(fcn,lims)** command to plot $f(x)$ as follows.

```
fplot('funczero01', [−1.5  1.5]); grid
```

This plot is shown in Figure A.15. As expected, this plot is identical to the plot of Figure A.14 that was obtained with the **plot(x,y)** command.

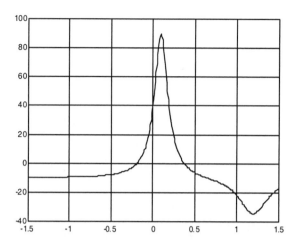

Figure A.15. Plot for Example A.16 using the fplot command

We will use the **fzero(f,x)** function to compute the roots of $f(x)$ in (A.20) more precisely. The code below must be saved with a file name, and then invoked with that file name.

```
x1= fzero('funczero01', -0.2);
x2= fzero('funczero01', 0.3);
fprintf('The roots (zeros) of this function are r1= %3.4f', x1);
fprintf(' and r2= %3.4f \n', x2)
```

MATLAB displays the following:

```
The roots (zeros) of this function are r1= -0.1919 and r2= 0.3788
```

Whenever we use the **fmin(f,x1,x2)** function, we must remember that this function searches for a minimum and it may display the values of local minima[*], if any, before displaying the function minimum. It is, therefore, advisable to plot the function with either the **plot(x,y)** or the **fplot(fcn,lims)** command to find the smallest possible interval within which the function minimum lies. For this example, we specify the range $0 \le x \le 1.5$ rather than the interval $-1.5 \le x \le 1.5$.

The minimum of $f(x)$ is found with the **fmin(f,x1,x2)** function as follows.

```
min_val=fmin('funczero01', 0, 1.5)
```

```
min_val = 1.2012
```

---

[*] *Local maxima or local minima, are the maximum or minimum values of a function within a restricted range of values in the independent variable. When the entire range is considered, the maxima and minima are considered be to the maximum and minimum values in the entire range in which the function is defined.*

This is the value of x at which $y = f(x)$ is minimum. To find the value of y corresponding to this value of x, we substitute it into $f(x)$, that is,

x=1.2012; y=1 / ((x–0.1) ^ 2 + 0.01) –1 / ((x–1.2) ^ 2 + 0.04) –10

```
y = -34.1812
```

To find the maximum value, we must first define a new function *m-file* that will produce $-f(x)$. We define it as follows:

function y=minusfunczero01(x)

% It is used to find maximum value from -f(x)

y=–(1/((x–0.1)^2+0.01)–1/((x–1.2)^2+0.04)–10);

We have placed the minus (–) sign in front of the right side of the last expression above, so that the maximum value will be displayed. Of course, this is equivalent to the negative of the **funczero01** function.

Now, we execute the following code to get the value of x where the maximum $y = f(x)$ occurs.

max_val=fmin('minusfunczero01', 0,1)

```
max_val = 0.0999
```

x=0.0999;% Using this value find the corresponding value of y
y=1 / ((x–0.1) ^ 2 + 0.01) –1 / ((x–1.2) ^ 2 + 0.04) –10

```
y = 89.2000
```

A.11 Display Formats

MATLAB displays the results on the screen in integer format without decimals if the result is an integer number, or in short floating point format with four decimals if it a fractional number. The format displayed has nothing to do with the accuracy in the computations. MATLAB performs all computations with accuracy up to 16 decimal places.

The output format can changed with the **format** command. The available formats can be displayed with the **help format** command as follows:

help format

FORMAT Set output format.

All computations in MATLAB are done in double precision.

FORMAT may be used to switch between different output display formats as follows:

FORMAT Default. Same as SHORT.

FORMAT SHORT Scaled fixed point format with 5 digits.

FORMAT LONG Scaled fixed point format with 15 digits.

FORMAT SHORT E Floating point format with 5 digits.

FORMAT LONG E Floating point format with 15 digits.

FORMAT SHORT G Best of fixed or floating point format with 5 digits.

FORMAT LONG G Best of fixed or floating point format with 15 digits.

FORMAT HEX   Hexadecimal format.

FORMAT +   The symbols +, - and blank are printed for positive, negative and zero elements.
                Imaginary parts are ignored.

FORMAT BANK   Fixed format for dollars and cents.

FORMAT RAT   Approximation by ratio of small integers.

Spacing:

FORMAT COMPACT Suppress extra line-feeds.

FORMAT LOOSE   Puts the extra line-feeds back in.

Some examples with different format displays age given below.

format short 33.3335 Four decimal digits (default)

format long 33.33333333333334 16 digits

format short e 3.3333e+01 Four decimal digits plus exponent

format short g 33.333 Better of format short or format short e

format bank 33.33 two decimal digits

format + only + or − or zero are printed

format rat 100/3 rational approximation

The **disp(X)** command displays the array **X** without printing the array name. If **X** is a string, the text is displayed.

The **fprintf(format,array)** command displays and prints both text and arrays. It uses specifiers to indicate where and in which format the values would be displayed and printed. Thus, if **%f** is used, the values will be displayed and printed in fixed decimal format, and if **%e** is used, the values will be displayed and printed in scientific notation format. With this command only the real part of each parameter is processed.

Appendix B

A Review of Complex Numbers

This appendix is a review of the algebra of complex numbers. The basic operations are defined and illustrated by several examples. Applications using Euler's identities are presented, and the exponential and polar forms are discussed and illustrated with examples.

B.1 Definition of a Complex Number

In the language of mathematics, the square root of minus one is denoted as i, that is, $i = \sqrt{-1}$. In the electrical engineering field, we denote i as j to avoid confusion with current i. Essentially, j is an operator that produces a 90-degree counterclockwise rotation to any vector to which it is applied as a multiplying factor. Thus, if it is given that a vector A has the direction along the right side of the x-axis as shown in Figure B.1, multiplication of this vector by the operator j will result in a new vector jA whose magnitude remains the same, but it has been rotated counterclockwise by $90°$. Also, another multiplication of the new vector jA by j will produce another $90°$ counterclockwise direction. In this case, the vector A has rotated $180°$ and its new value now is $-A$. When this vector is rotated by another $90°$ for a total of $270°$, its value becomes $j(-A) = -jA$. A fourth $90°$ rotation returns the vector to its original position, and thus its value is again A. Therefore, we conclude that $j^2 = -1$, $j^3 = -j$, and $j^4 = 1$.

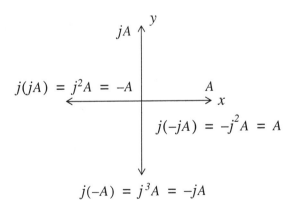

Figure B.1. The j operator

Note: In our subsequent discussion, we will designate the x-axis (abscissa) as the *real axis*, and the y-axis (ordinate) as the *imaginary axis* with the understanding that the "imaginary" axis is just as "real" as the real axis. In other words, the imaginary axis is just as important as the real axis.[*]

An *imaginary number* is the product of a real number, say r, by the operator j. Thus, r is a real number and jr is an imaginary number.

A *complex number* is the sum (or difference) of a real number and an imaginary number. For example, the number $A = a + jb$ where a and b are both real numbers, is a complex number. Then, $a = Re\{A\}$ and $b = Im\{A\}$ where $Re\{A\}$ denotes real part of A, and $b = Im\{A\}$ the imaginary part of A.

By definition, two complex numbers A and B where $A = a + jb$ and $B = c + jd$, are equal if and only if their real parts are equal, and also their imaginary parts are equal. Thus, $A = B$ if and only if $a = c$ and $b = d$.

B.2 Addition and Subtraction of Complex Numbers

The sum of two complex numbers has a real component equal to the sum of the real components, and an imaginary component equal to the sum of the imaginary components. For subtraction, we change the signs of the components of the subtrahend and we perform addition. Thus, if

$$A = a + jb \text{ and } B = c + jd$$

then

$$A + B = (a + c) + j(b + d)$$

and

$$A - B = (a - c) + j(b - d)$$

Example B.1

It is given that $A = 3 + j4$, and $B = 4 - j2$. Find $A + B$ and $A - B$

Solution:

$$A + B = (3 + j4) + (4 - j2) = (3 + 4) + j(4 - 2) = 7 + j2$$

and

$$A - B = (3 + j4) - (4 - j2) = (3 - 4) + j(4 + 2) = -1 + j6$$

---

[*] *We may think the real axis as the cosine axis and the imaginary axis as the sine axis.*

B.3 Multiplication of Complex Numbers

Complex numbers are multiplied using the rules of elementary algebra, and making use of the fact that $j^2 = -1$. Thus, if

$$A = a + jb \text{ and } B = c + jd$$

then

$$A \cdot B = (a + jb) \cdot (c + jd) = ac + jad + jbc + j^2bd$$

and since $j^2 = -1$, it follows that

$$A \cdot B = ac + jad + jbc - bd$$
$$= (ac - bd) + j(ad + bc) \tag{B.1}$$

Example B.2

It is given that $A = 3 + j4$ and $B = 4 - j2$. Find $A \cdot B$

Solution:

$$A \cdot B = (3 + j4) \cdot (4 - j2) = 12 - j6 + j16 - j^2 8 = 20 + j10$$

The *conjugate* of a complex number, denoted as $A*$, is another complex number with the same real component, and with an imaginary component of opposite sign. Thus, if $A = a + jb$, then $A* = a - jb$.

Example B.3

It is given that $A = 3 + j5$. Find $A*$

Solution:

The conjugate of the complex number A has the same real component, but the imaginary component has opposite sign. Then, $A* = 3 - j5$

If a complex number A is multiplied by its conjugate, the result is a real number. Thus, if $A = a + jb$, then

$$A \cdot A* = (a + jb)(a - jb) = a^2 - jab + jab - j^2b^2 = a^2 + b^2$$

Example B.4

It is given that $A = 3 + j5$. Find $A \cdot A*$

Solution:

$$A \cdot A* = (3 + j5)(3 - j5) = 3^2 + 5^2 = 9 + 25 = 34$$

B.4 Division of Complex Numbers

When performing division of complex numbers, it is desirable to obtain the quotient separated into a real part and an imaginary part. This procedure is called *rationalization of the quotient*, and it is done by multiplying the denominator by its conjugate. Thus, if $A = a + jb$ and $B = c + jd$, then,

$$\frac{A}{B} = \frac{a+jb}{c+jd} = \frac{(a+jb)(c-jd)}{(c+jd)(c-jd)} = \frac{A}{B} \cdot \frac{B^*}{B^*} = \frac{(ac+bd)+j(bc-ad)}{c^2+d^2}$$

$$= \frac{(ac+bd)}{c^2+d^2} + j\frac{(bc-ad)}{c^2+d^2} \tag{B.2}$$

In (B.2), we multiplied both the numerator and denominator by the conjugate of the denominator to eliminate the j operator from the denominator of the quotient. Using this procedure, we see that the quotient is easily separated into a real and an imaginary part.

Example B.5

It is given that $A = 3 + j4$, and $B = 4 + j3$. Find A/B

Solution:

Using the procedure of (B.2), we get

$$\frac{A}{B} = \frac{3+j4}{4+j3} = \frac{(3+j4)(4-j3)}{(4+j3)(4-j3)} = \frac{12-j9+j16+12}{4^2+3^2} = \frac{24+j7}{25} = \frac{24}{25} + j\frac{7}{25} = 0.96 + j0.28$$

B.5 Exponential and Polar Forms of Complex Numbers

The relations

$$e^{j\theta} = \cos\theta + j\sin\theta \tag{B.3}$$

and

$$e^{-j\theta} = \cos\theta - j\sin\theta \tag{B.4}$$

are known as the *Euler's identities*.

Multiplying (B.3) by the *real* positive constant C we get:

$$Ce^{j\theta} = C\cos\theta + jC\sin\theta \tag{B.5}$$

This expression represents a complex number, say $a + jb$, and thus

$$Ce^{j\theta} = a + jb \qquad \text{(B.6)}$$

where the left side of (B.6) is the *exponential form*, and the right side is the *rectangular form*.

Equating real and imaginary parts in (B.5) and (B.6), we get

$$a = C\cos\theta \quad \text{and} \quad b = C\sin\theta \qquad \text{(B.7)}$$

Squaring and adding the expressions in (B.7), we get

$$a^2 + b^2 = (C\cos\theta)^2 + (C\sin\theta)^2 = C^2(\cos^2\theta + \sin^2\theta) = C^2$$

Then,

$$C^2 = a^2 + b^2$$

or

$$\boxed{C = \sqrt{a^2 + b^2}} \qquad \text{(B.8)}$$

Also, from (B.7)

$$\frac{b}{a} = \frac{C\sin\theta}{C\cos\theta} = \tan\theta$$

or

$$\boxed{\theta = \tan^{-1}\left(\frac{b}{a}\right)} \qquad \text{(B.9)}$$

To convert a complex number from rectangular to exponential form, we use the expression

$$\boxed{a + jb = \sqrt{a^2 + b^2}\, e^{j\left(\tan^{-1}\frac{b}{a}\right)}} \qquad \text{(B.10)}$$

To convert a complex number from exponential to rectangular form, we use the expressions

$$\boxed{\begin{aligned} Ce^{j\theta} &= C\cos\theta + jC\sin\theta \\ Ce^{-j\theta} &= C\cos\theta - jC\sin\theta \end{aligned}} \qquad \text{(B.11)}$$

The *polar form* is essentially the same as the exponential form but the notation is different, that is,

$$\boxed{Ce^{j\theta} = C\angle\theta} \qquad \text{(B.12)}$$

where the left side of (B.12) is the exponential form, and the right side is the polar form.

We must remember that *the phase angle* θ *is always measured with respect to the positive real axis, and rotates in the counterclockwise direction.*

Example B.6

Convert the following complex numbers to exponential and polar forms:

a. $3 + j4$

b. $-1 + j2$

c. $-2 - j$

d. $4 - j3$

Solution:

a. The real and imaginary components of this complex number are shown in Figure B.2.

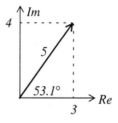

Figure B.2. The components of $3 + j4$

Then,

$$3 + j4 = \sqrt{3^2 + 4^2}\ e^{j\left(tan^{-1}\frac{4}{3}\right)} = 5e^{j53.1°} = 5\angle 53.1°$$

Check with MATLAB:

```
x=3+j*4; magx=abs(x); thetax=angle(x)*180/pi;  disp(magx); disp(thetax)
```

```
    5
  53.1301
```

b. The real and imaginary components of this complex number are shown in Figure B.3.

Then,

$$-1 + j2 = \sqrt{1^2 + 2^2}\ e^{j\left(tan^{-1}\frac{2}{-1}\right)} = \sqrt{5}e^{j116.6°} = \sqrt{5}\angle 116.6°$$

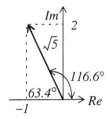

Figure B.3. The components of $-1 + j2$

Check with MATLAB:

y=−1+j*2; magy=abs(y); thetay=angle(y)*180/pi; disp(magy); disp(thetay)

```
    2.2361
  116.5651
```

c. The real and imaginary components of this complex number are shown in Figure B.4.

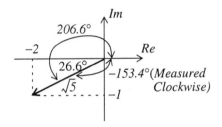

Figure B.4. The components of $-2 - j$

Then,

$$-2-j1 = \sqrt{2^2 + 1^2}\, e^{j\left(tan^{-1}\frac{-1}{-2}\right)} = \sqrt{5}e^{j206.6°} = \sqrt{5}\angle206.6° = \sqrt{5}e^{j(-153.4)°} = \sqrt{5}\angle-153.4°$$

Check with MATLAB:

v=−2−j*1; magv=abs(v); thetav=angle(v)*180/pi; disp(magv); disp(thetav)

```
    2.2361
 -153.4349
```

d. The real and imaginary components of this complex number are shown in Figure B.5.

Then,

$$4-j3 = \sqrt{4^2 + 3^2}\, e^{j\left(tan^{-1}\frac{-3}{4}\right)} = 5e^{j323.1°} = 5\angle323.1° = 5e^{-j36.9°} = 5\angle-36.9°$$

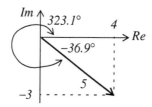

Figure B.5. The components of $4 - j3$

Check with MATLAB:

```
w=4–j*3; magw=abs(w); thetaw=angle(w)*180/pi;  disp(magw); disp(thetaw)
```

        5
     -36.8699

Example B.7

Express the complex number $-2\angle30°$ in exponential and in rectangular forms.

Solution:

We recall that $-1 = j^2$. Since each j rotates a vector by $90°$ counterclockwise, then $-2\angle30°$ is the same as $2\angle30°$ rotated counterclockwise by $180°$. Therefore,

$$-2\angle30° = 2\angle(30° + 180°) = 2\angle210° = 2\angle-150°$$

The components of this complex number are shown in Figure B.6.

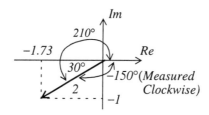

Figure B.6. The components of $2\angle-150°$

Then,

$$2\angle-150° = 2e^{-j150°} = 2(\cos150° - j\sin150°) = 2(-0.866 - j0.5) = -1.73 - j$$

Note: The rectangular form is most useful when we add or subtract complex numbers; however, the exponential and polar forms are most convenient when we multiply or divide complex numbers.

To multiply two complex numbers in exponential (or polar) form, we multiply the magnitudes and we add the phase angles, that is, if

$$A = M\angle\theta \quad \text{and} \quad B = N\angle\phi$$

then,

$$AB = MN\angle(\theta + \phi) = Me^{j\theta}Ne^{j\phi} = MNe^{j(\theta + \phi)} \qquad \text{(B.13)}$$

Example B.8

Multiply $A = 10\angle53.1°$ by $B = 5\angle-36.9°$

Solution:

Multiplication in polar form yields

$$AB = (10 \times 5)\angle[53.1° + (-36.9°)] = 50\angle16.2°$$

and multiplication in exponential form yields

$$AB = (10e^{j53.1°})(5e^{-j36.9°}) = 50e^{j(53.1° - 36.9°)} = 50e^{j16.2°}$$

To divide one complex number by another when both are expressed in exponential or polar form, we divide the magnitude of the dividend by the magnitude of the divisor, and we subtract the phase angle of the divisor from the phase angle of the dividend, that is, if

$$A = M\angle\theta \quad \text{and} \quad B = N\angle\phi$$

then,

$$\frac{A}{B} = \frac{M}{N}\angle(\theta - \phi) = \frac{Me^{j\theta}}{Ne^{j\phi}} = \frac{M}{N}e^{j(\theta - \phi)} \qquad \text{(B.14)}$$

Example B.9

Divide $A = 10\angle53.1°$ by $B = 5\angle-36.9°$

Solution:

Division in polar form yields

$$\frac{A}{B} = \frac{10\angle53.1°}{5\angle-36.9°} = 2\angle[53.1° - (-36.9°)] = 2\angle90°$$

Division in exponential form yields

$$\frac{A}{B} = \frac{10e^{j53.1°}}{5e^{-j36.9°}} = 2e^{j53.1°}e^{j36.9°} = 2e^{j90°}$$

NOTES

Appendix C

This chapter is an introduction to matrices and matrix operations. Determinants, Cramer's rule, and Gauss's elimination method are reviewed. Some definitions and examples are not applicable to subsequent material presented in this text, but are included for subject continuity, and reference to more advance topics in matrix theory. These are denoted with a dagger (†) and may be skipped.

C.1 Matrix Definition

A *matrix* is a rectangular array of numbers such as those shown below.

$$\begin{bmatrix} 2 & 3 & 7 \\ 1 & -1 & 5 \end{bmatrix} \quad or \quad \begin{bmatrix} 1 & 3 & 1 \\ -2 & 1 & -5 \\ 4 & -7 & 6 \end{bmatrix}$$

In general form, a matrix A is denoted as

$$A = \begin{bmatrix} a_{11} & a_{12} & a_{13} & \cdots & a_{1n} \\ a_{21} & a_{22} & a_{23} & \cdots & a_{2n} \\ a_{31} & a_{32} & a_{33} & \cdots & a_{3n} \\ \cdots & \cdots & \cdots & \cdots & \cdots \\ a_{m1} & a_{m2} & a_{m3} & \cdots & a_{mn} \end{bmatrix} \quad \text{(C.1)}$$

The numbers a_{ij} are the *elements* of the matrix where the index i indicates the row, and j indicates the column in which each element is positioned. Thus, a_{43} indicates the element positioned in the fourth row and third column.

A matrix of m rows and n columns is said to be of $m \times n$ *order matrix*.

If $m = n$, the matrix is said to be a *square matrix of order m* (or n). Thus, if a matrix has five rows and five columns, it is said to be a square matrix of order 5.

In a square matrix, the elements a_{11}, a_{22}, a_{33}, ..., a_{nn} are called the *main diagonal elements*. Alternately, we say that the matrix elements a_{11}, a_{22}, a_{33}, ..., a_{nn}, are located on the *main diagonal.*

† The sum of the diagonal elements of a square matrix A is called the *trace*[*] of A.

† A matrix in which every element is zero, is called a *zero matrix*.

C.2 Matrix Operations

Two matrices $A = \left[a_{ij} \right]$ and $B = \left[b_{ij} \right]$ are equal, that is, $A = B$, if and only if

$$a_{ij} = b_{ij} \qquad i = 1, 2, 3, ..., m \qquad j = 1, 2, 3, ..., n \qquad (C.2)$$

Two matrices are said to be *conformable for addition (subtraction)*, if they are of the same order $m \times n$.

If $A = \left[a_{ij} \right]$ and $B = \left[b_{ij} \right]$ are conformable for addition (subtraction), their sum (difference) will be another matrix C with the same order as A and B, where each element of C is the sum (difference) of the corresponding elements of A and B, that is,

$$C = A \pm B = [a_{ij} \pm b_{ij}] \qquad (C.3)$$

Example C.1

Compute $A + B$ and $A - B$ given that

$$A = \begin{bmatrix} 1 & 2 & 3 \\ 0 & 1 & 4 \end{bmatrix} \text{ and } B = \begin{bmatrix} 2 & 3 & 0 \\ -1 & 2 & 5 \end{bmatrix}$$

Solution:

$$A + B = \begin{bmatrix} 1+2 & 2+3 & 3+0 \\ 0-1 & 1+2 & 4+5 \end{bmatrix} = \begin{bmatrix} 3 & 5 & 3 \\ -1 & 3 & 9 \end{bmatrix}$$

and

$$A - B = \begin{bmatrix} 1-2 & 2-3 & 3-0 \\ 0+1 & 1-2 & 4-5 \end{bmatrix} = \begin{bmatrix} -1 & -1 & 3 \\ 1 & -1 & -1 \end{bmatrix}$$

Check with MATLAB:

```
A=[1 2 3;  0 1 4]; B=[2 3 0; –1 2 5];    % Define matrices A and B
A+B                                       % Add A and B
```

---

* *Henceforth, all paragraphs and topics preceded by a dagger (†) may be skipped. These are discussed in matrix theory textbooks.*

```
ans =
     3     5     3
    -1     3     9
```

A–B   % Subtract B from A

```
ans =
    -1    -1     3
     1    -1    -1
```

If k is any scalar (a positive or negative number), and not $[k]$ which is a 1×1 matrix, then multiplication of a matrix A by the scalar k is the multiplication of every element of A by k.

Example C.2

Multiply the matrix

$$A = \begin{bmatrix} 1 & -2 \\ 2 & 3 \end{bmatrix}$$

by

a. $k_1 = 5$

b. $k_2 = -3 + j2$

Solution:

a.

$$k_1 \cdot A = 5 \times \begin{bmatrix} 1 & -2 \\ 2 & 3 \end{bmatrix} = \begin{bmatrix} 5 \times 1 & 5 \times (-2) \\ 5 \times 2 & 5 \times 3 \end{bmatrix} = \begin{bmatrix} 5 & -10 \\ 10 & 15 \end{bmatrix}$$

b.

$$k_2 \cdot A = (-3 + j2) \times \begin{bmatrix} 1 & -2 \\ 2 & 3 \end{bmatrix} = \begin{bmatrix} (-3+j2) \times 1 & (-3+j2) \times (-2) \\ (-3+j2) \times 2 & (-3+j2) \times 3 \end{bmatrix} = \begin{bmatrix} -3+j2 & 6-j4 \\ -6+j4 & -9+j6 \end{bmatrix}$$

Check with MATLAB:

```
k1=5; k2=(-3 + 2*j);      % Define scalars k₁ and k₂
A=[1 -2; 2 3];            % Define matrix A
k1*A                      % Multiply matrix A by constant k₁
ans =
     5    -10
```

```
    10      15
```

k2*A   %Multiply matrix A by constant k$_2$

```
ans =
   -3.0000+ 2.0000i    6.0000-  4.0000i
   -6.0000+ 4.0000i   -9.0000+  6.0000i
```

Two matrices A and B are said to be *conformable for multiplication* $A \cdot B$ in that order, only when the number of columns of matrix A is equal to the number of rows of matrix B. That is, the product $A \cdot B$ (but not $B \cdot A$) is conformable for multiplication only if A is an $m \times p$ matrix and matrix B is an $p \times n$ matrix. The product $A \cdot B$ will then be an $m \times n$ matrix. A convenient way to determine if two matrices are conformable for multiplication is to write the dimensions of the two matrices side-by-side as shown below.

Shows that A and B are conformable for multiplication

$$A \quad\quad B$$
$$m \times p \quad p \times n$$

Indicates the dimension of the product $A \cdot B$

For the product $B \cdot A$ we have:

Here, B and A are not conformable for multiplication

$$B \quad\quad A$$
$$p \times n \quad m \times p$$

For matrix multiplication, the operation is row by column. Thus, to obtain the product $A \cdot B$, we multiply each element of a row of A by the corresponding element of a column of B; then, we add these products.

Example C.3

Matrices C and D are defined as

$$C = \begin{bmatrix} 2 & 3 & 4 \end{bmatrix} \text{ and } D = \begin{bmatrix} 1 \\ -1 \\ 2 \end{bmatrix}$$

Compute the products $C \cdot D$ and $D \cdot C$

Solution:

The dimensions of matrices C and D are respectively 1×3 3×1; therefore the product $C \cdot D$ is feasible, and will result in a 1×1, that is,

$$C \cdot D = \begin{bmatrix} 2 & 3 & 4 \end{bmatrix} \begin{bmatrix} 1 \\ -1 \\ 2 \end{bmatrix} = \begin{bmatrix} (2) \cdot (1) + (3) \cdot (-1) + (4) \cdot (2) \end{bmatrix} = \begin{bmatrix} 7 \end{bmatrix}$$

The dimensions for D and C are respectively 3×1 1×3 and therefore, the product $D \cdot C$ is also feasible. Multiplication of these will produce a 3×3 matrix as follows:

$$D \cdot C = \begin{bmatrix} 1 \\ -1 \\ 2 \end{bmatrix} \begin{bmatrix} 2 & 3 & 4 \end{bmatrix} = \begin{bmatrix} (1) \cdot (2) & (1) \cdot (3) & (1) \cdot (4) \\ (-1) \cdot (2) & (-1) \cdot (3) & (-1) \cdot (4) \\ (2) \cdot (2) & (2) \cdot (3) & (2) \cdot (4) \end{bmatrix} = \begin{bmatrix} 2 & 3 & 4 \\ -2 & -3 & -4 \\ 4 & 6 & 8 \end{bmatrix}$$

Check with MATLAB:

```
C=[2 3 4]; D=[1; -1; 2];        % Define matrices C and D
C*D                             % Multiply C by D

ans =
      7

D*C                             % Multiply D by C

ans =
      2      3      4
     -2     -3     -4
      4      6      8
```

Division of one matrix by another, is not defined. However, an equivalent operation exists, and it will become apparent later in this chapter, when we discuss the inverse of a matrix.

C.3 Special Forms of Matrices

† A square matrix is said to be *upper triangular* when all the elements below the diagonal are zero. The matrix A of (C.4) is an upper triangular matrix.

In an upper triangular matrix, not all elements above the diagonal need to be non-zero.

$$A = \begin{bmatrix} a_{11} & a_{12} & a_{13} & \dots & a_{1n} \\ 0 & a_{22} & a_{23} & \dots & a_{2n} \\ 0 & 0 & \ddots & \dots & \dots \\ \dots & \dots & 0 & \ddots & \dots \\ 0 & 0 & 0 & \dots & a_{mn} \end{bmatrix} \tag{C.4}$$

† A square matrix is said to be *lower triangular*, when all the elements above the diagonal are zero. The matrix B of (C.5) is a lower triangular matrix.

$$B = \begin{bmatrix} a_{11} & 0 & 0 & \dots & 0 \\ a_{21} & a_{22} & 0 & \dots & 0 \\ \dots & \dots & \ddots & 0 & 0 \\ \dots & \dots & \dots & \ddots & 0 \\ a_{m1} & a_{m2} & a_{m3} & \dots & a_{mn} \end{bmatrix} \tag{C.5}$$

In a lower triangular matrix, not all elements below the diagonal need to be non-zero.

† A square matrix is said to be *diagonal*, if all elements are zero, except those in the diagonal. The matrix C of (C.6) is a diagonal matrix.

$$C = \begin{bmatrix} a_{11} & 0 & 0 & \dots & 0 \\ 0 & a_{22} & 0 & \dots & 0 \\ 0 & 0 & \ddots & 0 & 0 \\ 0 & 0 & 0 & \ddots & 0 \\ 0 & 0 & 0 & \dots & a_{mn} \end{bmatrix} \tag{C.6}$$

† A diagonal matrix is called a *scalar matrix*, if $a_{11} = a_{22} = a_{33} = \dots = a_{nn} = k$ where k is a scalar. The matrix D of (C.7) is a scalar matrix with $k = 4$.

$$D = \begin{bmatrix} 4 & 0 & 0 & 0 \\ 0 & 4 & 0 & 0 \\ 0 & 0 & 4 & 0 \\ 0 & 0 & 0 & 4 \end{bmatrix} \tag{C.7}$$

A scalar matrix with $k = 1$, is called an *identity matrix* I. Shown below are 2×2, 3×3, and 4×4 identity matrices.

$$\begin{bmatrix} 1 & 0 \\ 0 & 1 \end{bmatrix} \qquad \begin{bmatrix} 1 & 0 & 0 \\ 0 & 1 & 0 \\ 0 & 0 & 1 \end{bmatrix} \qquad \begin{bmatrix} 1 & 0 & 0 & 0 \\ 0 & 1 & 0 & 0 \\ 0 & 0 & 1 & 0 \\ 0 & 0 & 0 & 1 \end{bmatrix} \qquad \text{(C.8)}$$

The MATLAB **eye(n)** function displays an $n \times n$ identity matrix. For example,

eye(4)   % Display a 4 by 4 identity matrix

ans =

```
    1        0        0        0
    0        1        0        0
    0        0        1        0
    0        0        0        1
```

Likewise, the **eye(size(A))** function, produces an identity matrix whose size is the same as matrix A. For example, let matrix A be defined as

A=[1 3 1; –2 1 –5; 4 –7 6]   % Define matrix A

A =

```
    1        3        1
   -2        1       -5
    4       -7        6
```

then,

eye(size(A))

displays

ans =

```
    1        0        0
    0        1        0
    0        0        1
```

† The *transpose of a matrix* A, denoted as A^T, is the matrix that is obtained when the rows and columns of matrix A are interchanged. For example, if

$$A = \begin{bmatrix} 1 & 2 & 3 \\ 4 & 5 & 6 \end{bmatrix} \text{ then } A^T = \begin{bmatrix} 1 & 4 \\ 2 & 5 \\ 3 & 6 \end{bmatrix} \qquad \text{(C.9)}$$

In MATLAB we use the apostrophe (') symbol to denote and obtain the transpose of a matrix. Thus, for the above example,

A=[1 2 3; 4 5 6]   % Define matrix A

```
A  =
       1        2        3
       4        5        6
```

A'   % Display the transpose of A

```
ans  =
       1        4
       2        5
       3        6
```

† A *symmetric matrix* A is a matrix such that $A^T = A$, that is, the transpose of a matrix A is the same as A. An example of a symmetric matrix is shown below.

$$A = \begin{bmatrix} 1 & 2 & 3 \\ 2 & 4 & -5 \\ 3 & -5 & 6 \end{bmatrix} \qquad A^T = \begin{bmatrix} 1 & 2 & 3 \\ 2 & 4 & -5 \\ 3 & -5 & 6 \end{bmatrix} = A \tag{C.10}$$

† If a matrix A has complex numbers as elements, the matrix obtained from A by replacing each element by its conjugate, is called the *conjugate of A*, and it is denoted as $A*$

An example is shown below.

$$A = \begin{bmatrix} 1+j2 & j \\ 3 & 2-j3 \end{bmatrix} \qquad A* = \begin{bmatrix} 1-j2 & -j \\ 3 & 2+j3 \end{bmatrix}$$

MATLAB has two built-in functions which compute the complex conjugate of a number. The first, **conj(x)**, computes the complex conjugate of any complex number, and the second, **conj(A)**, computes the conjugate of a matrix A. Using MATLAB with the matrix A defined as above, we get

A = [1+2j j; 3 2–3j]   % Define and display matrix A

```
A  =
     1.0000+  2.0000i            0+  1.0000i
     3.0000               2.0000-  3.0000i
```

conj_A=conj(A)   % Compute and display the conjugate of A

```
conj_A  =
```

```
1.0000- 2.0000i        0- 1.0000i
3.0000            2.0000+ 3.0000i
```

† A square matrix A such that $A^T = -A$ is called *skew-symmetric*. For example,

$$A = \begin{bmatrix} 0 & 2 & -3 \\ -2 & 0 & -4 \\ 3 & 4 & 0 \end{bmatrix} \quad A^T = \begin{bmatrix} 0 & -2 & 3 \\ 2 & 0 & 4 \\ -3 & -4 & 0 \end{bmatrix} = -A$$

Therefore, matrix A above is skew symmetric.

† A square matrix A such that $A^{T*} = A$ is called *Hermitian*. For example,

$$A = \begin{bmatrix} 1 & 1-j & 2 \\ 1+j & 3 & j \\ 2 & -j & 0 \end{bmatrix} \quad A^T = \begin{bmatrix} 1 & 1+j & 2 \\ 1-j & 3 & -j \\ 2 & j & 0 \end{bmatrix} \quad A^{T*} = \begin{bmatrix} 1 & 1+j & 2 \\ 1-j & 3 & -j \\ 2 & j & 0 \end{bmatrix} = A$$

Therefore, matrix A above is Hermitian.

† A square matrix A such that $A^{T*} = -A$ is called *skew–Hermitian*. For example,

$$A = \begin{bmatrix} j & 1-j & 2 \\ -1-j & 3j & j \\ -2 & j & 0 \end{bmatrix} \quad A^T = \begin{bmatrix} j & -1-j & -2 \\ 1-j & 3j & j \\ 2 & j & 0 \end{bmatrix} \quad A^{T*} = \begin{bmatrix} -j & -1+j & -2 \\ 1+j & -3j & -j \\ 2 & -j & 0 \end{bmatrix} = -A$$

Therefore, matrix A above is skew-Hermitian.

C.4 Determinants

Let matrix A be defined as the square matrix

$$A = \begin{bmatrix} a_{11} & a_{12} & a_{13} & \cdots & a_{1n} \\ a_{21} & a_{22} & a_{23} & \cdots & a_{2n} \\ a_{31} & a_{32} & a_{33} & \cdots & a_{3n} \\ \cdots & \cdots & \cdots & \cdots & \cdots \\ a_{n1} & a_{n2} & a_{n3} & \cdots & a_{nn} \end{bmatrix} \tag{C.11}$$

then, the *determinant of A*, denoted as $det A$, is defined as

$$detA = a_{11}a_{22}a_{33}\ldots a_{nn} + a_{12}a_{23}a_{34}\ldots a_{n1} + a_{13}a_{24}a_{35}\ldots a_{n2} + \ldots \tag{C.12}$$
$$-a_{n1}\ldots a_{22}a_{13}\ldots -a_{n2}\ldots a_{23}a_{14} - a_{n3}\ldots a_{24}a_{15} - \ldots$$

The determinant of a square matrix of order n is referred to as *determinant of order n*.

Let A be a *determinant of order 2*, that is,

$$A = \begin{bmatrix} a_{11} & a_{12} \\ a_{21} & a_{22} \end{bmatrix} \tag{C.13}$$

Then,

$$detA = a_{11}a_{22} - a_{21}a_{12} \tag{C.14}$$

Example C.4

Matrices A and B are defined as

$$A = \begin{bmatrix} 1 & 2 \\ 3 & 4 \end{bmatrix} \text{ and } B = \begin{bmatrix} 2 & -1 \\ 2 & 0 \end{bmatrix}$$

Compute $detA$ and $detB$.

Solution:

$$detA = 1 \cdot 4 - 3 \cdot 2 = 4 - 6 = -2$$
$$detB = 2 \cdot 0 - 2 \cdot (-1) = 0 - (-2) = 2$$

Check with MATLAB:

```
A=[1 2; 3 4]; B=[2 −1; 2 0];      % Define matrices A and B
det(A)                            % Compute the determinant of A

ans =
     -2

det(B)                            % Compute the determinant of B

ans =
      2
```

Let A be a matrix of order 3, that is,

$$A = \begin{bmatrix} a_{11} & a_{12} & a_{13} \\ a_{21} & a_{22} & a_{23} \\ a_{31} & a_{32} & a_{33} \end{bmatrix} \tag{C.15}$$

then, $detA$ is found from

$$detA = a_{11}a_{22}a_{33} + a_{12}a_{23}a_{31} + a_{11}a_{22}a_{33}$$
$$-a_{11}a_{22}a_{33} - a_{11}a_{22}a_{33} - a_{11}a_{22}a_{33}$$

(C.16)

A convenient method to evaluate the determinant of order *3*, is to write the first two columns to the right of the *3 × 3* matrix, and add the products formed by the diagonals from upper left to lower right; then subtract the products formed by the diagonals from lower left to upper right as shown on the diagram of the next page. When this is done properly, we obtain (C.16) above.

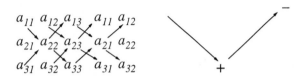

This method works only with second and third order determinants. To evaluate higher order determinants, we must first compute the *cofactors*; these will be defined shortly.

Example C.5

Compute $detA$ and $detB$ if matrices A and B are defined as

$$A = \begin{bmatrix} 2 & 3 & 5 \\ 1 & 0 & 1 \\ 2 & 1 & 0 \end{bmatrix} \text{ and } B = \begin{bmatrix} 2 & -3 & -4 \\ 1 & 0 & -2 \\ 0 & -5 & -6 \end{bmatrix}$$

Solution:

$$detA = \begin{array}{ccccc} 2 & 3 & 5 & 2 & 3 \\ 1 & 0 & 1 & 1 & 0 \\ 2 & 1 & 0 & 2 & 1 \end{array}$$

or

$$detA = (2 \times 0 \times 0) + (3 \times 1 \times 1) + (5 \times 1 \times 1)$$
$$- (2 \times 0 \times 5) - (1 \times 1 \times 2) - (0 \times 1 \times 3) = 11 - 2 = 9$$

Likewise,

$$detB = \begin{array}{ccccc} 2 & -3 & -4 & 2 & -3 \\ 1 & 0 & -2 & 1 & -2 \\ 0 & -5 & -6 & 2 & -6 \end{array}$$

or

$$detB = [2 \times 0 \times (-6)] + [(-3) \times (-2) \times 0] + [(-4) \times 1 \times (-5)]$$
$$- [0 \times 0 \times (-4)] - [(-5) \times (-2) \times 2] - [(-6) \times 1 \times (-3)] = 20 - 38 = -18$$

Check with MATLAB:

A=[2 3 5; 1 0 1; 2 1 0]; det(A)   % Define matrix A and compute detA

ans =

    9

B=[2 −3 −4; 1 0 −2; 0 −5 −6];det(B)   % Define matrix B and compute detB

ans =

    −18

C.5 Minors and Cofactors

Let matrix A be defined as the square matrix of order n as shown below.

$$A = \begin{bmatrix} a_{11} & a_{12} & a_{13} & \cdots & a_{1n} \\ a_{21} & a_{22} & a_{23} & \cdots & a_{2n} \\ a_{31} & a_{32} & a_{33} & \cdots & a_{3n} \\ \cdots & \cdots & \cdots & \cdots & \cdots \\ a_{n1} & a_{n2} & a_{n3} & \cdots & a_{nn} \end{bmatrix} \tag{C.17}$$

If we remove the elements of its *ith* row, and *jth* column, the remaining $n-1$ square matrix is called the *minor of A*, and it is denoted as $\left[M_{ij}\right]$.

The signed minor $(-1)^{i+j}\left[M_{ij}\right]$ is called the *cofactor* of a_{ij} and it is denoted as α_{ij}.

Example C.6

Matrix A is defined as

$$A = \begin{bmatrix} a_{11} & a_{12} & a_{13} \\ a_{21} & a_{22} & a_{23} \\ a_{31} & a_{32} & a_{33} \end{bmatrix} \tag{C.18}$$

Compute the minors $\left[M_{11}\right]$, $\left[M_{12}\right]$, $\left[M_{13}\right]$ and the cofactors α_{11}, α_{12} and α_{13}.

Solution:

$$[M_{11}] = \begin{bmatrix} a_{22} & a_{23} \\ a_{32} & a_{33} \end{bmatrix} \qquad [M_{12}] = \begin{bmatrix} a_{21} & a_{23} \\ a_{31} & a_{33} \end{bmatrix} \qquad [M_{11}] = \begin{bmatrix} a_{21} & a_{22} \\ a_{31} & a_{32} \end{bmatrix}$$

and

$$\alpha_{11} = (-1)^{1+1}[M_{11}] = [M_{11}] \qquad \alpha_{12} = (-1)^{1+2}[M_{12}] = -[M_{12}] \qquad \alpha_{13} = [M_{13}] = (-1)^{1+3}[M_{13}]$$

The remaining minors

$$[M_{21}], \quad [M_{22}], \quad [M_{23}], \quad [M_{31}], \quad [M_{32}], \quad [M_{33}]$$

and cofactors

$$\alpha_{21}, \alpha_{22}, \alpha_{23}, \alpha_{31}, \alpha_{32}, \text{ and } \alpha_{33}$$

are defined similarly.

Example C.7

Compute the cofactors of matrix A defined as

$$A = \begin{bmatrix} 1 & 2 & -3 \\ 2 & -4 & 2 \\ -1 & 2 & -6 \end{bmatrix} \tag{C.19}$$

Solution:

$$\alpha_{11} = (-1)^{1+1}\begin{bmatrix} -4 & 2 \\ 2 & -6 \end{bmatrix} = 20 \qquad \alpha_{12} = (-1)^{1+2}\begin{bmatrix} 2 & 2 \\ -1 & -6 \end{bmatrix} = 10 \tag{C.20}$$

$$\alpha_{13} = (-1)^{1+3}\begin{bmatrix} 2 & -4 \\ -1 & 2 \end{bmatrix} = 0 \qquad \alpha_{21} = (-1)^{2+1}\begin{bmatrix} 2 & -3 \\ 2 & -6 \end{bmatrix} = 6 \tag{C.21}$$

$$\alpha_{22} = (-1)^{2+2}\begin{bmatrix} 1 & -3 \\ -1 & -6 \end{bmatrix} = -9 \qquad \alpha_{23} = (-1)^{2+3}\begin{bmatrix} 1 & 2 \\ -1 & 2 \end{bmatrix} = -4 \tag{C.22}$$

$$\alpha_{31} = (-1)^{3+1}\begin{bmatrix} 2 & -3 \\ -4 & 2 \end{bmatrix} = -8, \qquad \alpha_{32} = (-1)^{3+2}\begin{bmatrix} 1 & -3 \\ 2 & 2 \end{bmatrix} = -8 \tag{C.23}$$

$$\alpha_{33} = (-1)^{3+3}\begin{bmatrix} 1 & 2 \\ 2 & -4 \end{bmatrix} = -8 \tag{C.24}$$

It is useful to remember that the signs of the cofactors follow the pattern

$$
\begin{matrix}
+ & - & + & - & + \\
- & + & - & + & - \\
+ & - & + & - & + \\
- & + & - & + & - \\
+ & - & + & - & +
\end{matrix}
$$

that is, the cofactors on the diagonals have the same sign as their minors.

Let A be a square matrix of any size; the value of the determinant of A is the sum of the products obtained by multiplying each element of *any* row or *any* column by its cofactor.

Example C.8

Matrix A is defined as

$$
A = \begin{bmatrix} 1 & 2 & -3 \\ 2 & -4 & 2 \\ -1 & 2 & -6 \end{bmatrix}
\tag{C.25}
$$

Compute the determinant of A using the elements of the first row.

Solution:

$$
detA = 1\begin{vmatrix} -4 & 2 \\ 2 & -6 \end{vmatrix} - 2\begin{vmatrix} 2 & 2 \\ -1 & -6 \end{vmatrix} - 3\begin{vmatrix} 2 & -4 \\ -1 & 2 \end{vmatrix} = 1 \times 20 - 2 \times (-10) - 3 \times 0 = 40
$$

Check with MATLAB:

A=[1 2 –3; 2 –4 2; –1 2 –6];det(A)   % Define matrix A and compute detA

ans =
    40

We must use the above procedure to find the determinant of a matrix A of order 4 or higher. Thus, a fourth-order determinant can first be expressed as the sum of the products of the elements of its first row by its cofactor as shown below.

$$A = \begin{bmatrix} a_{11} & a_{12} & a_{13} & a_{14} \\ a_{21} & a_{22} & a_{23} & a_{24} \\ a_{31} & a_{32} & a_{33} & a_{34} \\ a_{41} & a_{42} & a_{43} & a_{44} \end{bmatrix} = a_{11}\begin{vmatrix} a_{22} & a_{23} & a_{24} \\ a_{32} & a_{33} & a_{34} \\ a_{42} & a_{43} & a_{44} \end{vmatrix} - a_{21}\begin{vmatrix} a_{12} & a_{13} & a_{14} \\ a_{32} & a_{33} & a_{34} \\ a_{42} & a_{43} & a_{44} \end{vmatrix}$$

(C.26)

$$+ a_{31}\begin{vmatrix} a_{12} & a_{13} & a_{14} \\ a_{22} & a_{23} & a_{24} \\ a_{42} & a_{43} & a_{44} \end{vmatrix} - a_{41}\begin{vmatrix} a_{12} & a_{13} & a_{14} \\ a_{22} & a_{23} & a_{24} \\ a_{32} & a_{33} & a_{34} \end{vmatrix}$$

Determinants of order five or higher can be evaluated similarly.

Example C.9

Compute the value of the determinant of the matrix A defined as

$$A = \begin{bmatrix} 2 & -1 & 0 & -3 \\ -1 & 1 & 0 & -1 \\ 4 & 0 & 3 & -2 \\ -3 & 0 & 0 & 1 \end{bmatrix}$$

(C.27)

Solution:

Using the above procedure, we will multiply each element of the first column by its cofactor. Then,

$$A = \underbrace{2\begin{bmatrix} 1 & 0 & -1 \\ 0 & 3 & -2 \\ 0 & 0 & 1 \end{bmatrix}}_{[a]} \underbrace{-(-1)\begin{bmatrix} -1 & 0 & -3 \\ 0 & 3 & -2 \\ 0 & 0 & 1 \end{bmatrix}}_{[b]} \underbrace{+4\begin{bmatrix} -1 & 0 & -3 \\ 1 & 0 & -1 \\ 0 & 0 & 1 \end{bmatrix}}_{[c]} \underbrace{-(-3)\begin{bmatrix} -1 & 0 & -3 \\ 1 & 0 & -1 \\ 0 & 3 & -2 \end{bmatrix}}_{[d]}$$

Next, using the procedure of Example C.5 or Example C.8, we find

$$[a] = 6, \ [b] = -3, \ [c] = 0, \ [d] = -36$$

and thus

$$detA = [a] + [b] + [c] + [d] = 6 - 3 + 0 - 36 = -33$$

We can verify our answer with MATLAB as follows:

```
A=[2 −1 0 −3; −1 1 0 −1; 4 0 3 −2; −3 0 0 1]; delta = det(A)

delta =
    -33
```

Some useful properties of determinants are given below.

Property 1: *If all elements of one row or one column are zero, the determinant is zero.* An example of this is the determinant of the cofactor $[c]$ above.

Property 2: *If all the elements of one row or column are m times the corresponding elements of another row or column, the determinant is zero.* For example, if

$$A = \begin{bmatrix} 2 & 4 & 1 \\ 3 & 6 & 1 \\ 1 & 2 & 1 \end{bmatrix} \tag{C.28}$$

then,

$$detA = \begin{vmatrix} 2 & 4 & 1 \\ 3 & 6 & 1 \\ 1 & 2 & 1 \end{vmatrix} \begin{matrix} 2 & 4 \\ 3 & 6 \\ 1 & 2 \end{matrix} = 12 + 4 + 6 - 6 - 4 - 12 = 0 \tag{C.29}$$

Here, $detA$ is zero because the second column in A is 2 times the first column.

Check with MATLAB:

A=[2 4 1; 3 6 1; 1 2 1];det(A)

ans =
       0

Property 3: *If two rows or two columns of a matrix are identical, the determinant is zero.* This follows from Property 2 with $m = 1$.

C.6 Cramer's Rule

Let us consider the systems of the three equations below

$$\begin{aligned} a_{11}x + a_{12}y + a_{13}z &= A \\ a_{21}x + a_{22}y + a_{23}z &= B \\ a_{31}x + a_{32}y + a_{33}z &= C \end{aligned} \tag{C.30}$$

and let

$$\Delta = \begin{vmatrix} a_{11} & a_{12} & a_{13} \\ a_{21} & a_{22} & a_{23} \\ a_{31} & a_{32} & a_{33} \end{vmatrix} \quad D_1 = \begin{vmatrix} A & a_{11} & a_{13} \\ B & a_{21} & a_{23} \\ C & a_{31} & a_{33} \end{vmatrix} \quad D_2 = \begin{vmatrix} a_{11} & A & a_{13} \\ a_{21} & B & a_{23} \\ a_{31} & C & a_{33} \end{vmatrix} \quad D_3 = \begin{vmatrix} a_{11} & a_{12} & A \\ a_{21} & a_{22} & B \\ a_{31} & a_{32} & C \end{vmatrix}$$

Cramer's rule states that the unknowns x, y, and z can be found from the relations

$$x = \frac{D_1}{\Delta} \qquad y = \frac{D_2}{\Delta} \qquad z = \frac{D_3}{\Delta} \tag{C.31}$$

provided that the determinant Δ (delta) is not zero.

We observe that the numerators of (C.31) are determinants that are formed from Δ by the substitution of the known values A, B, and C, for the coefficients of the desired unknown.

Cramer's rule applies to systems of two or more equations.

If (C.30) is a homogeneous set of equations, that is, if $A = B = C = 0$, then, D_1, D_2, and D_3 are all zero as we found in Property 1 above. Then, $x = y = z = 0$ also.

Example C.10

Use Cramer's rule to find v_1, v_2, and v_3 if

$$\begin{aligned}
2v_1 - 5 - v_2 + 3v_3 &= 0 \\
-2v_3 - 3v_2 - 4v_1 &= 8 \\
v_2 + 3v_1 - 4 - v_3 &= 0
\end{aligned} \tag{C.32}$$

and verify your answers with MATLAB.

Solution:

Rearranging the unknowns v, and transferring known values to the right side, we get

$$\begin{aligned}
2v_1 - v_2 + 3v_3 &= 5 \\
-4v_1 - 3v_2 - 2v_3 &= 8 \\
3v_1 + v_2 - v_3 &= 4
\end{aligned} \tag{C.33}$$

Now, by Cramer's rule,

$$\Delta = \begin{vmatrix} 2 & -1 & 3 \\ -4 & -3 & -2 \\ 3 & 1 & -1 \end{vmatrix} \begin{matrix} 2 & -1 \\ -4 & -3 \\ 3 & 1 \end{matrix} = 6 + 6 - 12 + 27 + 4 + 4 = 35$$

$$D_1 = \begin{vmatrix} 5 & -1 & 3 \\ 8 & -3 & -2 \\ 4 & 1 & -1 \end{vmatrix} \begin{matrix} 5 & -1 \\ 8 & -3 \\ 4 & 1 \end{matrix} = 15 + 8 + 24 + 36 + 10 - 8 = 85$$

$$D_2 = \begin{vmatrix} 2 & 5 & 3 \\ -4 & 8 & -2 \\ 3 & 4 & -1 \end{vmatrix} \begin{matrix} 2 & 5 \\ -4 & 8 \\ 3 & 4 \end{matrix} = -16 - 30 - 48 - 72 + 16 - 20 = -170$$

$$D_3 = \begin{vmatrix} 2 & -1 & 5 \\ -4 & -3 & 8 \\ 3 & 1 & 4 \end{vmatrix} \begin{matrix} 2 & -1 \\ -4 & -3 \\ 3 & 1 \end{matrix} = -24 - 24 - 20 + 45 - 16 - 16 = -55$$

Then, using (C.31) we get

$$x_1 = \frac{D_1}{\Delta} = \frac{85}{35} = \frac{17}{7} \qquad x_2 = \frac{D_2}{\Delta} = -\frac{170}{35} = -\frac{34}{7} \qquad x_3 = \frac{D_3}{\Delta} = -\frac{55}{35} = -\frac{11}{7} \qquad \text{(C.34)}$$

We will verify with MATLAB as follows.

```
% The following code will compute and display the values of v1, v2 and v3.
format rat                          % Express answers in ratio form
B=[2 –1 3; –4 –3 –2; 3 1 –1];       % The elements of the determinant D of matrix B
delta=det(B);                       % Compute the determinant D of matrix B
d1=[5 -1 3; 8 -3 -2; 4 1 -1];       % The elements of D1
detd1=det(d1);                      % Compute the determinant of D1
d2=[2 5 3; -4 8 -2; 3 4 -1];        % The elements of D2
detd2=det(d2);                      % Compute the determinant of D2
d3=[2 -1 5; -4 -3 8; 3 1 4];        % The elements of D3
detd3=det(d3);                      % Compute he determinant of D3
v1=detd1/delta;                     % Compute the value of v1
v2=detd2/delta;                     % Compute the value of v2
v3=detd3/delta;                     % Compute the value of v3
%
disp('v1=');disp(v1);               % Display the value of v1
disp('v2=');disp(v2);               % Display the value of v2
disp('v3=');disp(v3);               % Display the value of v3

v1=
    17/7
v2=
    -34/7
v3=
    -11/7
```

These are the same values as in (C.34)

C.7 Gaussian Elimination Method

We can find the unknowns in a system of two or more equations also by the *Gaussian elimination method*. With this method, the objective is to eliminate one unknown at a time. This can be done by multiplying the terms of any of the equations of the system by a number such that we can add (or subtract) this equation to another equation in the system so that one of the unknowns will be eliminated. Then, by substitution to another equation with two unknowns, we can find the second unknown. Subsequently, substitution of the two values found can be made into an equation with three unknowns from which we can find the value of the third unknown. This procedure is repeated until all unknowns are found. This method is best illustrated with the following example which consists of the same equations as the previous example.

Example C.11

Use the Gaussian elimination method to find v_1, v_2, and v_3 of the system of equations

$$
\begin{align}
2v_1 - v_2 + 3v_3 &= 5 \\
-4v_1 - 3v_2 - 2v_3 &= 8 \\
3v_1 + v_2 - v_3 &= 4
\end{align}
\tag{C.35}
$$

Solution:

As a first step, we add the first equation of (C.35) with the third to eliminate the unknown v_2 and we obtain the following equation.

$$
5v_1 + 2v_3 = 9 \tag{C.36}
$$

Next, we multiply the third equation of (C.35) by 3, and we add it with the second to eliminate v_2. Then, we obtain the following equation.

$$
5v_1 - 5v_3 = 20 \tag{C.37}
$$

Subtraction of (C.37) from (C.36) yields

$$
7v_3 = -11 \ \text{ or } \ v_3 = -\frac{11}{7} \tag{C.38}
$$

Now, we can find the unknown v_1 from either (C.36) or (C.37). By substitution of (C.38) into (C.36) we get

$$
5v_1 + 2 \cdot \left(-\frac{11}{7}\right) = 9 \ \text{ or } \ v_1 = \frac{17}{7} \tag{C.39}
$$

Finally, we can find the last unknown v_2 from any of the three equations of (C.35). By substitution into the first equation we get

$$v_2 = 2v_1 + 3v_3 - 5 = \frac{34}{7} - \frac{33}{7} - \frac{35}{7} = -\frac{34}{7} \tag{C.40}$$

These are the same values as those we found in Example C.10.

The Gaussian elimination method works well if the coefficients of the unknowns are small integers, as in Example C.11. However, it becomes impractical if the coefficients are large or fractional numbers.

C.8 The Adjoint of a Matrix

Let us assume that A is an n square matrix and α_{ij} is the cofactor of a_{ij}. Then *the adjoint of A*, denoted as *adjA*, is defined as the n square matrix below.

$$adjA = \begin{bmatrix} \alpha_{11} & \alpha_{21} & \alpha_{31} & \cdots & \alpha_{n1} \\ \alpha_{12} & \alpha_{22} & \alpha_{32} & \cdots & \alpha_{n2} \\ \alpha_{13} & \alpha_{23} & \alpha_{33} & \cdots & \alpha_{n3} \\ \cdots & \cdots & \cdots & \cdots & \cdots \\ \alpha_{1n} & \alpha_{2n} & \alpha_{3n} & \cdots & \alpha_{nn} \end{bmatrix} \tag{C.41}$$

We observe that the cofactors of the elements of the ith row (column) of A are the elements of the ith column (row) of *adjA*.

Example C.12

Compute *adjA* if Matrix A is defined as

$$A = \begin{bmatrix} 1 & 2 & 3 \\ 1 & 3 & 4 \\ 1 & 4 & 3 \end{bmatrix} \tag{C.42}$$

Solution:

$$adjA = \begin{bmatrix} \begin{vmatrix} 3 & 4 \\ 4 & 3 \end{vmatrix} & -\begin{vmatrix} 2 & 3 \\ 4 & 3 \end{vmatrix} & \begin{vmatrix} 2 & 3 \\ 3 & 4 \end{vmatrix} \\ -\begin{vmatrix} 1 & 4 \\ 1 & 3 \end{vmatrix} & \begin{vmatrix} 1 & 3 \\ 1 & 3 \end{vmatrix} & -\begin{vmatrix} 2 & 3 \\ 3 & 4 \end{vmatrix} \\ \begin{vmatrix} 1 & 3 \\ 1 & 4 \end{vmatrix} & -\begin{vmatrix} 1 & 2 \\ 1 & 4 \end{vmatrix} & \begin{vmatrix} 1 & 2 \\ 1 & 3 \end{vmatrix} \end{bmatrix} = \begin{bmatrix} -7 & 6 & -1 \\ 1 & 0 & -1 \\ 1 & -2 & 1 \end{bmatrix}$$

C.9 Singular and Non-Singular Matrices

An n square matrix A is called *singular* if $detA = 0$; if $detA \neq 0$, A is called *non-singular*.

Example C.13

Matrix A is defined as

$$A = \begin{bmatrix} 1 & 2 & 3 \\ 2 & 3 & 4 \\ 3 & 5 & 7 \end{bmatrix} \qquad (C.43)$$

Determine whether this matrix is singular or non-singular.

Solution:

$$detA = \begin{vmatrix} 1 & 2 & 3 \\ 2 & 3 & 4 \\ 3 & 5 & 7 \end{vmatrix} \begin{matrix} 1 & 2 \\ 2 & 3 \\ 3 & 5 \end{matrix} = 21 + 24 + 30 - 27 - 20 - 28 = 0$$

Therefore, matrix A is singular.

C.10 The Inverse of a Matrix

If A and B are n square matrices such that $AB = BA = I$, where I is the identity matrix, B is called the *inverse* of A, denoted as $B = A^{-1}$, and likewise, A is called the *inverse* of B, that is, $A = B^{-1}$

If a matrix A is non-singular, we can compute its inverse A^{-1} from the relation

$$\boxed{A^{-1} = \frac{1}{detA} adjA} \qquad (C.44)$$

Example C.14

Matrix A is defined as

$$A = \begin{bmatrix} 1 & 2 & 3 \\ 1 & 3 & 4 \\ 1 & 4 & 3 \end{bmatrix} \qquad (C.45)$$

Compute its inverse, that is, find A^{-1}

Solution:

Here, $detA = 9 + 8 + 12 - 9 - 16 - 6 = -2$, and since this is a non-zero value, it is possible to compute the inverse of A using (C.44).

From Example C.12,

$$adjA = \begin{bmatrix} -7 & 6 & -1 \\ 1 & 0 & -1 \\ 1 & -2 & 1 \end{bmatrix}$$

Then,

$$A^{-1} = \frac{1}{detA} adjA = \frac{1}{-2} \begin{bmatrix} -7 & 6 & -1 \\ 1 & 0 & -1 \\ 1 & -2 & 1 \end{bmatrix} = \begin{bmatrix} 3.5 & -3 & 0.5 \\ -0.5 & 0 & 0.5 \\ -0.5 & 1 & -0.5 \end{bmatrix} \quad \text{(C.46)}$$

Check with MATLAB:

```
A=[1  2  3;  1  3  4;  1  4  3],  invA=inv(A)     % Define matrix A and compute its inverse
A =
      1       2       3
      1       3       4
      1       4       3
invA =
     3.5000    -3.0000     0.5000
    -0.5000          0     0.5000
    -0.5000     1.0000    -0.5000
```

Multiplication of a matrix A by its inverse A^{-1} produces the identity matrix I, that is,

$$AA^{-1} = I \quad or \quad A^{-1}A = I \quad \text{(C.47)}$$

Example C.15

Prove the validity of (C.47) for the Matrix A defined as

$$A = \begin{bmatrix} 4 & 3 \\ 2 & 2 \end{bmatrix}$$

Proof:

$$detA = 8 - 6 = 2 \quad and \quad adjA = \begin{bmatrix} 2 & -3 \\ -2 & 4 \end{bmatrix}$$

Then,

$$A^{-1} = \frac{1}{detA}adjA = \frac{1}{2}\begin{bmatrix} 2 & -3 \\ -2 & 4 \end{bmatrix} = \begin{bmatrix} 1 & -3/2 \\ -1 & 2 \end{bmatrix}$$

and

$$AA^{-1} = \begin{bmatrix} 4 & 3 \\ 2 & 2 \end{bmatrix}\begin{bmatrix} 1 & -3/2 \\ -1 & 2 \end{bmatrix} = \begin{bmatrix} 4-3 & -6+6 \\ 2-2 & -3+4 \end{bmatrix} = \begin{bmatrix} 1 & 0 \\ 0 & 1 \end{bmatrix} = I$$

C.11 Solution of Simultaneous Equations with Matrices

Consider the relation

$$AX = B \tag{C.48}$$

where A and B are matrices whose elements are known, and X is a matrix (a column vector) whose elements are the unknowns. We assume that A and X are conformable for multiplication. Multiplication of both sides of (C.48) by A^{-1} yields:

$$A^{-1}AX = A^{-1}B = IX = A^{-1}B \tag{C.49}$$

or

$$\boxed{X = A^{-1}B} \tag{C.50}$$

Therefore, we can use (C.50) to solve any set of simultaneous equations that have solutions. We will refer to this method as the *inverse matrix method of solution* of simultaneous equations.

Example C.16

For the system of the equations

$$\begin{cases} 2x_1 + 3x_2 + x_3 = 9 \\ x_1 + 2x_2 + 3x_3 = 6 \\ 3x_1 + x_2 + 2x_3 = 8 \end{cases} \tag{C.51}$$

compute the unknowns $x_1, x_2,$ and x_3 using the inverse matrix method.

Solution:

In matrix form, the given set of equations is $AX = B$ where

$$A = \begin{bmatrix} 2 & 3 & 1 \\ 1 & 2 & 3 \\ 3 & 1 & 2 \end{bmatrix}, \quad X = \begin{bmatrix} x_1 \\ x_2 \\ x_3 \end{bmatrix}, \quad B = \begin{bmatrix} 9 \\ 6 \\ 8 \end{bmatrix} \tag{C.52}$$

Then,

$$X = A^{-1}B \tag{C.53}$$

or

$$\begin{bmatrix} x_1 \\ x_2 \\ x_3 \end{bmatrix} = \begin{bmatrix} 2 & 3 & 1 \\ 1 & 2 & 3 \\ 3 & 1 & 2 \end{bmatrix}^{-1} \begin{bmatrix} 9 \\ 6 \\ 8 \end{bmatrix} \tag{C.54}$$

Next, we find the determinant $detA$, and the adjoint $adjA$

$$detA = 18 \quad and \quad adjA = \begin{bmatrix} 1 & -5 & 7 \\ 7 & 1 & -5 \\ -5 & 7 & 1 \end{bmatrix}$$

Therefore,

$$A^{-1} = \frac{1}{detA} \, adjA = \frac{1}{18} \begin{bmatrix} 1 & -5 & 7 \\ 7 & 1 & -5 \\ -5 & 7 & 1 \end{bmatrix}$$

and by (C.53) we obtain the solution as follows.

$$X = \begin{bmatrix} x_1 \\ x_2 \\ x_3 \end{bmatrix} = \frac{1}{18} \begin{bmatrix} 1 & -5 & 7 \\ 7 & 1 & -5 \\ -5 & 7 & 1 \end{bmatrix} \begin{bmatrix} 9 \\ 6 \\ 8 \end{bmatrix} = \frac{1}{18} \begin{bmatrix} 35 \\ 29 \\ 5 \end{bmatrix} = \begin{bmatrix} 35/18 \\ 29/18 \\ 5/18 \end{bmatrix} = \begin{bmatrix} 1.94 \\ 1.61 \\ 0.28 \end{bmatrix} \tag{C.55}$$

To verify our results, we could use the MATLAB's **inv(A)** function, and then multiply A^{-1} by B. However, it is easier to use the *matrix left division* operation $X = A \backslash B$; this is MATLAB's solution of $A^{-1}B$ for the matrix equation $A \cdot X = B$, where matrix X is the same size as matrix B. For this example,

```
A=[2 3 1; 1 2 3; 3 1 2]; B=[9 6 8]';
X=A \ B
```

X =
    1.9444
    1.6111
    0.2778

Example C.17

For the electric circuit of Figure C.1,

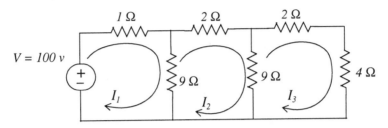

Figure C.1. Circuit for Example C.17

the loop equations are

$$10I_1 - 9I_2 \quad\quad = 100$$
$$-9I_1 + 20I_2 - 9I_3 = \quad 0 \tag{C.56}$$
$$-9I_2 + 15I_3 = \quad 0$$

Use the inverse matrix method to compute the values of the currents I_1, I_2, and I_3

Solution:

For this example, the matrix equation is $RI = V$ or $I = R^{-1}V$, where

$$R = \begin{bmatrix} 10 & -9 & 0 \\ -9 & 20 & -9 \\ 0 & -9 & 15 \end{bmatrix}, \quad V = \begin{bmatrix} 100 \\ 0 \\ 0 \end{bmatrix} \quad and \quad I = \begin{bmatrix} I_1 \\ I_2 \\ I_3 \end{bmatrix}$$

The next step is to find R^{-1}. This is found from the relation

$$R^{-1} = \frac{1}{detR} \, adjR \tag{C.57}$$

Therefore, we find the determinant and the adjoint of R. For this example, we find that

$$detR = 975, \quad adjR = \begin{bmatrix} 219 & 135 & 81 \\ 135 & 150 & 90 \\ 81 & 90 & 119 \end{bmatrix} \tag{C.58}$$

Then,

$$R^{-1} = \frac{1}{detR} adjR = \frac{1}{975} \begin{bmatrix} 219 & 135 & 81 \\ 135 & 150 & 90 \\ 81 & 90 & 119 \end{bmatrix}$$

and

$$I = \begin{bmatrix} I_1 \\ I_2 \\ I_3 \end{bmatrix} = \frac{1}{975} \begin{bmatrix} 219 & 135 & 81 \\ 135 & 150 & 90 \\ 81 & 90 & 119 \end{bmatrix} \begin{bmatrix} 100 \\ 0 \\ 0 \end{bmatrix} = \frac{100}{975} \begin{bmatrix} 219 \\ 135 \\ 81 \end{bmatrix} = \begin{bmatrix} 22.46 \\ 13.85 \\ 8.31 \end{bmatrix}$$

Check with MATLAB:

```
R=[10 -9 0; -9 20 -9; 0 -9 15]; V=[100 0 0]'; I=R\V

I =
   22.4615
   13.8462
    8.3077
```

We can also use subscripts to address the individual elements of the matrix. Accordingly, the above code could also have been written as:

```
R(1,1)=10; R(1,2)=-9;          % No need to make entry for A(1,3) since it is zero.
R(2,1)=-9; R(2,2)=20; R(2,3)=-9; R(3,2)=-9; R(3,3)=15; V=[100 0 0]'; I=R\V

I =
   22.4615
   13.8462
    8.3077
```

Spreadsheets also have the capability of solving simultaneous equations using the inverse matrix method. For instance, we can use Microsoft Excel's MINVERSE (Matrix Inversion) and MMULT (Matrix Multiplication) functions, to obtain the values of the three currents in Example C.17.

The procedure is as follows:

1. We start with a blank spreadsheet and in a block of cells, say B3:D5, we enter the elements of matrix R as shown in Figure C.2. Then, we enter the elements of matrix V in G3:G5.

2. Next, we compute and display the inverse of R, that is, R^{-1}. We choose B7:D9 for the elements of this inverted matrix. We format this block for number display with three decimal places. With this range highlighted and making sure that the cell marker is in B7, we type the formula

=MININVERSE(B3:D5)

and we press the *Crtl-Shift-Enter* keys simultaneously.

We observe that R^{-1} appears in these cells.

3. Now, we choose the block of cells G7:G9 for the values of the current *I*. As before, we highlight them, and with the cell marker positioned in G7, we type the formula

=MMULT(B7:D9,G3:G5)

and we press the Crtl-Shift-Enter keys simultaneously. The values of *I* then appear in G7:G9.

	A	B	C	D	E	F	G	H
1	Spreadsheet for Matrix Inversion and Matrix Multiplication							
2								
3		10	-9	0			100	
4	R=	-9	20	-9		V=	0	
5		0	-9	15			0	
6								
7		0.225	0.138	0.083			22.462	
8	R^{-1}=	0.138	0.154	0.092		I=	13.846	
9		0.083	0.092	0.122			8.3077	
10								

Figure C.2. Solution of Example C.17 with a spreadsheet

Example C.18

For the phasor circuit of Figure C.18

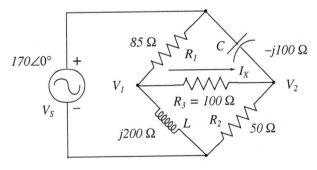

Figure C.3. Circuit for Example C.18

the current I_X can be found from the relation

$$I_X = \frac{V_1 - V_2}{R_3} \tag{C.59}$$

and the voltages V_1 and V_2 can be computed from the nodal equations

$$\frac{V_1 - 170\angle 0°}{85} + \frac{V_1 - V_2}{100} + \frac{V_1 - 0}{j200} = 0 \tag{C.60}$$

and

$$\frac{V_2 - 170\angle 0°}{-j100} + \frac{V_2 - V_1}{100} + \frac{V_2 - 0}{50} = 0 \tag{C.61}$$

Compute, and express the current I_x in both rectangular and polar forms by first simplifying like terms, collecting, and then writing the above relations in matrix form as $YV = I$, where $Y = Admittance$, $V = Voltage$, and $I = Current$

Solution:

The Y matrix elements are the coefficients of V_1 and V_2. Simplifying and rearranging the nodal equations of (C.60) and (C.61), we get

$$(0.0218 - j0.005)V_1 - 0.01V_2 = 2$$
$$-0.01V_1 + (0.03 + j0.01)V_2 = j1.7 \tag{C.62}$$

Next, we write (C.62) in matrix form as

$$\underbrace{\begin{bmatrix} 0.0218 - j0.005 & -0.01 \\ -0.01 & 0.03 + j0.01 \end{bmatrix}}_{Y} \underbrace{\begin{bmatrix} V_1 \\ V_2 \end{bmatrix}}_{V} = \underbrace{\begin{bmatrix} 2 \\ j1.7 \end{bmatrix}}_{I} \tag{C.63}$$

where the matrices Y, V, and I are as indicated.

We will use MATLAB to compute the voltages V_1 and V_2, and to do all other computations. The code is shown below.

```
Y=[0.0218–0.005j  –0.01;  –0.01  0.03+0.01j]; I=[2; 1.7j]; V=Y\I;% Define Y, I, and find V
fprintf('\n');                                          % Insert a line
disp('V1 = '); disp(V(1)); disp('V2 = '); disp(V(2));   % Display values of V1 and V2
```

```
V1 =
 1.0490e+002 + 4.9448e+001i

V2 =
  53.4162 + 55.3439i
```

Next, we find I_X from

R3=100; IX=(V(1)–V(2))/R3   % Compute the value of I_X

```
IX =
   0.5149- 0.0590i
```

This is the rectangular form of I_X. For the polar form we use

magIX=abs(IX)   % Compute the magnitude of I_X

```
magIX =
    0.5183
```

thetaIX=angle(IX)*180/pi   % Compute angle theta in degrees

```
thetaIX =
   -6.5326
```

Therefore, in polar form

$$I_X = 0.518\angle{-6.53°}$$

Spreadsheets have limited capabilities with complex numbers, and thus we cannot use them to compute matrices that include complex numbers in their elements as in Example C.18

C.12 Exercises

For Problems 1 through 3 below, the matrices A, B, C, and D are defined as:

$$A = \begin{bmatrix} 1 & -1 & -4 \\ 5 & 7 & -2 \\ 3 & -5 & 6 \end{bmatrix} \quad B = \begin{bmatrix} 5 & 9 & -3 \\ -2 & 8 & 2 \\ 7 & -4 & 6 \end{bmatrix} \quad C = \begin{bmatrix} 4 & 6 \\ -3 & 8 \\ 5 & -2 \end{bmatrix} \quad D = \begin{bmatrix} 1 & -2 & 3 \\ -3 & 6 & -4 \end{bmatrix}$$

1. Perform the following computations, if possible. Verify your answers with MATLAB.

   a. $A + B$ b. $A + C$ c. $B + D$ d. $C + D$

   e. $A - B$ f. $A - C$ g. $B - D$ h. $C - D$

2. Perform the following computations, if possible. Verify your answers with MATLAB.

   a. $A \cdot B$ b. $A \cdot C$ c. $B \cdot D$ d. $C \cdot D$

   e. $B \cdot A$ f. $C \cdot A$ g. $D \cdot A$ h. $\dot{D} \cdot C$

3. Perform the following computations, if possible. Verify your answers with MATLAB.

   a. $detA$ b. $detB$ c. $detC$ d. $detD$

   e. $det(A \cdot B)$ f. $det(A \cdot C)$

4. Solve the following systems of equations using Cramer's rule. Verify your answers with MATLAB.

   a.
   $$\begin{aligned} x_1 - 2x_2 + x_3 &= -4 \\ -2x_1 + 3x_2 + x_3 &= 9 \\ 3x_1 + 4x_2 - 5x_3 &= 0 \end{aligned}$$

   b.
   $$\begin{aligned} -x_1 + 2x_2 - 3x_3 + 5x_4 &= 14 \\ x_1 + 3x_2 + 2x_3 - x_4 &= 9 \\ 3x_1 - 3x_2 + 2x_3 + 4x_4 &= 19 \\ 4x_1 + 2x_2 + 5x_3 + x_4 &= 27 \end{aligned}$$

5. Repeat Exercise 4 using the Gaussian elimination method.

6. Solve the following systems of equations using the inverse matrix method. Verify your answers with MATLAB.

   a.
   $$\begin{bmatrix} 1 & 3 & 4 \\ 3 & 1 & -2 \\ 2 & 3 & 5 \end{bmatrix} \cdot \begin{bmatrix} x_1 \\ x_2 \\ x_3 \end{bmatrix} = \begin{bmatrix} -3 \\ -2 \\ 0 \end{bmatrix}$$

   b.
   $$\begin{bmatrix} 2 & 4 & 3 & -2 \\ 2 & -4 & 1 & 3 \\ -1 & 3 & -4 & 2 \\ 2 & -2 & 2 & 1 \end{bmatrix} \cdot \begin{bmatrix} x_1 \\ x_2 \\ x_3 \\ x_4 \end{bmatrix} = \begin{bmatrix} 1 \\ 10 \\ -14 \\ 7 \end{bmatrix}$$

Index